AF767576

Energy, Society, and Environment

Critical Introductions to Geography

Critical Introductions to Geography is a series of textbooks for undergraduate courses covering the key geographical subdisciplines and providing broad and introductory treatment with a critical edge. They are designed for the North American and international market and take a lively and engaging approach with a distinct geographical voice that distinguishes them from more traditional and outdated texts.

Prospective authors interested in the series should contact the series editor:

Deborah Dixon
University of Glasgow
Deborah.Dixon@glasgow.ac.uk

Published

Introduction to Cultural Geography: A Critical Approach
Jamie Winders and Declan Cullen

Migration: A Critical Introduction
Mary Gilmartin, Marlene H. Jacobsen, and Anna-Kaisa Kuusisto

Water: A Critical Introduction
Katie Meehan, Naho Mirumachi, Alex Loftus, and Majed Akhter

Environment and Society: A Critical Introduction, 3e
Paul Robbins, John Hintz, and Sarah A. Moore

Political Geography: A Critical Introduction
Sara Smith

Political Ecology: A Critical Introduction, 3e
Paul Robbins

Economic Geography: A Critical Introduction
Trevor J. Barnes and Brett Christophers

Health Geographies: A Critical Introduction
Tim Brown, Gavin J. Andrews, Steven Cummins, Beth Greenhough, Daniel Lewis, and Andrew Power

Urban Geography: A Critical Introduction
Andrew E.G. Jonas, Eugene McCann, and Mary Thomas

Geographic Thought: A Critical Introduction
Tim Cresswell

Mapping: A Critical Introduction to Cartography and GIS
Jeremy W. Crampton

Research Methods in Geography: A Critical Introduction
Basil Gomez and John Paul Jones III

Geographies of Media and Communication
Paul C. Adams

Social Geography: A Critical Introduction
Vincent J. Del Casino Jr.

Geographies of Globalization: A Critical Introduction
Andrew Herod

Geographic Thought: A Critical Introduction, 2e
Tim Cresswell

Energy, Society, and Environment

A Critical Introduction

Dustin Mulvaney

WILEY

This edition first published 2027

Registered Offices
John Wiley & Sons, Inc., 111 River Street, Hoboken, NJ 07030, USA
John Wiley & Sons Ltd, New Era House, 8 Oldlands Way, Bognor Regis, West Sussex, PO22 9NQ, UK

For details of our global editorial offices, customer services, and more information about Wiley products visit us at www.wiley.com.

The manufacturer's authorized representative according to the EU General Product Safety Regulation is Wiley-VCH GmbH, Boschstr. 12, 69469 Weinheim, Germany, e-mail: Product_Safety@wiley.com.

Wiley also publishes its books in a variety of electronic formats and by print-on-demand. Some content that appears in standard print versions of this book may not be available in other formats.

Library of Congress Cataloging-in-Publication Data

Names: Mulvaney, Dustin author
Title: Energy, society, and environment : a critical introduction / Dustin Mulvaney.
Description: First edition. | Hoboken, NJ : Wiley, 2027. | Series: Critical introductions to geography | Includes index.
Identifiers: LCCN 2025043211 (print) | LCCN 2025043212 (ebook) | ISBN 9781119734079 paperback | ISBN 9781119734093 adobe pdf | ISBN 9781119734086 epub
Subjects: LCSH: Power resources–Environmental aspects | Technology–Social aspects
Classification: LCC TD195.E49 M858 2027 (print) | LCC TD195.E49 (ebook) | DDC 333.79–dc23/eng/20260316
LC record available at https://lccn.loc.gov/2025043211
LC ebook record available at https://lccn.loc.gov/2025043212

Cover Design: Wiley
Cover Image: © Charles O'Rear/Getty Images

Set in 11/13.5pt Minion by Straive, Pondicherry, India
Printed and bound by CPI Group (UK) Ltd, Croydon, CR0 4YY

C9781119734079_080826

Contents

Chapter 1

Introduction: Energy, Society, and Environment

Technology is not neutral. We are inside of what we make, and it is inside all of us. We are living in a world of connections and it matters which are made and unmade.

—Donna Haraway, 2000, *The Cyborg Manifesto*

Energy connects us to the universe in a multitude of ways. As a force of nature and a social actor, it lights our world, powers the wind we harness for electricity, metabolizes the food we consume, and sustains food webs that underpin entire ecosystems. It forms the elements of the universe, provides warmth, and illuminates the stars we gaze upon. In short, energy makes our world go around. After all, energy is the ability to cause change—it is what sets things into motion or makes them shine or give off heat. Energy also is the stuff of life, powering the cells of living organisms. The conservation and utilization of energy have long been strategies in ecological evolution. Energy is not only everywhere—it is everything.

Energy is also profoundly social. As a natural resource transformed into a commodity, it powers the light, motion, and heat of human civilization. Societies construct massive infrastructures to extract, produce, transport, consume, and dispose of energy. Over the past 10,000 years of sedentary civilization, energy use has dramatically shaped the Earth—from agricultural land use to deforestation for wood and, later, charcoal used in metal smelting. The development of energy resources and technologies has connected distant regions of the world through intricate material relationships. Even earlier, humans were transforming ecosystems with fire. Early human societies burned grasslands and used fire to shape landscapes, enhancing hunting and foraging opportunities and aiding in tool-making. These practices influenced the development, spread, and persistence of certain ecosystems.

The discovery of denser, cheaper forms of energy—coal and petroleum—ushered in what many consider to be substantial improvements in quality of life. However, these same developments also accelerated global ecological change. Civilization's growing appetite for energy contributed to widespread deforestation, global human migrations, and the damming of major river systems. These transformations released vast amounts of carbon into the atmosphere, creating new risks—especially for the world's most vulnerable populations—by increasing exposure to disease and natural disasters. The energy economy today represents one of the largest natural resource sectors, responsible for significant material use and generating trillions of dollars in economic activity annually. Crucially, the benefits and harms of the global energy economy are not evenly distributed. Energy inequality persists as a defining issue of our time. The poorest populations often bear the brunt of air pollution, effluent discharges, water contamination, and land use changes driven by society's insatiable demand for energy. This chapter sets the context for understanding the depth and complexity of these challenges. It explores how deeply energy is embedded in our social systems and highlights the enduring struggle to address the environmental and social consequences of energy production and consumption.

What Is Energy? A Universe of Energy and Matter

Energy is all around us. The universe is made of space, matter, and energy. Our galaxy is where it is in space because energy propelled or guided it there. In our solar system, planets move around by Isaac Newton's kinetics, the mathematics that described the movement of physical objects from ballistics to planets. Our moon circles the Earth, and as it does, it pulls the liquid ocean by gravitational force, and the tides move. Other energy is delivered to Earth in the form of electromagnetic radiation from the sun and accumulates as matter of plants via photosynthesis. At the subatomic scale, energy behaves in ways that Einstein described 180 years after Newton, holding together the nuclei of atoms and shaping patterns of electrons that surround them. Energy works across these scales. Supernova explosions release energy in quantities to make all the elemental materials for life—the oxygen we breathe, the calcium in the bones of mammals, the iron in human blood, and the uranium that decays in our nuclear reactors. These stellar explosions started by combining particles in the Big Bang to convert hydrogen to helium, then helium to larger elements, and so on, giving us the balance of elements in the universe we have today. We live in a universe of energy and matter.

Energy is also the foundation of human civilization. Energy moves, warms, lights, clothes, and feeds societies. Extraction and use of energy can have major implications for the quality of life we live, even how long we live. Without heat for cooking, heating water, and keeping us warm, we would be less comfortable or worse. Data on human needs for energy suggest that there are significant improvements in life expectancy, education, and income with per capita energy use up to around 4 kW per person, though these relations also tend to be nonlinear (Lee and Yuan 2024). Even before the Neolithic revolution—when humans began more widespread use of agriculture and fire and shifted from a species on the move to one based in sedentary civilization—humans depended on energy to survive. Humans need energy for food, warmth, efficient (and inefficient) movements of goods and people, and to power cooling devices for our buildings, which incidentally use up to 40% of our end-use energy. Energy runs throughout the ecosphere and human civilization, transforming it into waste and emissions. One study of the entire natural resource throughput of the global economy estimates that since 1900 humanity has released 2,500 gigatons of waste and emissions to Earth's water, air, subsurface, and land (Krausmann et al. 2018).

Energy is political. Where and how humans extract and produce energy to power our devices and make our products is shaped by political forces, even geopolitical ones. These forces can ensure certain benefits accrue to owners, communities, or nations, or they might represent efforts to protect consumers from the owners of energy. Wars are fought over energy. Energy is also politicized. Access to and costs of energy can be motivations that organize communities to press for ownership over energy systems or foundational to democracy. In the United States for example, the early development of electricity was dominated by private interests, entrepreneurs, and inventors like Edison and Tesla, General Electric, and Westinghouse. That is, until a movement for public ownership facilitated just that in the New Deal of the 1930s with the Tennessee Valley Authority—a new publicly owned utility created by the federal government to bring power to rural and impoverished Appalachia (Vaheesan 2024). The tradeoffs of energy development and the different types of ownership models are topics of great dispute and controversy; private, public, or cooperative are a few models. But energy and political power will always be intimately interlinked.

In everyday life, we encounter energy as fuels or electric power for light, for heat, for movement, or any kind. But there can be so many steps between natural resources and how they are processed, used, or consumed that it is easy to lose sight of where the energy we use comes from. We know we get electricity for our phones from the electrical outlet. We probably know there is a wire that enters our home or apartment from the street. But where did it come from before it arrived in the neighborhood? Following the lines, we find an interconnected electric grid with millions of poles, millions of miles of wires, step-up and step-down transformers, substations, tall metal structures, and power plants. The electricity grids in developed economies are incredibly complicated machines that bring us power, something we too often take for granted considering the number of people on the planet without electricity. Similarly, many people fuel their vehicles at gasoline stations. But where did that fuel come from? Where was it processed and mixed? Where was it extracted from the Earth? How does access to fuel impact the life of someone who needs to use it to fuel a tractor to raise crops to sell to feed their family or community? Our understanding of how we are connected through both our material connections and also the availability of the services (light, heat) turns out to be complicated.

Energy powering human societies come from natural resources and causes pollution. Whether taken from the breeze or a surface mine, energy resources and infrastructures to transform it have material substance. Whatever the source of fuels, electricity, heat, and the devices they power, energy has a material dimension that affects Earth somewhere. Resources are taken, transformed, and then the residuals are left behind. Energy extraction or generation and production usually is accompanied by toxic or noxious air pollutants, wastewater effluents, and/or hazardous materials disposal. In fact, many early environmental laws were specifically intended to reduce environmental impacts from air pollution from energy combustion. Fossil fuel energy resources, when combusted, result in air pollution that impacts human health, add greenhouse gases to Earth's atmosphere, and can require solid waste disposal. Fossil fuel energy comes with environmental impacts to the planet we share with other people and other species, from mountain top removal for coal to crude oil spills in sensitive ecosystems or fisheries. Renewable energy industries too can have important impacts for consideration, especially with land. Consider the deforestation for palm oil or soybeans used for animal feed, food stuffs, or to make biodiesel. Once heralded as a low-carbon fuel, biodiesel from soy and palm is acknowledged to have significant greenhouse gas emissions from land use change where the issue is not monitored and managed.

Energy is a resource also at the center of many geopolitical controversies and disputes, if not war and armed conflict (Le Billon 2004). Petroleum, for example, has been at the center of violent conflict and efforts to control the resource since it became an important commodity in the nineteenth century. Oil fields in Iraq were lit on fire in the Gulf War on the orders of Iraqi political leadership as they retreated, causing one of the worst oil pollution disasters to air quality we have seen. In the Niger Delta region of West Africa, a globally import area of oil exploitation and extraction, the Ogoni people have struggled to gain some of the benefits from a multinational corporation that makes a profitable commodity that results in the contamination of their land, air, and water. The Russian invasion of Ukraine in 2022 put European reliance on natural gas imports in the spotlight, leading to new policies in Europe to phase out some energy imports and ramp up other imports from places considered more allied. Energy is entangled in questions of war and peace.

Energy resources through history have also been linked time and again to stories about corruption, remarkable wealth generation, and frontier exploitation where land, ecosystems, and people are dispossessed, perpetuating systems that reproduce global inequality. What is it about any commodity or product or infrastructure that causes this particular material to be imbued with such agency—the ability to act (or be reacted to by our interests and desires)? Such agency to shape lives and destinies of everyone around them? Energy strategies and their importance to economic activity shape decisions to wage war, to make violence, and to protest infrastructures; they feed grand narratives, capture national imaginations, and influence who we are and where we live and go.

This textbook aims to situate the many dimensions in which energy influences our lives and the complex tradeoffs and choices that lie ahead as humans come to understand the impacts of civilization's energy choices. Important decisions about the future are beckoning us today. Look at the headlines, and there are news stories about climate change and energy transitions every day. Look more closely, and you see emerging conflicts over places to get metals and minerals for renewable energy devices. New frontiers include deep seabed mining far offshore, but also more familiar conflicts, such as extraction on or near sacred sites, or in wilderness areas. These will shift the petro- and fossil fuel conflicts today to conflicts about strategic metals and minerals. It's the authors' contention that more familiarity with these issues can allow us to better engage with the questions that are raised about tradeoffs, sacrifices, and the specter of new possibilities in a low-carbon future.

Energy Transitions Are Underway

Today human activities use more renewable and solar electricity than at any other time since the fossil fuel era began. Every week headlines announce new records set for solar and wind in places that are leading the sustainable energy transition. By 2033, according to the International Energy Agency (2024), worldwide production of solar electricity is expected to pass coal power generation, a thought that once seemed unimaginable. While plenty of evidence suggests the progress away from fossil fuels is too slow—80% of all primary energy is still from fossil fuels—renewable and solar energy have continued sustained growth and are contributing to efforts to decarbonize electricity. The search to develop and install innovative energy solutions to climate and air pollution while increasing the overall supply to electrify homes and bring energy to those without energy access will be an ongoing, multi-faceted, and complex endeavor. But whatever suite of technologies or organizational forms decarbonization efforts take, renewable and solar energy will play a leading role in the energy future. How did the solar and renewable energy industries achieve such rapid and widespread success in recent years? How might this success be sustained? Will it result in backlash or setbacks?

Some of the biggest energy challenges facing civilization, communities, and our planet will require difficult decisions and tradeoffs. See Box 1.1 for example, the conflicts over critical metals, minerals, and materials that are needed for electrification and decarbonization of transportation and the electricity grid. Cobalt, manganese, lithium, graphite, and a handful of others are needed for the batteries to electrify vehicles and for grid energy storage, but they are in limited supply or suffer from production bottlenecks. Batteries offer solutions to the widespread use of the internal combustion engine, on say matters of air quality and certainly greenhouse gas emissions. But the reality is the development of new sources for these novel and emerging materials will result in new impacts that will require us to plan for them and be wise about how we go about energy transitions (Sovacool et al. 2020). Scholars of science and technology studies Clark Miller and Jennifer Richter describe something similar when they talk about social planning for energy transitions (Miller and Richter 2014). They argue that the intentional planning to prepare for the impacts of energy transitions is crucial to ensuring equity and sustainability. Without careful planning and thoughtful collaboration with communities on the ground, it is hard to see how these new developments do not just reproduce the inequalities they have produced before.

The topic of energy transitions is of increasing public interest because of growing interest in deep decarbonization of the global economy to stave off the worst of global climate change, processes that are already well underway in some parts of the world. There are emerging debates about how to best move along decarbonization pathways,

Box 1.1 Energy Sources Are in Transition. What Energy Systems Are Changing Around You?

The scale and pace of a clean energy transition needed may be unparalleled in global history. Different minerals and metals will be needed to substitute for fossil energy. Different energy landscapes will emerge; some will be contested, others will be championed. New devices will have to enter our homes and businesses. In some cases, we just simply have to change the way we do and make things altogether (plastics, steel, etc.) and build new infrastructures (pedestrian- and bike-safe streets, more low-income housing, better and more convenient public transportation). Note things changing around you. If it is not immediately apparent, take a look at information about your town or state or nation. What is changing with the energy system? Have any power plants been closed or upgraded? What is new? Are there new nearby power lines or wind farms or solar farms? Describe or sketch the change and how it might be impacting your community, region, or landscape.

and there are many experts offering valuable perspectives on important decisions that will shape future energy systems.

- Should resources be invested in deploying technologies that are available today, or should those funds be invested in research and development for the next generation of technologies?
- What are the key economic and technical obstacles to overcome with energy transitions? What tools do we have to overcome them?
- Transitions energy systems mean building a lot of new infrastructure. How do we improve public acceptance of green infrastructures?
- How can we improve collaboration in energy transitions with Indigenous communities and First Nations?
- How can we prevent green housing and high-quality public transit infrastructure from being unevenly realized or leading to displacement of marginalized groups?
- Do we need more or less energy per capita? What is the future composition of sustainable energy systems?

A rapid transition to renewable and clean energy seemed unlikely not long ago, but conventional energy systems have become more expensive and less and less a part of the energy mix. How did these changes happen? Scaling up deployment and driving down costs is part of that story. Just look to the rapid cost drops in solar and wind power, followed by energy storage (see Figure 1.1). Public policy commitments are another driver, as commitments to renewables in the future send the signals to producers that there will be future customers. Others are looking to avoid commitments to infrastructures that may no longer be viable or too costly to maintain in the

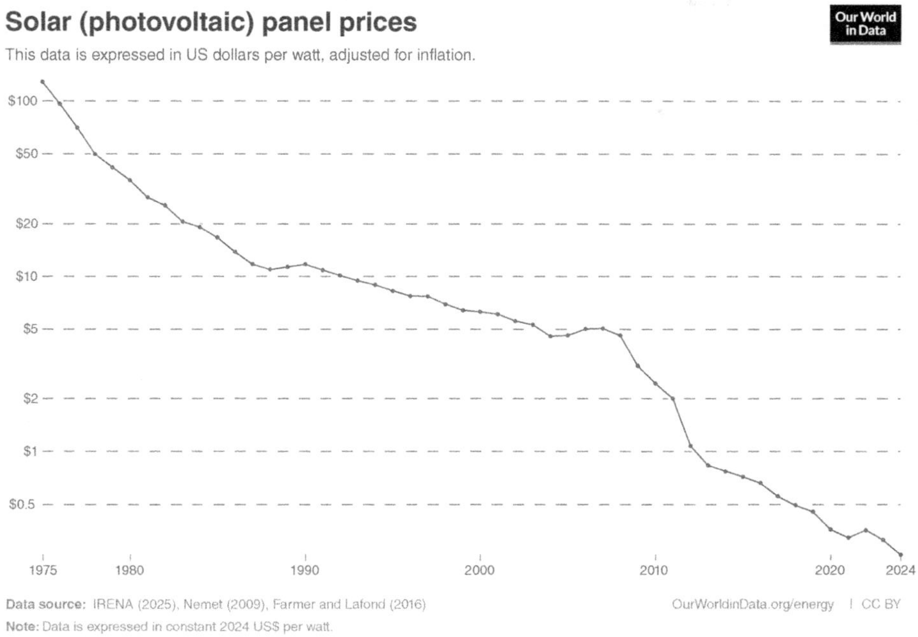

Figure 1.1 Price declines in solar energy. Source: Our world in Data, 2024/CC BY 4.0.

future, so are avoiding upkeep and investments in fossil- and fissile-fueled infrastructures that may be costly to fuel, at the limits of air pollution permits, or too inflexible to operate.

Energy transition researchers Michaël Aklin and Johannes Urpelainen argue that external shocks like price swings or supply disruptions expose the costs of carbon lock-in and, thus, open windows of opportunity for alternative policies to succeed. Carbon lock-in is when an energy system becomes reliant on older technologies instead of newer ones and commits the owner to using it into the future because of this invested capital. The authors investigate how renewables transitioned from expensive, marginal, niche technologies in the 1970s to widespread, inexpensive, leading sources of electricity generation within 50 years. They outline the processes of external shocks, politicization, and lock-in that explain why some energy technologies take off and then stall or slow. The core thesis is that external shocks to the system act as major impetuses for change. Shocks cause energy systems to become politicized, and this opens political systems up to incentives that favor alternative technologies. In the 1970s policy architecture that led to the early development of renewables in various leading places like Denmark, Germany, and California, revisiting a time when California Governor Jerry Brown was advising then-U.S. President Jimmy Carter, who in turn was consulting Amory Lovins, arguing that much of the response was because of concerns about how prices for energy imports. The major shocks were those that occurred with rapid, spiking prices for energy, even energy shortages like those seen with petroleum in 1973 and 1979. They suggest other mechanisms related to the stall in nuclear power. Alkin and Urpelainen also point at elite ideology as a major element of making this shift happen in some places, implicating this as a major driver in California. The national shock of an aggressive nuclear weapons deployment by an armed neighbor led those in Finland to see the benefits of nuclear power development if it helped illustrate the country's ability to make electricity using an important knowledge base. Theories they propose attend to key questions of energy transitions, most revolving around the core contention that political obstacles are the primary things blocking energy transitions, except in developing countries, where capital constraints and wider political economic issues remain the major limiting factors.

While energy is deeply intertwined with social and political issues, it is also clearly delimited in the material world, which means the geographies of extraction, production, and use shift with energy transitions. Energy is something "out there" and moving from one resource to another literally means enrolling in new places, infrastructures, ecologies, and communities. Energy is a resource, technology, and commodity that people explore for, develop, produce, sell, and use, but it is still a biophysical phenomenon that is a driver of the material world from the movement of galaxies to the food webs of ecosystems to the metabolism of cells and effects of subatomic particles. The physicist, chemist, and mathematician Alfred James Lotka is one of the originators of the research field of population ecology—how ecosystems evolve over space and time (Kingsbury 2015). His research on age-structured populations noted that population dynamics could be reduced to exchanges of matter and energy in an ecosystem (Bacaër 2011). Howard T. Odum is another important contributor to early understandings of the significance of power and the role it plays in society. In a widely read and referenced scientific paper "Environment, Power, and Society for the Twenty-First Century: The Hierarchy of Energy," Odum develops an eloquent theoretical basis for how energy moves through ecosystems and how that shapes the material basis for our world, the ecologies we live among, and basis for the organization of society. Howard Odum and Eugene Odum later would write about how the valuation of an ecosystem and its energetic flows is by way of the concept of emergy—energy memory—which was introduced as a concept to represent the usefulness of energy (Odum 2007). Emergy traces energy through an ecosystem from its transformation from solar energy through ecosystems and its dissipation into the environment usually as low-level radiation, but also as wildfire and other events (Brown and Cohen 2019).

As for understanding the development of society, anthropologists have sought to understand how energy shapes cultural evolution and development. "White's law" named for anthropologist Leslie White suggested that the materialist development of technology and advancement of society was predicated on obtaining increasing amounts of energy. The idea asserted that technological development benefited the advancement of society, and more specifically that more per capita energy use led to more desirable social outcomes. White also argued that increasing the amount of energy used by a society led each to spend that energy more efficiently. The quest for more energy would be inevitable in White's view, "Unless he [sic] can harness additional quantities of energy-by tapping new sources-cultural development will come to an end" (White 1943, p. 342). As anthropologist Dominic Boyer (2014) has noted on reflection, White wrote about peak coal and oil and raised concerns about the dwindling resource base for humans and its implications for social and human evolution.

A generation later, energy scholar Laura Nadar was a leading voice pushing back on some of these precepts about energy use and human evolution. Narratives about energy systems changes ignored import dimensions from gender dynamics to corporate power that were crucially undermining efforts (Nadar 2004). Importantly, Nadar pointed out that the "crisis model" of system change, where a moment of crisis is used to justify new policies and directions for energy supplies favors corporations who can use government to serve their ends. To Nadar, democratic control over energy systems and future pathways is crucial to ensuring an energy system that provides for people without degrading the environment.

To pull another thread from Leslie White, is increasing energy demand in human civilization inevitable? White wasn't the only one to find himself in the camp of a future that requires more energy. The contemporary expression of this idea comes from so-called "ecomoderns" a group of thinkers, futurists, and policy advocates on energy-climate policy issues. They echo the sentiments that attempts to limit future energy use—a central tenet they say is held by activists and scholars of degrowth (more on this later)—are futile because developing countries need access to clean, reliable energy and changing consumption habits in the developed world is futile. Energy scholar Vaclav Smil has made similar points to the ecomoderns about the benefits of more energy for increased well-being (Smil 2003). Smil also has long challenged views on energy systems and contends their propensity to change is glacially slow (2019). The widely read energy author's critique is based on the engagement with the history of energy and is spirited by a focus on energy density, something provided by fossil fuels and nuclear power. These trends suggest to the ecomoderns that it is best to pursue more or less all of the above energy technologies and resources, focusing on cleaning them up and innovating toward generation solutions, whether that is future nuclear reactors, carbon capture technologies from natural gas, or renewable energy.

Renewable energy is human civilization's most rapidly growing form of energy. This gets it laudatory attention and much praise in the press or from activists or regular people concerned about what they hear could happen because of changes to our climate. But renewable energy itself deserves greater scrutiny and critical analysis. New energy sources, fuels, and generation devices have different materialities—resource bases and supply chains that differ from conventional energy sources, raising new questions about environmental impact and justice considerations when scaling up renewable energy deployment (see Box 1.2). The differential ways that these production systems evolve (or do not) offer opportunities to comparatively evaluate critical environmental and social issues and to ask what factors make these issues vary as they do. Supply chains offer one way to organize production systems to compare one versus another but also to compare processes within and along the chains.

Box 1.2 Common but Differentiated Responsibility

The concept of **Common but Differentiated Responsibilities** was enshrined as Principle 7 of the Rio Convention of 1992 (Harris 1999). The key idea is that everyone has common responsibilities to protect the planet and its resources, but given the historic patterns of development and underdevelopment owing to colonialism, extractivism, and other rampant or sanctioned activities over the past few centuries in particular, that some nations bear more responsibility than others. The framework is often drawn on to provoke investments by developed economies into developing ones to help them obtain the cleanest possible energy sources. These can be direct investments in the power plants and infrastructures that could lower carbon emissions. But they could also suggest research and development investments. Some suggest this idea should apply between nations, especially where climate refugees might be displaced by rising seas, dangerous storm events, or landslides induced by climate change.

Key Energy Themes for Environment and Society

Clearly, we as a species trying to coexist with other species on Earth are going to be in trouble if we do not confront these energy dilemmas. Environmental degradation, species and habitat loss, and public health are all impacted severely by energy choices, necessitating energy transitions. But how will this play out? Who will take risks and who benefits? These are just a few of the options before us. There will be several reoccurring themes throughout this textbook, and this section aims to introduce some of those briefly to prepare the reader for more detailed case studies or concepts later (Figure 1.2).

Less energy poverty and more energy justice

Tackling energy poverty is a leading theme for environment and society because of how important access to energy is for well-being and quality of life (Lippert and Sareen 2023). Entire communities can be uplifted with access to energy as they are now able to participate in other pursuits which also raise their well-being. Hence, when we speak of energy justice, often the context is ensuring that people have access to electricity and modern fuels. This notion extends to industrialized nations where electricity shutoffs for nonpayment by electric and gas utilities can make households vulnerable to weather extremes and have no way to store cold or frozen food (Baker et al. 2023) (Figure 1.2).

Today energy justice is becoming a mainstream policy concept used in government agencies and decision-making. In 1994, U.S. President Clinton signed an Executive

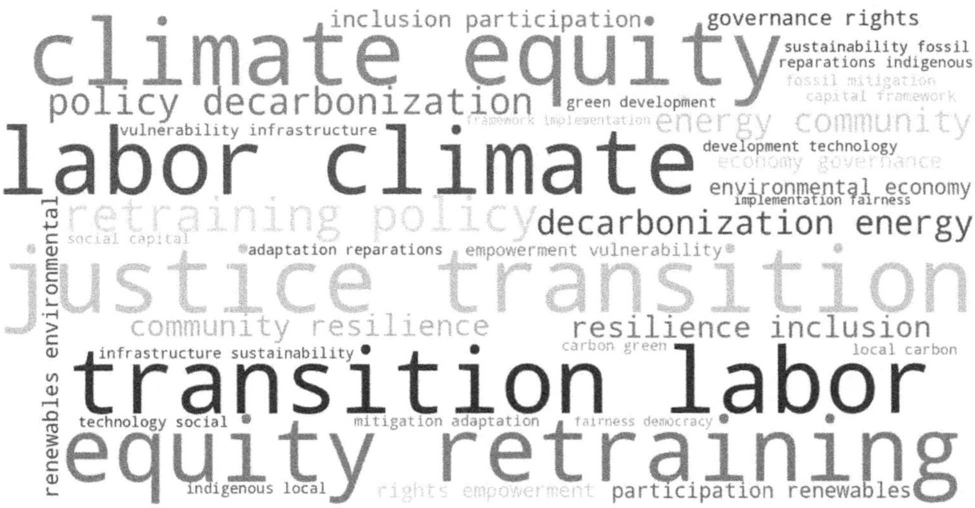

Figure 1.2 Keywords in just transitions research.

Order for Environmental Justice, only a decade after environmental justice became a more common term in understanding the distributional impact of pollution on society. But environmental justice as a concept mobilized to change policy was rarely used in a meaningful way. Fast forward almost 30 years and there is now a position in the U.S. Department of Energy to understand and develop policy around ensuring environmental justice, a manor policy in the United States requires spending a large percentage (40%) in environmental justice communities. In 2021, Shalanda Baker was named deputy director for energy justice at the U.S. Department of Energy bringing someone familiar with and expert in solutions to the key issues that were critical to communities overburdened with industrial and energy pollution. U.S. President Biden also created other senior positions in government to support environmental justice efforts. California created an office of environmental justice under the Attorney General, and by 2022 the federal government created a similar post with considerable power and influence.

Communities and workers that receive the greatest exposure to air and water pollution are fenceline communities near industrial production. From refineries to recycling facilities, pollution from these "locally unwanted land uses" and operations can be deadly for communities. Breast cancer rates and other public health afflictions have been documented to be far more severe in communities living near coal or oil burning power plants and refineries. People living near refineries frequently receive shelter-in-place warning when facility operations require things like "flaring" to burn off excess fuel accumulated in part of the refinery to resume safe operations. Recycling facilities can cause very local heavy metal contamination in soils or in the air. These unequal exposures are often at the center of environmental justice campaigns by fenceline communities (Figure 1.3).

Figure 1.3 The movement for climate justice in Sacramento, California. Source: Lorie Shaull / Flickr.

Greenhouse gases, global warming, and climate justice

Leading most public and scientific concern about fossil fuel use over the past few decades is how greenhouse gas pollution is impacting our planet's temperature. Human settlements and activities today have severe and unequal vulnerabilities to extreme weather. The world's most vulnerable are also usually the poorest or not able to mobilize resources, while the least able to adapt and reduce their exposure to hazard. Prior to the elevation of climate change as a key concern, the primary questions about fossil fuels were different—air pollution like acid rain and smog or the overall loss of fossil fuels due to resource degradation, so-called peak oil. Peak oil never materialized and is less a concern as discoveries further afield became cost-competitive with the higher prices of oil frequently registering in global petroleum markets. But today climate is first and foremost in the mind of those engaged with the fossil fuel industry with the exception of fenceline communities dealing with air pollution from routine flaring, "normal accidents," and persistent pollution from industrial facilities.

There are many researchers that contend that extractivism itself is prone to violence, and the inherent aspects of violence are not just limited to the global flows of oil, gas, and coal (Dunlap et al. 2024). The nexus of extractivism and violence becomes particularly problematic as there are calls to speed up and deregulate extraction as the economy and energy systems respond to the specter of climate change. In fact, as there are increasing interests pushing to open up more areas to extraction and mining for clean energy metals and minerals, raising concern that this too will result in unequal distributions of violence related to extractivism as developing countries and

Figure 1.4 Long a centerpiece of the local economy, the Navajo Generating Station was closed and imploded. Source: Alan Stark / Flickr.

marginalized communities and landscapes will bear the brunt of these critical materials for decarbonization. Geographer Meredith DeBoom coined the term "climate necropolitics" to help us see and understand "the multiscalar processes, practices, discourses, and logics through which anthropocentric (human-centered) imaginaries can be used to render extractive violence legitimate in the name of climate change response" (DeBoom 2020, p. 900) (Figure 1.4).

Energy as a socio-technical problem

The approach used by researchers who study socio-technical systems takes as a core assumption that the human and technical are constitutive or coproduced. This means the interactions that humans hope to make and the materials they work with result in outcomes that are not always easily predictable or capable of control but instead are often subject to contingencies that are hard to predict or unknown altogether. Take, for example the idea that the government should invest money to develop a cleaner technology; it is not known that the technologies will pan out or might do so in surprising ways. This is similar to what people mean when they say that nature is an actor in shaping social outcomes. It is to emphasize this relationship between our desires and understandings and "willingness" of the social and biophysical worlds to oblige to those outcomes. Sometimes nature bites back at our knowledge limits. Sometimes these knowledge limits are imposed by a failure to characterize the problem, but other times they are imposed by competing urgencies or conflicts with values that seem unresolvable. The idea described as "post-normal science" aims to situate the scientific process in a context to understand that science is not value-free and can have implications for things that are outside of the scientific process (Funtowicz and Ravetz 1993).

We are learning more about how our gendered, racialized, colonial, and anthropocentric logics operate and develop energy resources within capitalist value relations that do not only find economic value in nature (Baruah 2017; Huber 2009). The uneven spatio-temporal dynamics, such as where people live, work, where pollution goes, and how benefits from those systems circulate, will continue to be pressing topics for energy and society. Moving away from the subterranean energy regime could mean more land use conflicts, new sites of extraction, and socio-ecological impacts where they haven't been made before. As we will explore further in Chapter 2, these dynamics necessitate a critical approach.

Energy as a spatial problem

Bringing natural resources for energy from sites of extraction to where they are generated and ultimately used has always presented spatial challenges. Think about oil tanker ships, pipelines, and transmission lines as examples. Furthermore, the justice dimensions too are spatially distributed. Some communities live near industrial energy generation facilities, while others go unburdened by the cost of energy. The energy spatial problem continues with the challenges of scaling up renewables, from spatial extent of wind or solar farms, but also through the ways that electricity needs to be moved from rural or offshore areas to urban areas. The spatial challenges of energy development, supply chains, emissions, and disposal are inherent to the extraction of natural resources for energy.

After two centuries of society's economic activities built on carboniferous energy (Mumford 1936), atomic power, and mega-dam hydropower projects (Nixon 2010), renewable energy is making inroads in the twenty-first century. Prompted by public policies and technical advances, bioenergy, wind, solar, smaller-scale hydro, and geothermal power are increasingly planned, advocated for, and adopted as a means to reduce greenhouse gas emissions, reduce air pollution, and improve water efficiency from energy use.

One way to organize our thinking about these spatial challenges is to map the material flows through which energy systems pass. Systems of production, sometimes called commodity chains or value chains, are complicated assemblages, and energy systems that supply the chains to make or generate energy are no different. Every system of production has a supply chain; this refers to a sequence of production from extraction to end use or disposal. Production systems can reach to the most remote ends of our planet because components for a product could come from any part of the world before reaching a consumer. Cobalt mined in the Democratic Republic of Congo might be refined into cobalt chemicals in China, manufactured into lithium-ion battery parts in Japan, assembled in Germany, and installed in the United States.

This geographic variation of production resources and manufacturing offers opportunities to understand how renewable energy supply chains are shaping and are shaped by political circumstances and governance. Situating renewable energy supply chains in this way allows researchers to see comparative dimensions to understand what drives similarities and differences in production across space and time to better understand topics as varied as environmental inequality, international relations, and innovation policy.

For understanding the role of the spatial, it is helpful to look to geography, where space and its network of socio-ecological relations have been studied and theorized for a long time. In human geography, space is the socially produced realm where interactions occur that constitute social and material relationships. Objects are arranged in space, for example people, firms, states, and non-human nature. Modifying these spatial arrangements through action and meaning requires operating through relationships of power. Those with political power can transform particular dimensions of spatial interaction to serve particular ends, for example, creating property rights or regulating the practices of firms. As geographer Lefebvre (1991) suggests, all politics is at its roots spatial, and spatial practices are political. Edward Soja argued that a spatial turn is needed in geography as space constitutes nature–society relationships and is marked by power. These themes have been carried over to the study of energy studies more broadly, to understand the relational space through which energy provisions are made (Bridge 2018; Castán Broto and Baker 2018). Space, as a socio-ecological relationship, "emerges as a negotiating ground not just to secure energy access and move towards sustainable energy systems, but also to make explicit how we know and understand these problems" (Castán Broto and Baker 2018, p. 3). This view sees struggles over future energy pathways as discursive struggles over the use of space, the appropriate scale of regulation, and the consequences of energy systems, technologies, and devices to particular places. It may not seem like much, but moving from thinking about space as distances and containers in this way to a terrain across which relationships are made and negotiated allows us to see more of the totality of processes that make a place, instead of restricting our analysis of a problem's causes and effects.

Energy as a terrain of struggle

As opposed to technology- and economics-based arguments for change, viewing energy and its resources as a terrain of struggle centers social movements and political contestation as key drivers of spatial–temporal configurations of life and labor (Hess 2018; Huber and McCarthy 2017; McCarthy 2015). To characterize the delimitation of political spaces, geographers use the notion of scale, which can refer to the multiple intersecting levels of spatial representation or organization from the global, national, and regional to the local, community, household to the body and genome. Scale is important to think about in the context of energy and society because of our interactions with energy and the systems by which energy is governed. For example, in the United States, individual states regulate the proportion of clean or renewable energy that must be bought by electric utilities. This means those decision-makers at the state scale are the ones who control the situation with renewable energy deployment. Yet also in the United States, the federal government manages the programs that offer incentives to electricity customers to make that system cheaper. Local government might be in charge of building permits and electrical inspections for rooftop solar, which might involve government actors from the local up through the federal government to claim rebates offered by the Internal Revenue Service. Understanding the different scales of governance that co-occur can help organize our thinking about energy systems strategically. Where can we push on the levers for change?

Geographers like Sally Marston (2000) argued that concepts of scale—federal, state, regional, local, etc.—are socially constructed, that is, produced through social struggle and stabilized through laws, institutions, and every days practices (Smith 1984). Scale also can imply characteristics about size or extent. The social aspects of scale can be reorganized or changed by law or other social practice. To say scale is also "socially constructed" is to acknowledge the role of human decision making in arbitrating political processes often shaped by inequitable power asymmetries. Some have more influence in local government, in their households, or over their bodies, while others find more influence at the state, national, or even global scales.

Energy production and its pollution are also spatial in how the distribution of impacts occurs. The term "fenceline community" is used to describe environmental justice. Power asymmetries can be the reasons that actors strategically "jump scale" (Smith 1992) to shop new political venues or arenas that are more favorable to their interests. The successful production or reproduction of scale can empower or contain. For example, proponents of neoliberal economic policies rescale authority to the global, where multinational corporations can transcend the autonomy of national interests and harmonize the rules of investment through the jurisdiction of global trade agreements. In contrast, rescaling to the local devolves authority to where place-based interests dominate state, national, and global jurisdictions. Scales instead of boundaries might be described as relational constellations of actors. There are always actors outside of the bounds of scale seeking to redraw and extend the boundaries to reorder power relations. Think of the idea of gerrymandering voting districts itself as emblematic of those in the very literal sense. Likewise, there are actors inside the bounds of scale trying to delimit or secede themselves from others. The global touches down in the local, and the local is embedded in the global, and far apart locales can be networked together through these socio-ecological relations. This is how much of our global economy is built today and will continue to develop as we shift to a low-carbon economy (Box 1.3).

Box 1.3 The Strait of Hormuz: Why War in the Gulf Reaches Fuel, Food, and Plastics

The 2026 war involving Iran shows why the Strait of Hormuz is one of the most consequential chokepoints in the world economy. The strait links the Persian Gulf to the open ocean, and in 2024 and early 2025 it carried more than one-quarter of global seaborne oil trade, about one-fifth of global oil and petroleum consumption, and roughly one-fifth of global liquefied natural gas trade. Because there are only limited alternatives for moving those volumes, conflict in or around Hormuz quickly becomes a global energy-security problem. As of late March 2026, the war has sharply disrupted traffic through the strait, illustrating how a regional conflict can reverberate through fuel prices, shipping costs, and state strategy worldwide.

Hormuz matters not only because it affects gasoline, diesel, and jet fuel. Oil and gas are also major feedstocks for petrochemicals, which are used to make plastics, packaging, detergents, synthetic fibers, and many other everyday materials. Natural gas is also the key feedstock for most ammonia production, and about 70% of ammonia is used to make fertilizer. That means insecurity in Hormuz can move well beyond the energy sector, disrupting fertilizer supply, raising food-production costs, and destabilizing the material supply chains that support modern agriculture, manufacturing, and household consumption. Seen this way, the security dimension of Hormuz extends far beyond oil markets, reaching into agriculture, manufacturing, and everyday consumption. Significant volumes of helium for semiconductor manufacturing, sulfur and urea for agricultural fertilizers, and other industrial inputs must pass through the water body.

Hormuz is not just naval or geopolitical. It is about the vulnerability created by fossil-fuel dependence itself: a narrow maritime passage can shape not only oil markets, but also the cost of food, the availability of fertilizer, and the petrochemical base of modern life. The strait is a reminder that energy systems are inseparable from broader questions of infrastructure, trade, and everyday survival.

Toward a Just Transition to a Low-Carbon Economy

Look around and you will see that there is an energy transition to low-carbon energy futures underway. Several key renewable and clean energy devices and infrastructures and associated supply chains must be scaled and manufactured at an unprecedented rate for renewable energy to become a dominant source of electricity to satisfy global climate objectives. Photovoltaics, wind turbines, and lithium-ion batteries for electric vehicles and stationary electric storage will be key technologies for clean power conversion long into the foreseeable future, but these require new and expanded extractive industries (Sovacool et al. 2020). This means that the impacts of clean and renewable energy devices shift impacts from embodied impacts away from emissions as is the primary impact of combustion (Grubert and Zacarias 2022). The global nature of these production systems means that we need to broaden what we mean by justice in climate adaptation. Renewable industries are already embroiled in geopolitical disputes (Klinger 2015), land use conflict from India to the North American West (Mulvaney 2017; Stock and Birkenholtz 2021), and end-of-life waste management challenges for wind turbine blades and solar panels.

Clean energy and the materials to make them are also central to the imaginaries of statecraft and nation building as many governments aspire to exploit their own resources to develop and industrialize, as with the ambitions of Bolivia or Chile or Afghanistan or Nevada, USA, to supply lithium for batteries. Similarly, talk about global competition on the frontiers of energy innovation involves building supply chains to gain geopolitical advantage. Finally, digging into natural energy resources and commodity chains reveals a more complicated entanglement of petrochemical industries and electronics companies than the typical clean and renewable energy versus extractives and fossil fuel industries discourse commonly represented in energy transitions conversations. We have a lot of work ahead of us.

Takeaways/Highlights

- Changes to our energy system are underway
- The direction that energy resources and technologies evolve is not inevitable but will be the outcome of a social effort and material limitations or discoveries
- Energy problems are multi-faceted spatial and socio-technical problems that defy simple economic or technological explanations
- A transition to clean and renewable energy will not automatically be just, but will require special attention to detail for how low-carbon energy sources are developed.

Questions and Concepts for Review

1. Energy transitions happen when societies change the resources they use for energy or the ways they use energy in new technologies or behaviors. What are key energy transitions that have occurred in the past two decades?
2. The impacts of energy generation and extraction are severely unequal including who is burdened by pollution and who receives benefits. Think about the variations in air quality in different parts of your home town or where you live. Where is the air pollution at its worst? What types of industries or infrastructures are there? Are there any known or persistent issues with water or air quality there? What is the plan to clean it up?
3. What does space and scale help us understand in changing energy systems?
4. Energy transitions will require new natural resources that we currently do not use. What guardrails can be put in place to keep communities from being exploited for natural resources?

Exercise 1.1

Think about and track your own energy uses every day. What kinds of energy does your household use throughout the day? Where does it likely come from? What are the indirect uses? What are the different types of energy you encounter?

Exercise 1.2

Search for the words "energy transition" and "climate change" in an internet search engine. Skim the first five articles that you see that look relevant. What are the key themes? Make a list of words that resurface across the articles. What are the key themes of the energy transition?

Exercise 1.3

Search for an academic research article that uses scale or space in the keywords in a paper about energy systems. How are they using it and why? How does it help the authors understand what they are studying?

References

Bacaër, N. (2011). *A Short History of Mathematical Population Dynamics*. London: Springer.

Baker, E., Carley, S., Castellanos, S. et al. (2023). Metrics for decision-making in energy justice. *Annual Review of Environment and Resources* 48 (1): 737–760.

Baruah, B. (2017). Renewable inequity? Women's employment in clean energy in industrialized, emerging and developing economies. *Natural Resources Forum* 41 (1): 18–29.

Boyer, D. (2014). Energopower: an introduction. *Anthropological Quarterly* 87 (2): 309–333.

Bridge, G. (2018). The map is not the territory: a sympathetic critique of energy research's spatial turn. *Energy Research & Social Science* 36: 11–20.

Brown, M. and Cohen, M. (2019). Energy ecosystems and network analysis. In: *Encyclopedia of Ecology* (ed. B. Fath). https://www.sciencedirect.com/referencework/9780444641304/encyclopedia-of-ecology.

Castán Broto, V. and Baker, L. (2018). Spatial adventures in energy studies: an introduction to the special issue. *Energy Research & Social Science* 36: 1–10.

DeBoom, M.J. (2020). Climate necropolitics: Ecological civilization and the distributive geographies of extractive violence in the Anthropocene. *Annals of the American Association of Geographers* 111 (3): 900–912.

Dunlap, A., Verweijen, J., and Tornel, C. (2024). The political ecologies of "green" extractivism(s): an introduction. *Journal of Political Ecology* 31 (1): 436–463.

Farmer, J.D. and Lafond, F. (2016). How predictable is technological progress? *Research Policy 45* (3): 647–665.

Funtowicz, S.O. and Ravetz, J.R. (1993). Science for the post-normal age. *Futures* 25 (7): 739–755.

Grubert, E. and Zacarias, M. (2022). Paradigm shifts for environmental assessment of decarbonizing energy systems: emerging dominance of embodied impacts and design-oriented decision support needs. *Renewable and Sustainable Energy Reviews* 159: 112208.

Haraway, D.J. (2000). A cyborg manifesto: science, technology, and socialist-feminism in the late twentieth century. In: *Posthumanism* (ed. L.J. Nicholson), 69–84. London: Palgrave.

Harris, P.G. (1999). Common but differentiated responsibility: the Kyoto protocol and United States policy. *NYU Environmental Law Journal* **7**: 27.

Hess, D.J. (2018). Energy democracy and social movements: a multi-coalition perspective on the politics of sustainability transitions. *Energy Research & Social Science* 40: 177–189.

Huber, M. (2009). Energizing historical materialism: fossil fuels, space and the capitalist mode of production. *Geoforum* 40 (1): 105–115. https://doi.org/10.1016/j.geoforum.2008.08.004.

Huber, M. and McCarthy, J. (2017). Beyond the subterranean energy regime? Fuel, land use and the production of space. *Transactions of the Institute of British Geographers* 42 (4): 655–668. https://doi.org/10.1111/tran.12182.

International Energy Agency (2024). World Energy Outlook 2024. https://www.iea.org/reports/world-energy-outlook-2024 (accessed 24 October 2025).

IRENA (2023). Renewable Energy Statistics. https://www.irena.org/Publications/2023/Jul/Renewable-energy-statistics-2023.

Kingsland, S. (2015). Alfred J. Lotka and the origins of theoretical population ecology. *Proceedings of the National Academy of Sciences of the United States of America* 112 (31): 9493–9495.

Klinger, J.M. (2015). A historical geography of rare earth elements: from discovery to the atomic age. *The Extractive Industries and Society* **2** (3): 572–580.

Krausmann, F., Lauk, C., Haas, W., and Wiedenhofer, D. (2018). From resource extraction to outflows of wastes and emissions: the socioeconomic metabolism of the global economy, 1900–2015. *Global Environmental Change* 52: 131–140.

Le Billon, P. (2004). *The Geopolitics of Resource Wars*. Routledge.

Lee, C.C. and Yuan, Z. (2024). Impact of energy poverty on public health: a non-linear study from an international perspective. *World Development* 174: 106444.

Lefebvre, H. (1991). The production of space. In: *The People, Place and Space Reader* (ed. J.J. Gieseking, W. Mangold, C. Katz, et al.). New York: Routledge.

Lippert, I. and Sareen, S. (2023). Alleviation of energy poverty through transitions to low-carbon energy infrastructure. *Energy Research & Social Science* 100: 103087.

Marston, S.A. (2000). The social construction of scale. *Progress in Human Geography* 24 (2): 219–242.

McCarthy, J. (2015). A socioecological fix to capitalist crisis and climate change? The possibilities and limits of renewable energy. *Environment and Planning A* 47 (12): 2485–2502. https://doi.org/10.1177/0308518X15602491.

Miller, C.A. and Richter, J. (2014). Social planning for energy transitions. *Current Sustainable/Renewable Energy Reports* **1** (3): 77–84.
Mulvaney, D. (2017). Identifying the roots of Green Civil War over utility-scale solar energy projects on public lands across the American Southwest. *Journal of Land Use Science* 12 (6): 493–515.
Mumford, L. (1936). *Technics and Civilization*. University of Chicago Press.
Nadar, L. (2004). The harder path—shifting gears. *Anthropological Quarterly* 77 (4): 771–791.
Nemet, G.F. (2009). Demand-pull, technology-push, and government-led incentives for non-incremental technical change. *Research Policy* 38 (5): 700–709.
Nixon, R. (2010). Unimagined communities: Developmental refugees, megadams and monumental modernity. *New Formations* 69 (69): 62–80.
Odum, H.T. (2007). *Environment, Power, and Society for the Twenty-First Century: The Hierarchy of Energy*. Columbia University Press.
Smil, V. (2019). *Energy in Civilization*. New York: Routledge.
Smith, N. (1984). *Uneven Development: The Production of Nature Under Capitalism*. The Johns Hopkins University.
Smith, N. (1992). Contours of a spatialized politics: homeless vehicles and the production of geographical scale. *Social Text* 33: 55–81.
Sovacool, B.K., Ali, S.H., Bazilian, M. et al. (2020). Sustainable minerals and metals for a low-carbon future. *Science* 367 (6473): 30–33.
Stock, R. and Birkenholtz, T. (2021). The sun and the scythe: energy dispossessions and the agrarian question of labor in solar parks. *The Journal of Peasant Studies* 48 (5): 984–1007.
Vaheesan, S. (2024). *Democracy in Power*. University of Chicago Press.
White, L.A. (1943). Energy and the evolution of culture. *American Anthropologist* 45 (3): 335–356.

Further Reading

Aronoff, K., Battistoni, A., Cohen, D.A. et al. (2019). *A Planet To Win: Why We Need A Green New Deal*. London: Verso.
Bridge, G., Barr, S., and Bouzarovski, S. (2018). *Energy and Society: A Critical Perspective*. London: Routledge. https://doi.org/10.4324/9781351019026.
Calvert, K. (2016). From 'energy geography' to 'energy geographies': perspectives on a fertile academic borderland. *Progress in Human Geography* 40 (1): 105–125. https://doi.org/10.1177/0309132514566343.
Miller, C.A., Richter, J., and O'Leary, J. (2015). Socio-energy systems design: a policy framework for energy transitions. *Energy Research & Social Science* 6: 29–40.
Rhodes, R. (2018). *Energy: A Human History*. Simon and Schuster.
Smil, V. (2003). *Energy at the Crossroads: Global Perspectives and Uncertainties*. Cambridge, MA: MIT Press.
Webber, M.E. (2016). *Thirst for Power*. Yale University Press.

Chapter 2

Critical Energy Studies

Social science must be critical of its object. In order to be able to explain and understand social phenomena we have to evaluate them critically.

—Andrew Sayer, *Method in Social Science*

What Does It Mean to Engage with Energy Critically?

To engage with energy critically means to examine the systems, institutions, and relationships that shape how energy is produced, distributed, and consumed. *Critical energy studies* encompass a broad range of research, practice, and scholarship focused on understanding how and why energy systems change—or, in some cases, why they remain unchanged. Sites of inquiry may include places undergoing significant investment or technological innovation, as well as those caught in stasis or experiencing economic or environmental decline. These sites are shaped by power structures and embedded within complex social and cultural dynamics that affect people, other organisms, and the environment. A critical approach to energy prioritizes explanation and understanding, often by centering questions about power, inequality, and the socio-political forces that underlie energy transitions and stagnations. Power dynamics can manifest through force or more subtly through coercion via language, discourse, policy, and the many ways energy shapes politics, economics, and the distribution of society's benefits and harms. The "structures" in the phrase *power structures* refer to the social and environmental processes that constrain possibilities and make certain outcomes more likely. For instance, markets have power structures. So do families. Race, class, and gender are all expressed through power structures in which some individuals and groups hold more power than others. In *critical studies*, the emphasis on critique reflects a deep interest in uncovering the causal mechanisms behind inequity, land use change, and environmental degradation—and in analyzing how asymmetrical power structures are created, maintained, and reproduced. Research in this field often centers on the oppressed, the marginalized, and the subaltern—on places and peoples that are either valued or devalued by capital and financial interests or considered expendable. The goals of such scholarship may include influencing policy, shaping strategy, or serving as a check on power imbalances by identifying injustices and proposing alternatives.

Critical Energy Studies

To be critical is also to be reflexive about the categories we have inherited to describe and make sense of the world. This chapter introduces several themes common across various approaches to critical energy studies. We begin by distinguishing between *critical* and *acritical* studies of energy—an idea I owe to my editor, who suggested it as a useful explanatory device. We will explore both critical and acritical approaches to the study of energy, highlighting how each framework offers a distinct lens for interpreting the world. Through selected case studies, we will demonstrate the limitations of acritical approaches, particularly those that reduce social and technological change to merely economic or technical variables. As we will argue, adopting a critical approach is also a question of *ontology*—that is, of how different parts of the world are related to one another. The ways in which we organize concepts and categories in our explanations shape our understanding of the world, our methods of inquiry, and our visions for change.

What about an *acritical* study then? An acritical study, at its core, avoids the kinds of questions that reveal power dynamics or interrogate the causes of inequality, racism, or colonial legacies. These approaches often relegate such concerns to the political realm, seeking to draw a boundary between what is technical and what is political. Acritical investigations may assert that their focus is on technological innovation or cost reduction—and claim that these are neutral, apolitical endeavors. Critical analyses challenge this premise. The critical approach, in contrast, places both the technical and the political at the center of analysis, exploring how they are interconnected. It reveals how political decisions shape material outcomes and how technological systems carry embedded political values and effects. Rather than separating the political from the technical, critical analyses engage with *all* relevant processes to build more comprehensive explanations. This holistic approach enables a deeper understanding of the social, environmental, and economic forces that shape energy systems—and the power relations that sustain them.

Critical and *Acritical* Approaches to Energy, Society, and the Environment

The word critical stems from the Greek *kritikos* which means to "judge or discern." Critical approaches to understanding society and the environment aim to describe and explain power and inequity. You may have heard of "critical theory." One of the basics tenets of critical theory is to question taken-for-granted assumptions about the world and society. In the world of environmental ethics for example, we might question the continued domination of Earth's ecosystem by human activities and interests (Leiss 2023). Understanding it was not possible to dominate nature was something many cultures in human histories around the world embraced, but was lost in the Western world and by reductionist approaches to science until mainstreamed again by the likes of Murray Bookchin and Rachel Carson and many other post-World War environmental writers and philosophers (White 2003). Writers like Murray Bookchin debated the root causes of humanity's domination of the Earth. Bookchin, who identified as a social ecologist, argued that the domination of nature stemmed from the domination of humans by other humans. Others attributed it to a more general human dominance over nature, while still others pointed to patriarchy as the underlying driver (Bookchin 2022; Plumwood 1986).

At a broad level, critical studies are defined by their reflexive engagement with commonly held assumptions. For much of human history, it was widely accepted that mountains existed to be mined, and that fisheries, wetlands, and forests were merely sources of food, energy, or commodities (Smith 1996). Natural resources were seen as having value only in terms of human use or economic utility; nature was something to be dominated and exploited (Leiss 2023). It was only as humans began to develop a deeper understanding of their interactions with the natural world that this perspective began to shift. Initial conservation efforts focused on preserving resources for continued human benefit, but over time, protections were extended to species and ecosystems themselves. Today, some countries have even gone so far as to enshrine the rights of nature in their national constitutions (Laastad 2020).

Not long ago certain classes and races of humans too were treated like commodities, and in some instances still are today. A critical race theory approach centers the perspective of those who experience racism or other inhumane treatment of others to both reflect on how that history was sustained and how it still manifests in human civilization and neighborhoods, social patterns, and interactions today. Redlining would be an example of how a historic pattern of exclusion—not allowing people to buy homes, get loans, or obtain home insurance in an area because of the color of their skin—led to segregated parts of cities often near to industrial facilities, airports, or highways (Nardone et al. 2020). Many of these patterns remain with us today. What makes it *critical* is that it starts from the assumption that things like racism exists and the critique is of the structures that maintain it, contributing to anti-racist and decolonial scholarship (Puttick 2023).

Geographer Timothy Creswell describes the history of critical thought in human geography, spanning feminism to anarchism to postmodernist and black geography perspectives (Cresswell 2024). Each area is evaluating what they see against a historical context that is crucial for understanding socio-ecological processes and relations. Understanding the implications for housing affordability or gentrification in an area about to receive green infrastructure investments could be deeply rooted in historic processes, such as with redlining in Southern California and other parts of the United States not too long ago (Solórzano and Vélez 2017). Feminist perspectives are important to bring to critical studies since they put things in the context of the social relations between genders. The famed energy guru Amory Lovins called the energy path to renewable energy the "soft path," and anthropologist Margaret Mead warned Lovins that the frame of being "soft" would not be accepted by the deep currents of masculinity in American society. In this instance, taking the perspective that the world is dominated by misogyny and masculinity allows Mead to see the uptake of language and the symbols they are connected to in a more informed way than Lovins, who was more focused on the ideas. Mead was right about the influence of language in energy strategies.

Some areas of research in critical studies explore the relations between humans and nonhumans. Critical animal studies say analogous themes with the emergence of humans rethinking their interactions with other organisms and nonhumans and the slow move towards some kinds of animal rights (Weitzenfeld and Joy 2014). Some vegans (who do not eat living organisms) and vegetarians (who do not eat animals) say we should not eat meat because of the welfare of animals and their sentience (Singer 1972). In fact, some call for animal rights and liberation (Adams et al. 2011). Many places reflect social preferences to protect animals and have laws to ensure animal welfare from how we treat our pets to animals that are used in medical research. Not long ago, it was not common to believe that animals needed to be treated for their welfare; these were once ideas outside of the mainstream. Animals were once just for humans to use. But our understanding of our relations and place in the world evolved and continues to. Critical analyses are meant to be self-reflexive in this respect.

Today many question the "take-make-waste" approach to natural resource use, where a metal might be mined in one part of the world, processed in another place, and used in still another, only to eventually be disposed of in a landfill. This linear view of natural resource use raises numerous questions. Might we run out of certain metals and minerals? The common view at the time was that more stuff is better, and more consumption is more happiness. But that view was questioned, and now today we think of words like sustainability to describe the highest environmental ideal. The critique of a wasteful society led many to look for alternatives in more circular approaches to materials and energy use. But as phrases like sustainability and circular economy become mainstream, the focus of the critical work could shift. In fact, we already see in critical waste studies the questioning of these rosier depictions of circular economy, instead asking questions about social processes of valorization and value (Rathore 2020). This critical turn here will be important because wastes can be toxic and usually imply the need for worker health and safety protections.

If there were a normative dimension or ethics to critical theory, it would be the liberation of beings (humans, other species, or ecosystems) from power structures. There are even scholars that see critical theory itself in need to be "decolonized" given its roots in European traditions (Bhambra 2021). For nonhuman subjects, similar ideas apply for biocentric and ecocentric perspectives that see living organisms as more central in moral consideration.

Part of what this chapter aims to accomplish is to develop a framework for exploring key questions at the intersection of energy and society, as much as to describe how critical and social sciences understand these relationships. The social sciences play a crucial role in organizing human relationships, creating categories and classifications, and explaining how and why the world unfolds as it does. A critical social science approach actively questions every dimension of inquiry—not only the outcomes of research designs, but also the very categories and assumptions used to explain phenomena. For example, researchers have long studied inequity using metrics such as wealth or income. What critical analyses do particularly well is interrogate received ideas—challenging assumptions about objects, subjects, and the connections between them. This means unpacking terms like *renewable energy*, *coal industry*, and *solar power*, revealing their many layers and dimensions (see Box 2.1). Sometimes, critical analysis involves deconstructing categories we take for granted. Consider the categories *renewable energy* and *fossil fuels*. These classifications are useful when distinguishing between energy sources that are continuously replenished and those derived from carbon-rich materials extracted from beneath the Earth's surface. In some discourses, fossil fuels are cast as the villain, while renewable energy is positioned as the hero. This narrative is helpful in many cases—but how do we understand an energy company that simultaneously invests in both fossil fuels and renewable energy? Such cases disrupt our prior frameworks and demand an exploration of other sets of relationships. When we think about energy resources or technologies, they evoke specific images and associations. Critical inquiry challenges us to reflect on those associations, question the categories themselves, and understand the broader political, economic, and ecological contexts in which they are embedded (Jasanoff and Kim 2015).

Early interventions into critical studies sought also to counter the "sociological imperialism" of the social sciences, some of which reduced all explanations to theories of three white Western scholars, Karl Marx, Emile Durkheim, and Max Weber (Sayer 1992). Many social scientists at the time were interested in grand theories to explain the social world, but this have evolved to a more complex understanding, with the help of many people across genders, languages, and cultures. The development of more postmodern thinking, where multiple truths can exist across a situation simultaneously, started to challenge many of the tenets.

Box 2.1 Critical Versus Acritical Analyses of Solar Power

One contemporary example of an *acritical* versus critical analysis is how different people think about solar power. Solar power is the most popular technology in the world. Solar power does not generate fumes, carbon dioxide, or other pollution while generating electricity; in fact, it is one of the only stationary sources of electricity generation with no moving parts. It is also the only energy technology we can live under or integrate into our built environment. Given its attributes of being clean and renewable, solar power is questionably something we want more of. Society broadly all benefits from having more solar power, and this makes a lot of sense when we think about climate change and strategies to lower our greenhouse gas emissions. Solar power often ranks as one of the top strategies to displace fossil pollution. At the same time, solar is land-use intensive; it sometimes requires disturbance to extensive areas. Solar projects can impact wildlife, rural areas, and farmland. The critical analysis would look from the impact side or the worker perspective and try to improve. Whereas, a technology booster may shy away from critique, for fear it feeds the opposition to renewable energy. Solar technology also competes with other types of power plants in the utility-scale market, so the industry, because of competition they also want to produce a product at low cost and have a reputation to protect. "Solar PV as ethical green products are not subject to enough critical examination. Consumers, the public and even environmentally conscious minds may reproduce a commodity fetish with photovoltaics that mask socio economic relations, while crowning them with a green Halo" (Mulvaney 2019).

Critical research on energy might look to understand the working conditions at a coal mining projects, with the reasoning that exploitation of workers is widespread and common at certain types of facilities or in some parts of the world. It might ask why poverty is so pervasive in areas with so much coal resource wealth. With the rise of renewable energy, critical energy studies might look at proposed infrastructure or development projects that most everyone thinks is a good idea and supports in general. Consider a solar farm, photovoltaics are the most popular technology in the world, so we would expect most everyone likes green power from a solar farm. But a critical analysis of a project or region might point out some aspect in a case study that may undermine its claims to be green or good for communities. The project may have displaced landless people (Stock and Birkenholtz 2021).

In *Carbon Democracy*, Timothy Mitchell (2009) describes how the social and material organization of energy systems like coal in the early twentieth century made them vulnerable to labor union protest. Mitchell's analysis shows how democracy was advanced through the empowerment of labor unions alongside the development of the resource. Coal enabled labor unions to become more powerful actors and negotiate better livelihoods for the people in coal communities over time as the same industries became powerful entities in their own right. This approach differs from the analysis that sees workers simply as exploited by coal companies or the evolution of these systems as simply economic or technological.

In questions facing areas undergoing energy transitions, can we assume the transition from fossil fuels to renewable energy will involve a move away from exploitation of labor or degradation of land? If no, then what will it take to get there? When the focus is on power dynamics, there may be less change in labor relations or materials/mining relations than the technologies shift towards renewables. But these categories of renewables, fossil fuels, and nuclear power are more interconnected and tangled. The categories make sense in some context, for example efforts to move away from combustion fuels or mitigate climate change. But how do we map these onto the embodied fossil fuels that might be in renewable energy—in the plastics, semiconductors, glass, metals, and resins? How do we tell a story about an organization that may develop both renewables and fossil fuels? How do we reconcile our categories when confronted with corporations that develop renewable energy and natural gas companies? Our abstract categories need to be refined.

We are starting to see this reconsideration of the boundaries between fossil fuel industries and renewable industries emerge in the research. Instead of looking at them as distinct domains, geographers for example are using the concept of "stacked energy geographies" to explain these entanglements (Harlan and Baka 2024). Political scientists too are suggesting moving beyond viewing the energy transition as a contest between incumbent fossil fuel interests and renewables or clean energy industries and instead suggest a framework that sees these organizations as interlinked. This advance is important because it points to pathways to resolve some of the tensions with firms that see low-carbon energy transitions as existential threats with financial futures at stake, and instead opens the possibilities for a "decarbonization bargain" in public policy (Kupzok and Nahm 2024).

Critical Realism

The "critical" in critical studies also may refer to a specific orientation in the philosophy of science. Andrew Sayer is one well-known critical realist who developed ideas that inform the methods of social theory, causality—why things happen, and the multidimensionality of complicated social relations (Sayer 1992). Chaotic conception is a term popularized by Sayer by that from comes from Karl Marx in a book called the *Gundrisse* (Marx 1973) to describe how we construct ideas and hold them to be true, while in reality there are much more fuzzy and possibly fraught categories. When Marx used the term, they were referring to the concept of population, the numbers of workers needed to support surplus extraction, a key process in capitalism.

> For example, capital is nothing without wage labour, without value, money, price etc. Thus, if I were to begin with the population, this would be a chaotic conception [*Vorstellung*] of the whole, and I would then, by means of further determination, move analytically towards ever more simple concepts [*Begriff*], from the imagined concrete towards ever thinner abstractions until I had arrived at the simplest determinations. From there the journey would have to be retraced until I had finally arrived at the population again, but this time not as the chaotic conception of a whole, but as a rich totality of many determinations and relations.

From this, we can see how population is being used as an analytical device to understand the relations of labor by breaking it into smaller processes. Sayer explains the concepts just by thinking about a lecture in a classroom. We could think of the lecture as some particular event with everybody there to learn and take notes, but the reality is the lecture is an assemblage. But in reality, that is part of the reality. The lecture turns out to be a complex mixture of people and activities, some watching the lecturer, maybe taking notes; the rest are looking at their phones—maybe the professor accidentally left their ringer on and drifted to whatever the call could be about—others are looking at the clock, talking to their classmates, leaving class early, coming into class late, or thinking about what they are going to have for lunch, or more important things going on in life. The idea that there is one coherent "lecture" as we think of it abstract, a speech to be delivered in this context, and all the complexity connecting all the different actors that are involved in this particular event we are calling lecture is lost. That is just the human side; there's the computer, the projector, the lights, the microphone (it fits a big hall), the mouse under the chair in the far corner, and so on.

Sayer asks to consider the abstract and concrete to allow for more flexible understandings of circumstances. He suggests we make abstractions—the lecture—sensitive to the structure of their objects. Sayer calls this rational abstraction because it rationally focuses on causal connections. That does not deny the existence of the lecture, but allows us to see it as much more, with entanglements that stretch outside of the lecture hall. In the social world, we create our own categories. What scholars of science and technology studies tell us is that we should be self-aware and critical of even these categories. In this sense, critical realism is contrary to views of scientism, positivistic, and technocratic thinking, which are built upon systems of classification, ordering, and pattern finding. Critical realism embraces idiosyncrasy and the messiness of the real world. These topics get pretty deep into questions of ontology—how our world is connected—and epistemology, the study of how we know truth (Figure 2.1).

Another key thread in the critical analysis of energy systems is the focus on questions about technology and its role in dominating nature or other people. Andrew Feenberg's critical analysis of technology has a more nefarious understanding of the role in society, with roots in the philosophy of technology and constructivism. Feenberg's work is sometimes referred to as techno-dystopian. According to Feenberg (p. 19), "Technology is a two-sided phenomenon: on the one hand the operator, on the

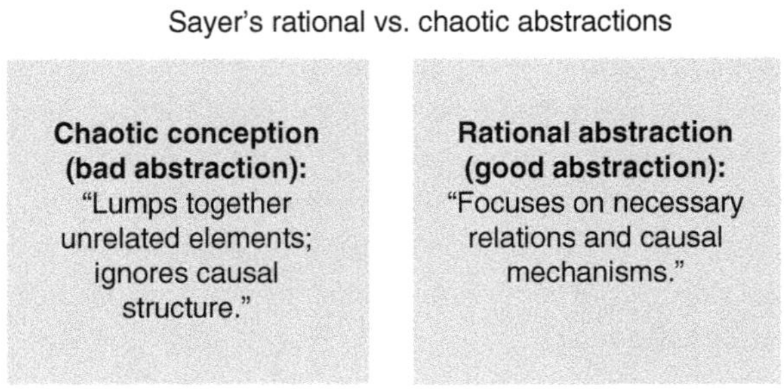

Figure 2.1 Sayer's chaotic versus rational abstractions.

other the object. Where both operator and object are human beings, technical action is an exercise of power." That is a rather different orientation than one that sees the evolution of technology inherently as progress, but it does not deny it either. It shows that these dynamics can be complicated, and reflecting on our inherited ideas can open up our understandings of stasis and change.

Philosopher Herbert Marcuse was one of the first European philosophers to introduce the idea that proposing technical solutions to social problems was morally and politically significant. It is so because they reduce the challenges, which are embedded across multiple aspects of social life to only its technological dimensions. Marcuse questions analyses of technologies that divorce them from the social relations, the socio-ecological relations, and cultural dimensions. "Moreover, it is certainly true that many philosophical concepts are mere 'foggy ideas' arising out of the domination of existence by an uncontrolled economy and, accordingly, are to be explained precisely by the material conditions of life" (Marcuse 2020).

The most instrumental accounts of technology change treat efficiency as the primary driver of technology in this or that way. But the history of technological change in energy shows these reasons to be far more complicated with factors such as social acceptance, existing relations with other technologies or mediating that evolution not just the prices alone. Instead, technologies are a part of our culture, our infrastructures, and our daily lives, which make it hard to generalize about the role of prices along an arc of technological development. Take the issue of cookstoves in developing countries to improve indoor quality. For decades programs have been designed to bring cookstoves to communities that use traditional biomass that causes significant indoor air pollution.

Tim Forsyth popularized much of the philosophy of critical realism for political ecology, bringing it into important conversations bridging political ecology and science and technology studies (Forsyth 2004). These conversations are still active sites of debates about methodology and interpretation, resting on critical realisms distinctions between the "real" and "observable" world. Critical realist approaches that underlie political ecology have, "a crucial role in critically analyzing environmental science and truth claims about ecology. It can also engage with these claims to make

Box 2.2 Science and Technology Studies—What Is Inside Black Boxes

The black box features in many case studies in the sociology of science because it powerfully exhibits the metaphor of complex mechanics not needed in detail to understand the general process of things. Take this exchange from Gregory Bateson's, Steps to An Ecology of Mind.

> F: It's a word that comes from the engineers. When they draw a diagram of a complicated machine, they use a sort of shorthand. Instead of drawing all the details, they put a box to stand for a whole bunch of parts, and they label the box with what that bunch of parts is supposed to *do*.
> D: So a "black box" is a label for what a bunch of things are supposed to do...
> F: That's right. But it is not an explanation of *how* the bunch works.

them more accurate, diverse, and socially inclusive, even if it cannot explain underlying ontological structures" (Forsyth 2023, p. 196).

The black box described in Box 2.2 hides the messy political and cultural or other dynamics. That may be fine to describe some topics, but will not be for others. The inner workings can sometimes hide social influence on decisions about how to build models, assumptions about processes, or other tasks that seem technical just through the act of "black-boxing."

Technoscience and Philosophies of Science

Critical studies of science and technology coined a term technoscience—a term taken to mean the mixing of or blurring of boundaries between science and technology, its cultural aspects, and political economy. Popularized by Donna Haraway and others in the subfield of anthropology called Science and Technology Studies, the critical nature of this concept starts from the perspective that there are power dynamics in science and technology. This is very different than analyses in say economics or business or any *acritical* analysis, which treat technology as neutral and independent. Just like a scientist must be skeptical of claims until they can be verified, the social scientist must question the very categories that are created to understand the complexity of the social world. The questions that these scholars are concerned with circulate around democracy in science. Going back to John Dewey, who noted the technical aspects of modern technological civilization are in tension with democracy because making public policy or engaging with other aspects of daily life increasingly needed to be conducted with a significant amount of expertise. Scholars such as Haraway and Sheila Jasanoff (Jasanoff and Kim 2015) bring attention to these tensions by coupling this public distance between the public and experts with the influence of political economy on matters of science and technology. These views on democracy and science do not aim to cede scientific credibility to nonscience, but instead to daylight scientific practice so it operates out of the shadows, especially on matters of importance to the public. Haraway described the need for critical analysis this way, "such a democracy does require a critical *science* politics at the national level, 'Critical' means evaluative, public, multiactor, mutiagenda, oriented to equality and heterogenous well-being" (Haraway 1997, p. 95). This is contrary to an ahistoric science that champions its narrative without questioning certain aspects. Take, for example, a technology we might celebrate because of its invention and adoption. A critical analysis might start from the positionality that the subaltern or most vulnerable are more likely to bear the consequences of these transitions. To paraphrase Haraway, for whom does [insert technology of choice] live or die? Those dynamics might be key to understanding the most important issues where policy or practice could intervene. Science and technology studies scholar Steve Fuller, on the other hand, argues that scholars have never truly been critical of science because it does not level radical enough critiques and instead focuses on description (Fuller 2000).

Bruno Latour is another analyst who has brought new thinking to our understanding of science, technology, and society. This work looks to a critical lens to emancipate to democratize the sciences. The scholar pushes back on the notion that science can be depoliticized. In fact, the effort to separate the technical from the political is a political strategy or at least has political consequences as certain kinds of claims are pushed outside the scope of consideration (Latour 2004). Research looking to overcome these challenges must look toward where their work can make contributions towards emancipatory movements in scientific practice and industry. To bring science into politics and vice versa, Latour is committed to this idea that we can reconstruct objects and subjects, humans and nonhumans, false dualisms in this view into sociotechnical networks that reveal how we might best understand socio-ecological processes to bring attention to issues central to critical studies, power asymmetries, environmental and social inequities.

The Critical Turn in Energy Studies

The growth of interest in energy studies has come alongside the growth in each of the technologies. We can point to critical studies of biofuels in the 2010s that observed how lands were being labeled wastelands and made available to biofuels companies (Baka 2011). At the time the biofuels were celebrated for their clean energy potential, and this type of analysis was key to drawing attention to many of the problematic aspects of biofuels. Today there are emerging critical energy analyses of wind power (Avila-Calero 2017), critical minerals mining (Riofrancos 2023), solar power (Stock 2023), and carbon capture (Buck 2022). These studies lead to questions about the received views towards these technologies. In the case of wind power, we hear stories of communities dispossessed of land or living in the shadows of renewable facilities with no access to electricity (Scheidel et al. 2023). With solar power, an acritical approach may assume many of the claims associated with it are all positive. We might hear of an investment from an international climate agreement and assume it to be good because it helps the cause of climate action, and shows global cooperation on major challenge that impacts countries around the world. What happens when climate finance directed by international climate accords results in the taking of agrarian lands used by peasants for their livelihoods? Unless directly looking for these considerations, who are the losers in energy transitions? It may never arise as a consideration. But clearly, if we take environmental justice seriously, these are very important observations (see Box 2.3). We need to acknowledge these consequences and develop responses that protect the most vulnerable (Sovacool et al. 2021).

Box 2.3 What Is Critical Environmental Justice?

David Naguib Pellow uses the phrase critical environmental justice to address shortcomings in environmental justice scholarship (Pellow 2025). Pellow summarized these as oversimplified analyses of race or class, single-scale or multiscalar research, questions of when to embrace versus resist dominant institutions, and why some populations continue to be expendable in the face of socio-ecological and political economic forces. A critical approach to environmental justice would bring more complexity to environmental racism or inequity by bringing to it questions of gender and expanding consideration to the nonhuman world. He argues it further requires attention to scale—something long argued by others as well—to understand the justice and multi-scalar dimensions of the global economy, to connect disparate sites and capture global and local processes simultaneously. There is also the role of the state in implementing and enforcing environmental justice, and whether other parties should be engaged in helping communities achieve environmental justice. The development of community benefits agreements or good neighbor agreements are emerging examples of a triangulation of the state, private actors, and communities (which are often not straightforward to define) in negotiation. Finally, the questions are, who is expendable? Who is to be a sacrifice zone? Pellow asks critical environmental justice research, scholarship, and activism to confront narratives of environmental justice research, scholarship, and activism to confront narratives of expendability to a framing around indispensability, which pushes back on the implicit idea of inevitability where someone says there is a necessary tradeoff that requires making another sacrifice zone.

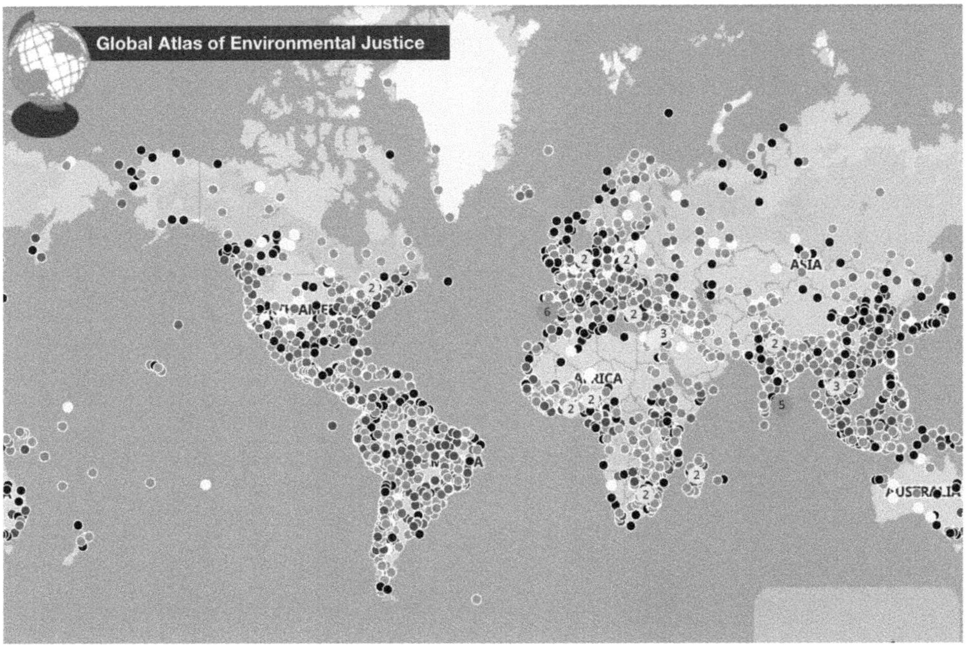

Figure 2.2 The Global EJ Atlas shows there are environmental conflicts around the world worthy of critical analyses. Source: The Global Environmental Justice Atlas/EJAtlas.

Critical studies of energy, and specifically clean and renewable energy, will continue to get more popular alongside growth in interest and advocacy on climate change and the development of low-carbon devices and clean energy infrastructures. More communities have stakes in how energy and natural resource projects develop, more profits are to be made, and the expectation is the labor conditions and environmental performance generally get better with time. Yet, there are many proposed solutions and pathways to get to a low-carbon economy. Which promise to reshape the world in various ways, with different outcomes, winners, and losers. This is why this is a "critical" introduction, not just an introduction (Figure 2.2).

Critical is meant to convey a close introspection into the convention ways we think about environmental and natural resource problems. Too often we describe environmental problems or natural resource conflicts as inevitable or as a result of indeterminate economic forces—think about how some describe hurricanes as "natural disasters." But the reality is many really tough problems that are situated, not only in politics, but in socio-material relations. Political economies, cultures, behaviors, subjectivities, prices, and views from those towards the state to the best uses of ecosystems, water, and land all play a role in explanation for phenomenon. Some of these challenges are deeply important and also require attention to relations of power. Many of the acritical analyses turn away from power relations, arguing it is "too political" or makes the situation too messy. Unfortunately, the world is a messy place. Even how we language can complicate our understandings of the world (see Box 2.4).

Box 2.4 Other Uses of "Critical" in Energy Studies

There are a few other uses of the word *critical* in the broader energy studies space that deserve a note of explanation before we move forward in the book. First, there are types of reviews we often see in the engineering literature called critical reviews. It is a type of thorough review of a life cycle assessment project. Similar to a peer-review, critical reviewers assess the data quality, assumptions, methods, analysis, and interpretation used in life cycle assessments, and there are different formats for their assembly and use. Second, you will read a lot about critical metals and minerals in the energy space because there are some energy technologies that are dependent on them, or at least they are cheaper than substitutes. Some refer to these as transition minerals, pointing to the agency implied in the adjective critical itself. These definitions are of course political economic in nature. In the United States, there are two separate assessments of critical minerals done by the USGS and the Department of Energy. One taking a global perspective on the overall availability of certain metal and mineral supplies, and the other taking a more national security focus that looks at domestic supplies and U.S. dependence on overseas supplies. Both regularly have these lists updated based on market conditions and changes in demand for minerals and metals. Criticality studies are evaluations of the availability of certain metals, minerals, or other resources needed as inputs to other processes. Finally, critical mass is the amount of nuclear material needed to sustain a nuclear reaction. Critical turns out to be a critical word to many in many different contexts!

Takeaways/Highlights

- The philosophical and methodological roots of critical studies trace back to critical realism, Marxism, and social constructionism
- A key purpose of critical studies is to critique power structures and underlying forces that drive inequity and environmental degradation
- A critical analytical approach can reshape how we classify and problems in society, these are important if we want to understand why change happens
- Critical means a few different things to people across the broader field of energy and natural resource studies

Questions and Concepts for Review

1. What is the difference between ontology and epistemology?
2. Reflect on the core tenets of critical realism, particularly its emphasis on the existence of real structures and mechanisms that may not be immediately observable but still shape observable events. How does this philosophical approach help us understand the relationship between agency and structure in social and

environmental systems? In your response, consider how critical realism differs from positivist or constructivist approaches, and discuss its relevance or usefulness in studying energy, society, or environmental change.

3. A critical approach at a very basic level aims to bring attention to taken-for-granted ideas that may be worth questioning. What are some things that society generally accepts that you think future societies might question about today's? Try to think of examples relevant to energy and natural resources.

Exercise 2.1

Take an object that represents energy—an oil well, furnace, solar panel—any object. Write down the object in the center of a piece of paper and draw lines radiating out for each social or material relationship you can think of. For example, start with a solar panel. What materials are needed for it? We might list copper, silicon, and glass. What are the social relationships? What do people think about the object? We might show low-carbon electricity, or jobs, or climate action. See how many ideas you can link to the object.

Exercise 2.2

Find a peer-reviewed journal article from the references to this chapter that looks at some place or energy source. See if your library has access to it and download the article. Read it and tell us what is "critical" about the paper. What would be the acritical version of it?

Exercise 2.3

Skim the headlines of a periodical like a newspaper or blog. Find a newspaper article that describes a current event in energy. As you read it, are there any assumptions that the author might take for granted? Who are the main antagonists in the story? What are their roles? Is there an alternative explanation?

References

Adams, C.J., Bell, A., Benton, T. et al. (2011). *Critical Theory and Animal Liberation.* Rowman & Littlefield Publishers.

Avila-Calero, S. (2017). Contesting energy transitions: wind power and conflicts in the isthmus of tehuantepec. *Journal of Political Ecology* 24 (1): 992–1012.

Baka, J. (2011). Biofuels and wasteland grabbing: how India's biofuel policy is facilitating land grabs in Tamil Nadu, India. *In Institute of Development Studies*, 6–8.

Bhambra, G.K. (2021). Decolonizing critical theory? Epistemological justice, progress, reparations. *Critical Times* 4 (1): 73–89.

Bookchin, M. (2022). *The Philosophy of Social Ecology: Essays on Dialectical Naturalism*. AK Press.

Buck, H.J. (2022). Mining the air: political ecologies of the circular carbon economy. *Environment and Planning E: Nature and Space* 5 (3): 1086–1105.

Cresswell, T. (2024). *Geographic Thought: A Critical Introduction*. Wiley.

Forsyth, T. (2004). *Critical Political Ecology: The Politics of Environmental Science*. Routledge.

Forsyth, T. (2023). Political ecology and ontology: is literal critical realism the answer? A response to Knudsen. *Journal of Political Ecology* 30 (1): 191–197.

Fuller, S. (2000). Why science studies has never been critical of science: some recent lessons on how to be a helpful nuisance and a harmless radical. *Philosophy of the Social Sciences* 30 (1): 5–32.

Haraway, D. (1997). *Modest_Witness@Second_Millennium.FemaleMan©_Meets_OncoMouse™: Feminism and Technoscience*. New York: Routledge.

Harlan, T. and Baka, J. (2024). Stacked energyscapes: conceptualizing fossil fuel and renewable energy entanglements in low-carbon transitions. *Energy Research & Social Science* 115: 103648.

Jasanoff, S. and Kim, S.-H. (2015). *Dreamscapes of Modernity: Sociotechnical Imaginaries and the Fabrication of Power*. University of Chicago Press.

Kupzok, N. and Nahm, J. (2024). The decarbonization bargain: how the decarbonizable sector shapes climate politics. *Perspectives on Politics* 22 (4): 1203–1223.

Laastad, S.G. (2020). Nature as a subject of rights? National discourses on Ecuador's constitutional rights of nature. *Forum for Development Studies* 47 (3): 401–425.

Latour, B. (2004). *Politics of Nature*. Harvard University Press.

Leiss, W. (2023). *89 The Domination of Nature: New Edition*. McGill-Queen's Press-MQUP.

Marcuse, H. (2020). *Critical Theory and Society*. Routledge.

Marx, K. (1973). *Gundrisse*. New York: Penguin Books.

Mitchell, T. (2009). Carbon democracy. *Economy and Society* 38 (3): 399–432.

Mulvaney, D. (2019). *Solar Power: Innovation, Sustainability, and Environmental Justice*. University of California Press.

Nardone, A., Chiang, J., and Corburn, J. (2020). Historic redlining and urban health today in US cities. *Environmental Justice* 13 (4): 109–119.

Pellow, D.N. (2025). *What is Critical Environmental Justice*, 2e. Medford, MA: Polity.

Plumwood, V. (1986). Ecofeminism: an overview and discussion of positions and arguments. *Australasian Journal of Philosophy* 64 (Suppl 1): 120–138.

Puttick, S. (2023). Geographical education II: anti-racist, decolonial futures. *Progress in Human Geography* 47 (6): 850–858.

Rathore, G.J.S. (2020). Circulating waste, circulating bodies? A critical review of e-waste trade. *Geoforum* 110: 180–182.

Riofrancos, T. (2023). The security–sustainability nexus: lithium onshoring in the global north. *Global Environmental Politics* 23 (1): 20–41. https://doi.org/10.1162/glep_a_00668.

Sayer, A. (1992). *Method in Social Science: A Realist Approach*. Routledge.

Scheidel, A., Sorman, A.H., Avila, S. et al. (2023). Renewables grabbing: land and resource appropriations in the global energy transition. In: *Routledge Handbook of Global Land and Resource Grabbing* (eds. A. Neef, C. Ngin, T. Moreda, and S. Mollett), 189–204. Routledge.

Singer, P. (1972). Famine, affluence, and morality. *Philosophy & Public Affairs* 1 (3): 229–243.

Smith, N. (1996). The production of nature. In: *Future Natural* (eds. J. Bird, B. Curtis, M. Mash, T. Putnam, G. Robertson, L. Tickner), 47–66. Routledge.

Solórzano, D.G. and Vélez, V.N. (2017). Using critical race spatial analysis to examine redlining in Southern California communities of color, circa 1939. In: *Critical Race Spatial Analysis* (eds. D.D. Jackson, D. Morrison, S.A. Annamma), 91–108. Routledge.

Sovacool, B.K., Turnheim, B., Hook, A. et al. (2021). Dispossessed by decarbonisation: reducing vulnerability, injustice, and inequality in the lived experience of low-carbon pathways. *World Development* 137: 105116.

Stock, R. (2023). Power for the plantationocene: solar parks as the colonial form of an energy plantation. *The Journal of Peasant Studies* 50 (1): 162–184.

Stock, R. and Birkenholtz, T. (2021). The sun and the scythe: energy dispossessions and the agrarian question of labor in solar parks. *The Journal of Peasant Studies* 48 (5): 984–1007.

Weitzenfeld, A. and Joy, M. (2014). An overview of anthropocentrism, humanism, and speciesism in critical animal theory. *Counterpoints* 448: 3–27.

White, D.F. (2003). Hierarchy, domination, nature: considering Bookchin's critical social theory. *Organization & Environment* 16 (1): 34–65.

Part 1

Resources and Society

Part 1 introduces energy not as an abstract "fuel mix," but as a set of physical resources and socio-technical systems that tie geology, land use, labor, corporate power, and state policy together. Energy infrastructures are more than technologies; they are rooted in material and political realities, showing that every major energy source comes with distinctive infrastructures, risks, and unevenly distributed harms and benefits. Across these chapters, we will learn to read energy systems as both biophysical flows (energy density, extraction rates, emissions) and social arrangements (ownership, regulation, coercion, dispossession, and environmental justice). The section builds a foundation for understanding why energy transitions are contested: they rearrange landscapes, employment, public health, and geopolitical relationships, even when they reduce greenhouse gas emissions.

Chapter 3 (Coal) frames coal as the archetypal "industrial" energy resource: abundant, transportable, and historically central to electrification, steelmaking, and the rise of high-throughput economies. The chapter highlights coal's deep entanglement with labor history and the politics of extraction—company towns, union struggles, occupational disease, and the persistent danger of underground mining. Coal also anchors a discussion of the "externalities" that define modern energy systems: air pollution, acid mine drainage, coal ash disposal, land disturbance, and cumulative health burdens often borne by working-class and marginalized communities. Coal becomes a key lens for climate politics because it is both carbon-intensive and capital-intensive: plants, rail networks, and mines lock societies into long-lived investments. That lock-in sets up end-of-era dilemmas—closures, unemployment, stranded assets, and the ethical questions of who pays for cleanup and transition support.

Chapter 4 (Oil and Gas) shifts from solid fuels to liquids and gases—and with that, to a globalized infrastructure of wells, refineries, pipelines, ports, tankers, and petrochemical supply chains. Oil is presented as more than transportation fuel: it is a foundation of modern mobility systems and material life, from asphalt and lubricants to plastics and chemical feedstocks. Gas appears as both a heating and power-generation fuel and as a contested "bridge fuel," complicated by methane leakage, local air pollution, and long-lived infrastructure investments. The chapter emphasizes geopolitical dimensions: petrostates, security doctrines, price shocks, and the ways energy abundance and dependency shape state power. It also foregrounds the social geography of refining and petrochemical production—fenceline communities, cumulative exposures, and the political struggle over regulation, monitoring, and accountability.

Chapter 5 (Bioenergy) introduces an energy category that is deceptively familiar—wood, crops, residues—yet politically complex because it sits at the boundary of energy, land, and food systems. Students are introduced to why bioenergy is controversial: carbon accounting debates (especially about time, regrowth, and indirect land-use change), biodiversity impacts, and "food vs. fuel" dynamics when arable land is diverted into energy crops. The chapter explores multiple bioenergy pathways (solid biomass, ethanol, biodiesel, biogas) and shows how each has different supply chains, labor conditions, and ecological implications. Importantly, bioenergy becomes a bridge to later concepts like carbon removal and BECCS (bioenergy with carbon capture and storage): a climate strategy that raises ethical questions about land requirements, whose land is used, and whether promised negative emissions could reinforce extractive models under a new banner.

Chapter 6 (Hydropower) treats dams as quintessential infrastructures of modernity, producing electricity while simultaneously reorganizing rivers, ecosystems, and governance. Hydropower is presented as low-carbon in operational emissions, but high-impact in social and ecological terms: displacement, altered fisheries, sediment trapping, downstream livelihood impacts, and the reconfiguration of water rights and sovereignty. The chapter draws attention to the political economy of "big dams," including how promises of development and electrification can clash with local costs and contested consent. It also develops a key analytic skill: separating *withdrawals* from *consumption* and recognizing that energy systems interact with water systems in different ways. Hydropower's dependence on hydrology makes it newly vulnerable to climate change, turning reliability and long-term planning into central concerns. Students also encounter newer framings—run-of-river projects, upgrades to existing dams, and pumped storage—as ways hydropower may evolve in a renewables-heavy grid.

Chapter 7 (Nuclear & Geothermal) pairs two low-carbon resources with very different public imaginaries. Nuclear is framed through its paradoxes: exceptional energy density and reliable generation alongside catastrophic risk perception, accident history, waste storage dilemmas, decommissioning burdens, proliferation concerns, and the legacy of uranium mining impacts on Indigenous and rural communities. The chapter invites students to see nuclear power as an institutional system as much as a technology, defined by regulation, insurance and liability structures, security protocols, and political trust. Geothermal, by contrast, is treated as a locally bounded but promising heat resource: steady generation where geology permits, with challenges including high upfront exploration risk, induced seismicity in enhanced geothermal systems, and potential water and brine management issues. Together, these resources are used to interrogate what "clean" means: low carbon does not automatically mean low impact, and risk, governance capacity, and social acceptance shape viability as much as technical potential.

Chapter 8 (Solar, Wind, Waves, and Tides) introduces renewable energy as a set of *flows*—intermittent, spatially distributed, and often modular—rather than fuel stocks that can be stored and transported easily. The chapter emphasizes how renewables shift the problem from fuel supply to infrastructure: siting, transmission, storage, balancing, and land (or ocean) use. Solar and wind are treated as mature and scaling technologies with meaningful tradeoffs: habitat disruption, visual and cultural landscape impacts, bird and bat mortality for wind, supply-chain and waste questions for panels and turbines, and the politics of permitting and community benefit. Marine renewables (waves and tides) appear as emergent—high potential in some regions, but constrained by engineering challenges, cost, ecological uncertainty, and harsh operating environments. A key throughline is that renewables do not remove politics from energy; they often intensify it by requiring new spatial negotiations over land, seabeds, and transmission corridors.

Chapter 9 (Metals and Minerals) completes Part 1 by showing that energy transition is also a materials transition. As societies electrify transportation, buildings, and industry, they shift from moving molecules (coal, oil, gas) to moving electrons—and electrons ride on copper, aluminum, silicon, steel, and a wider set of battery and magnet materials. The chapter frames the transition's "mineral intensity" as both a planning challenge and a justice challenge: more mining, more processing, more tailings and overburden, and new "resource frontiers" that bring ecosystems and communities into extractive zones. Students learn why critical minerals debates are never only technical: they involve labor conditions (including artisanal mining), state power, corporate consolidation, geopolitics, and the limits of traceability once materials enter smelters and global supply chains. The chapter also introduces mitigation pathways like recycling, substitution, design for disassembly, and certification schemes, all while emphasizing that best practices are not automatic and that "tradeoff" narratives can erase real alternatives, local rights, and consent.

Part 1 provides the foundation for explaining why energy resources are inseparable from questions of land, labor, infrastructure, and justice. The chapters collectively show that decarbonization is necessary but not sufficient: energy systems must also be evaluated through lifecycle impacts, governance capacity, and distributional consequences. Part 1 sets up the core tension that the rest of the book builds on—how to transform energy systems quickly enough to address climate change while avoiding the reproduction of extractive harm, sacrifice zones, and unequal development in new forms.

Chapter 3
Coal

If energy security means having reliable access to ample quantities of energy at an affordable price, no resource has had more impact

—Thurber, 2019

There may be no more important rock or mineral in human history than coal. As a natural resource, the materiality of coal derives from the remains of organisms living on Earth tens to hundreds of millions of years ago that left energy stored between the chemical bonds of fossil fuels. Coal has been widely used in electric power and heat in homes, power plants, and industries across history, but also in metallurgy. As an energy source, coal can be combusted, gasified, and burned, or it can even be made into liquid diesel fuels. The biophysical processes that make plant matter into coal are the composition of the carbon-containing organisms, heat, pressure, and ultimately time. These factors result in coal resources that have different qualities—lignite, subbituminous, bituminous, and anthracite, and peat that are found in only certain locations. Types of coal are geographically situated among the communities they are found in, and these qualities can affect how much communities benefit from coal development and use. 85% of coal is in Asia and North America, and about 80% of the world's coal is to make steam or coke materials, while about 20% is lignite and peat (IEA 2024a).

Coal is also important economically, socially, and culturally. The rock has provided many with livelihoods, helped strengthen labor unions, enhanced the pace of development in many regions, and is a part of the cultural history of place; think about how many songs are written about coal miners. But coal extraction has also devastated mountains, been the site of many deadly accidents with sometimes thousands of casualties, and its combustion is overloading our atmosphere with carbon. The chapter will describe coal as a resource and how these industries touch down in communities using examples from Appalachia, Wyoming's Powder River Basin, China, and India, drawing on research in geography, sociology, anthropology, and cology. The pollution from coal use—acid mine drainage, heavy metals and fly ash, and combustion emissions that cause acid rain, mercury deposition, smog, and climate change—is presented as well as policies and practices to reduce them. The chapter closes with a description of syngas, carbon capture technologies, and coalbed methane as other related energy sources.

Energy from a Shiny, Black Rock

The use of coal by humans as an energy source goes back thousands of years. People in the area that is today China used coal to make heat to stay warm over two thousand years ago. Coal is used in metallurgy too, not just as a source of heat but also as a source of carbon, say for silicon or steel. Coal resources were used to help shape iron and to burn lime by the Saxons as early as the thirteenth century. Coal came into wider use for heating in the hearths in homes in Britain in the fourteenth century as new designs for fireplaces and stoves to keep smoke out of homes became common. The steam engine and similar engine devices brought the extensive use of coal power. When coal and steam were harnessed by inventors, the pace and magnitude of Industrial Revolution rapidly intensified. Up until then industries and production processes relied on charcoal, wood, hydropower, and animal and human power to drive machines with the occasional wind gristmill or water wheel or pump. Coal was abundant in some places, literally exposed at the surface. The shiny black rock would soon become the main driver of global economic development for well over a century.

The black rock speaks to so much of human history, particularly the turn to the Industrial Revolution but even in recent years up to today (Smil 2016). Most coal deposits used for electricity production are a product of geochemical processes that occurred with carbon-containing organisms and sedimentary rock formation from between 360 and 280 million years ago, when the Earth's climate was warmer and more humid, and ecosystems were more productive and transforming solar energy into carbon-based life forms. Vegetation and other sources of carbon are sedimented upon, and after years of geotechnics and weather, starved of oxygen and decay under pressure, at high heat, and over time, to form coal. The substance that becomes coal is produced from the cellulose and lignin in dead and decaying plants. When these materials are buried in sediments, they decay to a class of materials called vitrinites, a type of kerogen that is derived from the lignin-cellulose walls of plant cells. The youngest form of coal is peat, a material produced from the compression of sedimentary layers of rock, soil, and sand that is only a few million years old or younger. Many parts of the world still use peat as a heating fuel, though it has moisture and other impurities that reduce its effectiveness and cause it to burn incompletely with high levels of pollution. Eventually, under more heat, pressure, and the patience of a few million years, peat forms lignite, often called "brown coal." Most lignite or brown coal reserves come from the Quaternary period—starting over two million years ago. The lignite is usually burned in power plants near to the mines because it is not as valuable to transport and trade in coal markets.

Further compression and time form subbituminous and bituminous coal, the most common coals. Where pressures are greatest, a very carbon-rich form of coal called anthracite is formed. Anthracite is an extremely uncommon coal type that forms where there are significant plate collisions and carbon-rich ecosystems under sediments. The anthracite mines of eastern Pennsylvania in the United States, for example, were formed when the plate that carries present-day Africa collided with the east coast of North America. The collision occurred a long time after a sedimentary basin with carbon-rich ecosystem was deposited, covered by sediments, and subject to the heat and pressures of Earth.

As coal ages and is pressurized and heated, the fixed carbon content of coal increases. Peat has the lowest carbon content per unit energy, followed by lignite, subbituminous, bituminous, and anthracite. The volatile matter and moisture content follow the opposite trend with anthracite having very little of those impurities. Anthracite with the highest carbon content is called meta-anthracite, while semi-anthracite is a bit lower in fixed carbon content. These grades also predict the energy content or calorific value of coal as bituminous coal and anthracite have the highest energy content. Energy density is a measure of how much energy is contained by some unit of mass or volume. As the carbon content increases, the heating value or energy density is reduced slightly. So high volatile bituminous coal and semi-anthracite have similar energy densities compared to the renewables discussed later in this chapter; coal has a very high energy density. Lignite resources can have densities that compare to dried dung or air-dried wood at 10–20 megajoules per kilogram (MJ/kg); bituminous coal can range from 20 to 26 MJ/kg, whereas anthracite ranges from 27 to 30 MJ/kg (Smil 1991).

Today coal is in steep decline in some parts of the world, while it continues to be widely used in others. The birthplace of coal-fired electricity, the United Kingdom, said goodbye to coal in October 2024, with the retirement of the coal-fired power station called Ratcliffe-on-Soar. Meanwhile, coal use in China and other places across the globe continues. In 2023, China accounted for 95% of new global coal-fired power plant construction activity (Lempriere 2024). Malaysia, Indonesia, Veitnam, and Thailand have targets to retire coal-fired power generation between 2045 and 2056, an area that will also see growth in per capita energy use in the coming decades (EIA 2024a). Some countries are co-firing coal with fuels such as biomass or ammonia in an effort to reduce the carbon intensity of coal combustion for electricity and heat. While most industrialized countries have plans to phase out coal, there are outliers in the global south with dependencies between countries like Turkey and Colombia that are expected to continue at least to 2050 (Cardoso and Turhan 2018). China also is increasingly committing to coal-to-chemicals to compete with other sources of materials for lucrative plastics, which could have consequences for climate action targets. 94% of China's fossil fuel resources are coal (Chengcheng and Schape 2024).

The Labor of Coal

More than just an energy source, the increased use of coal for steam-powered industry and rail also figured in the cultural imagination and the structuring of social relations and working conditions. The book *Germinal* by Emile Zola depicted the class differences and workers' organizing in the mines of Belgium in the eighteenth century. Coal mining and its products helped make a middle class. But it also benefited the most powerful interests in society. Mines, mine infrastructures, mining equipment, and machine patents enriched wealthy elites; at the same time, coal created a working class that could be exploited for their labor and time. It also accelerated the destructive forces of colonialism and natural resource depletion with railroads, lumber mills, hydraulic mining, and other tools of extraction that liberated human activities from weather-dependent resources like hydropower and wind.

With such large portions of coal fueling communities and bringing them economic prosperity, coal became part of the national imagination in many places, the subject of art, poetry, song, and language. The idiom "coals to Newcastle" of British origin describes a pointless action. At the time, the overproduction of coal and lost markets over time meant that coal could sometimes be had for next to nothing. As the phrase suggests, it would not make sense to bring coal to Newcastle because it had little to no value there. Newcastle at the time was the largest coal exporting port in the nineteenth century.

In the United States, coal is deeply woven into the history of the struggle to unionize labor. Labor and the coal industry were part of the first union contract dispute negotiated by the federal government, by President Theodore Roosevelt's administration in 1904. This significant enrollment among the unionized ranks would become a political block that would vote with labor, but in some cases would block reforms needed in the coal industry from mine safety to carbon pollution. All to say, these natural resources have constituencies that speak for them when needed, and they can be powerful political actors and effective at getting what they want.

Much of the early use of coal for steam power was from the machines developed first by Thomas Newcomen and improved upon by James Watt and others in the

Figure 3.1 Vincent Van Gough's Coal Barges, 1888. Source: Vincent van Gogh / Wikimedia Commons / Public domain.

eighteenth century. First used to pump water out of mines, coal was used to make steam and ironwork. The coal-guzzling Newcomen engine pulled steam into a cylinder, creating a partial vacuum when contacted by cold water, which allowed the higher atmospheric pressure to expand and push a piston into the cylinder. The Newcomen engine gained popularity not just because of its novelty, but because its efficiency gains made it economical to use. The Newcomen engines could do the work of 50 horses, and horses were expensive to feed and care for. The steam engine types advanced by James Watt used a separate condenser to improve the overall efficiency of the coal fuel and work done by the devices. The engine, formerly shared by Watt and Boulton, had four times better fuel efficiency, allowing the steam engine to leave the mines and begin to power facilities like textile mills. Paradoxically, Watt's focus on invention shifted to patent defense, and some argue the 25-year monopoly delayed more innovative developments in the technology.

Coal ushered in new capabilities for transportation (Figure 3.1). Coal use expanded to soon power industrial enterprises, including steam locomotives and steamships, literally changing transportation patterns by allowing ships to move beyond where they were limited by wind and railways connecting far-off places for the first time. The first locomotives were made from machines that shifted a piston back-and-forth but converted energy into rotary motion. The locomotive at the time was the first major prime mover to power machines since the windmill was developed and used to gristmill grains.

The development of coal resources transformed landscapes, economies, and social relations. Coal collieries or mines in Europe and later in the United States would be sites of some of the earliest labor organizing. Coal mining is dangerous work, requiring a mix of skilled and unskilled labor. Early mining operations employed children because they could nimbly fit into smaller places. In the U.S. anthracite mines of Pennsylvania, child laborers were referred to as "breaker boys" and they separated slate and other non-coal materials from the coal while breaking the coal into smaller pieces.

Among the many dangers of mining, including the collapse of the mine itself, are many "damps" found deep in coal mines. Choke damp or black damp is found in closed spaces like coal mines, and it is the result of oxygen being absorbed by the mines' walls and the release of water vapor and CO_2. Most deaths from the tragedy were from asphyxiation from black damp. The Hartley Colliery disaster in Northumberland killed 204 people including children when the pumping engine fell off its beam and into the mineshaft. The accident led to reforms which required two entrances to every mine shaft. A fire at a mine in Soma, Turkey in 2014 killed over 300 people from an undetected fire damp. It is common that after investigations of such accidents some sort of negligence or labor safety violation played a role in mining disasters. China in the late twentieth century had some of the worst mining disasters ever, some accidents killing thousands.

Underground mixtures of gases can be fatal under some circumstances. White damp is another dangerous gaseous mixture of mostly carbon monoxide, a blood poison. The "canary in the coal mine" expression/colloquialism was used because the animal was used to be on the lookout for white damp, because it was odorless and could cause death in the bird at very low concentrations. Fire damp, as mines got deeper, miners faced a dangerously high levels of methane. Fire damp must be burned off to ensure that catastrophic explosion did not kill or injure workers. Stink damp is mostly hydrogen sulfide, which has a characteristic rotten egg odor, sometimes experienced near hot springs or places with high levels of dissolved sulfur in groundwater. Afterdamp is the mixture of gases in a mine after a fire or explosion, which can include dust and hazardous fumes.

Collieries is another term for coal mines and in addition to worker safety, there are often safety issues posing threats to communities. These relations are made complicated often because these industries are also the economic foundation of the local economy often, making it challenging to get the appropriate level of safety best practices because of political ties between the companies and the elected or appointed officials. At older collieries, mines had piles of what might be thought of as tailings but were called colliery spoils, basically rocks and other debris that are not coal except for small coal fragments. In Wales, a colliery spill known as the Aberfan disaster in 1966 occurred when a spot that received significant amounts of rainfall suddenly turned the colliery tailings into a slurry of coal, which slid down the hillside and destroyed a school and homes, killing 116 children and 28 adults. The tragedy still affects that community over 55 years later through psychological trauma.

While coal companies pushed to weaken labor laws and regulations, if not prevent them altogether, labor unions were often leading the response to mine safety and better working conditions. The infamous Molly Maguires were an alleged secret labor society in Ireland that later emigrated to the anthracite coal region of eastern Pennsylvania. They were antagonists of labor disputes at coal mining sites where Irish immigrants were often employed. They were purportedly involved in violent acts of resistance at coal mining strikes, particularly a strike in 1875, and 20 were rounded up by private police with ties to the Reading Railroad who controlled the anthracite mines, tried, and hung in 1878 and 1879 on terrorism charges. The trails and executions led by the railroad's private army, the Pinkerton's would inhibit unionization in the anthracite fields for another 20 years or more. The Pinkerton violence against workers would be an infamous theme of capitalist social relations with its private police force from around the late 1870s through the early twentieth century.

The United States role in labor disputes changes with the coal strike in 1902. Previous U.S. Presidents sided with industry and at times used force to break up labor disputes. But in 1902, U.S. President and federal government, when faced with a strike in Pennsylvania's important anthracite mines, instead played the role of mediator in a dispute where workers demanded shorter hours and higher pay. In October 1902, the 163-day strike ended. The labor union led by major industries like coal and steel would continue to become more powerful in the next few decades before beginning a long decline in power for labor into the next century.

Coal was so important in World War II that the U.S. President Franklin Roosevelt asked the U.S. Interior Secretary Harold Ickes to seize coal mines to head off a potential labor dispute in 1943. The production of coal for steel and energy could not be disrupted given rates of military equipment and machinery destroyed in that war. Coal was an important natural resource for cheap electricity, for steam locomotives, but also critical for coke for steel, for iron, and for heating homes in many regions. A half-million coal miners went on strike in April 1943. The miners eventually won pay increases and pay for travel from the shaft to the coal face.

In part because many widespread coal mining jobs are in some places, coal mining has also been a place to see social class and racial conflicts. Dr. Rebecca Scott is a researcher of Appalachian coal mining community, and found these places to be "where white hegemony is challenged and reaffirmed" by its position both as a marginalized but idealized identity. In other words, there is some level of recognition about being marginalized by the distribution of economic benefits, but this is read through some identities in a way that reinforces racialized stereotypes about the causes of uneven development by viewing some racial identities as superior or more deserving than others (Scott 2009).

After Steam, Electricity

After steam, electricity would be the next major energy carrier put to work by coal. The idea of conduction was demonstrated by Stephen Gray in the 1740s. But it took a better understanding of how electric charges could be utilized, which started to be experimented on by scientists such as Luigi Galvani in the 1780s, with the infamous experiments passing electric current through frogs' legs, hoping to understand the origins of the electricity. Alessandro Volta in 1799 built the first real prototype of a battery, a stack known as the Volta pile. The Volta pile set off a series of scientific discoveries, including the decomposition of water into hydrogen and oxygen that we refer today as electrolysis. Volta piles were later used to isolate several elements for the first time, including barium, magnesium, strontium, boron, and calcium. These early batteries all produced direct current.

It took a better understanding of the relation between magnetism and electricity—electromagnetism induction—to see great potential for electricity use in machines. Hans Christian Ørsted, a chemist first showed that electric current produces a circular magnetic field, beginning what would become a unified concept of energy. He noticed a compass needle was deflected when electric current was passed through a Volta pile; the needle usually pointed in the direction of Earth's magnetic field, but the presence of electricity altered the needle to point in a different direction. André-Marie Ampère a French mathematician and physicist, for whom the unit of electric current is named, began to inquire about a field he termed electrodynamics. Ampere took two wires in parallel and showed the wires were attracted or deflected based on whether the currents were in the same directions or flowing in opposite directions. Ampere's law states that magnetic fields are produced by electrical currents, and the magnetic field is proportional to the electric current that coproduces it. The direction of the magnetic field is perpendicular to the flow of electric current, both of which have direction and magnitude. Later, English scientist Michael Faraday hypothesized the existence of electromagnetic fields, and popularized the terms anode, cathode, and electrode. Scottish physicist James Clerk Maxwell is widely known as one of the key players in our understandings of a unified theory of energy. This line of inquiry eventually yields our understanding of the interchangeability of matter and energy (and the eventual explanation for the release of heat with nuclear decay).

These are the basic principles behind the operation of all sorts of equipment, from transformers at electricity substations, solenoids, induction hotplates and stoves, and phone chargers. Ohm's law is a scientific idea that explains the conductive properties of materials. The law states that the current through a material across two points is directly proportional to its voltage. Investigated by Georg Ohm, the research involved trying to understand how and why an electromotive force decays across the length of a wire. These led to the designs of resistors so integral to modern electronics.

By 1870 the inventor Zenobe Gramme built the first dynamo able to power motors. Thomas Edison's Pearl Street central power station would be operating by 1882. Electricity generation from coal started to become widespread by the 1880s. Sir Charles Parsons is credited with building the first steam turbine capable of generating electricity in 1884. The basic version of an alternating current (AC) dynamo came by 1932 from Hippolyte Pixii.

By the turn of the century coal-fired power plants were on the order of 1–10 MW. 50 years later they would be proposed on the order of GWs, with the largest ever

Box 3.1 Thermal Coal and Metallurgical Coal

Thermal coal is the type of coal that is used for electricity generation and heat. Metallurgical coal is a distinct type of coal used to make pig iron, steel, and silicon metals. 87% of metallurgical coal is coking coal used in blast furnaces as a reducing agent; coal is more than just energy. It is an important raw material for the production of steel and silicon metals, which can be used to make photovoltaics. So even though many see a bleak future for coal-powered electricity generation, it will continue to be mined for making other products. As we will discover in later chapters, there are industries that depend on it today for material hardening or other reactions, including steel and silicon metals. This means their demand can be subject to broader economic dynamics such as growth and contractions in real estate development that might affect things like demand for steel

Source: Adapted from IEA (2024b).

proposed in the United States, on the Kaiparowits Plateau at 6 GW at one point. The largest coal-fired power plant ever built at 6.7 GW is the Tuoketuo Power Station located in the Inner Mongolia Autonomous Zone, China (Box 3.1).

Carboniferous Society and Climate Change

The historian Lewis Mumford noted that "carboniferous capitalism" relies on the "accumulated wealth" of carbon-based energy that took millennia to establish. The fossil fuels we rely on today are the products of long ecological and geological processes, and our use diminishes a stock of finite resources (see Box 3.2). He distinguished carboniferous capitalism from other era(s) where human civilization acquired flows of energy from the sun, mainly from wood and agricultural biomass. Mumford argued that carboniferous capitalism's "habituation to wreckage and debris" with "disregard for a balanced mode of production and consumption" would strip human civilization of the energy and other natural resources upon which it is built (Figure 3.2). For these and other reasons including its outsized greenhouse gas impact relative to energy generation, so advocate to keep coal in the ground and use other sources for electricity (see Box 3.3).

In the late nineteenth century, Swedish scientist Svante Arrhenius suggested burning fossil fuels could possibly heat the Earth's climate system through a greenhouse gas effect. In the philosophical magazine and *Journal of Science* in 1896, he wrote on the influence of carbonic acid in the air on the temperature on the ground. The release of carbon dioxide to the atmosphere goes more than into the air, including exchanges between the sea, rocks, and soils. According to the Natural Resources Defense Council,

Box 3.2 Earth's Climate System Due to Human Interventions in the Carbon Cycle

Among other landscapes polluted by coal, emissions of greenhouse gases from coal and other fossil fuels are causing severe impacts to Earth's climate system. A Scripps scientist, Charles Keeling started measuring CO_2 levels in the atmosphere on Mona Loa a volcano in Hawaii in 1958. Since the dawn of the industrial era, the carbon dioxide levels in the atmosphere have risen by 45%, meaning there are more gases capable of trapping radiation that would otherwise be remitted into space. Since 1990, the heating influence of human-produced greenhouse gases has increased 47%. Ocean heat has increased 0.58–0.78 watts per square meter from This warming is causing shifts in some ecosystems such as the Arctic, where since 1979 sea-ice cover extent has been reduced by 40%. The rate of mountain glacier loss is roughly doubling each decade, posing risks to species and communities that depend on that water. Humans will have to adapt to these circumstances, although it is not clear what will happen to ecosystems.

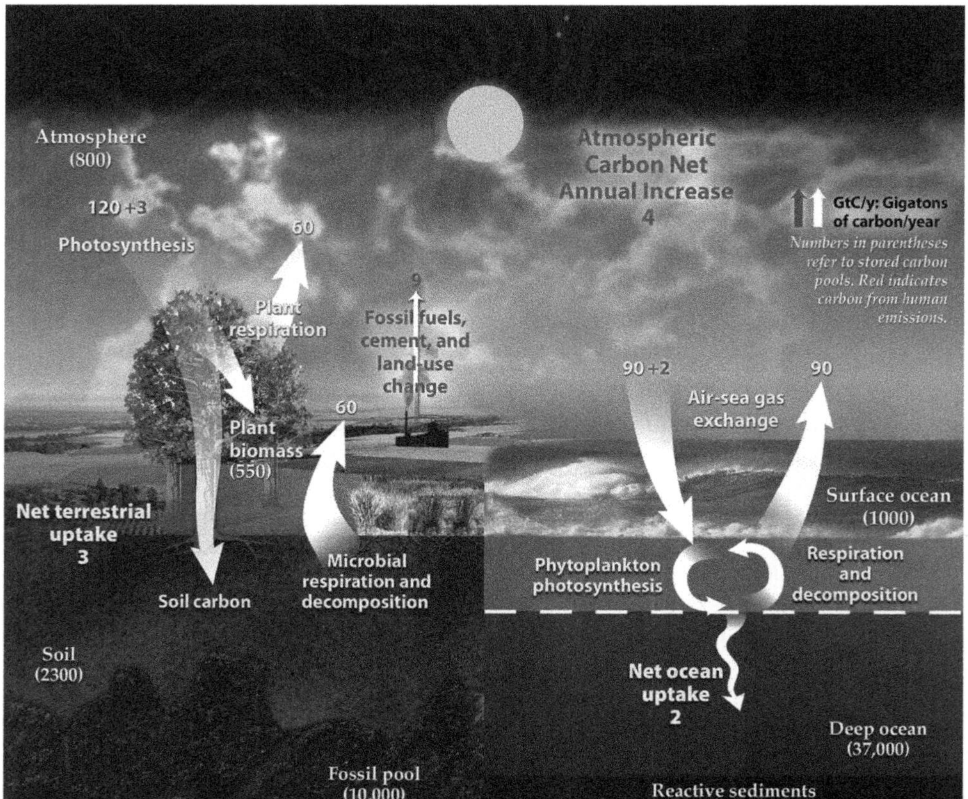

Figure 3.2 The carbon cycle. Source: United States Department of Energy / Wikimedia Commons / Public domain.

Box 3.3 Do We Need Fossil Fuel Non-Proliferation Treaty to Keep It in the Ground?

The idea for a Fossil Fuel Non-Proliferation Treaty stems from the nuclear weapons treaties signed through the 1960s, where countries agreed to stop making nuclear weapons. Given the amount of fossil fuels that countries plan to develop and that private companies plan to produce, some think it is about time we considered an agreement to phase out fossil fuel investments in infrastructures and reserve development (Newell 2021). The idea would be to create a global registry of fossil fuel reserves and infrastructures, followed by agreements to keep reserves in the ground or retire existing investments.

our oceans are 30% more acidic since the beginning of fossil fuel combustion at the beginning of the Industrial Revolution (Denchak 2018). Part of the reason is direct deposits of acid rain from nitric and sulfuric acid resulting from combustion. But the other reason is the other product of combustion, carbon dioxide dissolved in seawater, is acidic. A large portion of the increased emissions since the industrial revolution has already been absorbed.

As energy crises come and go and countries seek out ways to become more secure in terms of energy supply, there have been several efforts, some of them sustained, to make liquid transportation fuels from coal. Fischer-Tropsch coal-to-liquids was once touted as a solution to some countries' dependence on oil imports from other nations. In the 1970s, advances in the development of coal-to-liquids were seen as a national priority in the United States because coal can be converted to gasoline and diesel fuels. By the 2000s, China also dependent on petroleum imports was planning to build out coal-to-liquids projects. These projects were expensive and water intensive, so soon many of these efforts were abandoned. The process is amenable in theory to carbon capture when processed this way, but economics will ultimately dictate these possibilities. Many coal-bearing regions of the world are arid and face periodic water shortages, which also does not bode well for the water-intensive coal-to-liquids process.

Coal gasification is the process of taking coal and converting it into syngas. Syngas is short for synthetic gas and was an early fuel used in some metropolitan areas in the late nineteenth century. Gasification and coal-to-liquids are very likely the only way that coal could continue to provide electricity if carbon capture technologies could be deployed. Syngas will likely be one of the key pathways for petrochemicals, methanol, synthetic natural gas, and ammonia production in parts of the world where natural gas and petroleum are lacking. Very valuable fertilizers, polyesters, and olefins can be produced from these intermediaries. As the profitability in the fossil fuel sectors begins to shift, these will be important areas to follow in the political economy as the terrain of struggle will shift from protecting profits in fossil fuels to petrochemicals

Box 3.4 Stranded Assets and Climate Debts

Climate mitigation science suggests that 90% of proven coal and 60% of oil and gas reserves must stay in the ground in order to meet climate targets and stay within 1.5 or 2° of warming. This means that major global investments in mines, power plants, and all the pieces associated and in between may become what are called stranded assets (Lewis et al. 2014). These are investments that become obsolete before they yield a return. This includes the physical assets themselves like LNG terminals and pipelines, but also the financial assets—stocks, bonds, and pension funds tied to these industries. As energy systems evolve towards decarbonization some assets may rapidly devalue and trigger cascading financial instability alongside unemployment. The exposure to risks from stranded assets could be significant, especially for investors that fossil fuel debt and assets, including pension funds intended for retirees. This situation creates a dilemma where communities dependent on fossil fuel jobs and revenues face the loss of livelihoods due to stranded assets, while others face worsening climate harm with no compensation or support. A "climate-debt trap," where low-income countries are forced to borrow money often from wealthy nations to adapt to climate crises caused by emissions they didn't produce. Addressing both stranded assets and climate debt is essential for a **just global energy transition**. That means not only phasing out carbon-intensive assets, but also creating financial mechanisms to compensate affected workers and regions, investing in alternative livelihoods, and fulfilling climate finance commitments to the Global South for adaptation, mitigation, and loss and damage.

Source: Adapted from Semieniuk et al. (2022).

versus leaving the public with debt, stranded investments, and cleanups (see Box 3.4). Industries will tend to protect their interests so sometimes legal or policy interventions are needed to advance collective goals.

The Geography of Coal

Geographically, coal production is concentrated in a few places, but it is spread throughout the world wherever the carbon-based rocks of viable age and content are present. Despite progress towards renewable energy in 2024, a record 8.77 billion tons of coal were produced globally (IEA 2024a). The predominance of coal is produced in emerging and developing economies led by China, India, and Indonesia. 2024 coal production by major consumers include China (4,883 million tons (Mt)), India (1,245 Mt), Indonesia (775 Mt), Australia (459 Mt), the United States (386 Mt), and Australia (493 Mt) (IEA 2024b). Only a fraction of global coal production is traded globally. Regions and countries that import significant amounts of coal include China, India, Japan, and Korea, with major exporters including Indonesia, Australia, and Russia. The global distribution of coal is highly uneven with over three-quarters of coal reserves in the USA, China, Russia, India, and Australia.

In the US, there are several important coal-bearing regions, including the (1) Intermountain West, which includes the Rocky Mountain and Colorado Plateau regions, and the states of Wyoming, Montana, Colorado, Utah, New Mexico, and Arizona; (2) Appalachia, including the states of West Virginia, Kentucky, Pennsylvania, Tennessee, and Mississippi; and the (3) Interior region, where coal is scattered from the Gulf Coast to the Illinois Basin states, including Texas, Louisiana, Illinois, Michigan, Indiana, and other states to a lesser extent. The Powder River Basin surface mines in Wyoming supply up to 40% of United States coal production annually, mainly from 13 large strip mines near Gillette. With 40 billion tons of extractable coal remaining, the Powder River Basin will be the major supplier of coal to the United States for decades to come. Appalachia, with the high mine-mouth costs of mountain-top removal (MTR), and the high sulfur content in their coal, has seen a dramatic decline over the past four decades.

Coal usually requires transport from where it is mined to its site of use—the spatial question for energy we raised in Chapter 1. Throughout history humans built canals, railroads, and highways to move coal, at substantial public and private investment. The top coal power electricity-generating states in the United States are Texas, Indiana, Ohio, Illinois, and Pennsylvania. Interestingly, few of these states get much of their coal needs from their own states, as many of these plants are fed with coal from the American Interior West. Over 90% of coal is used in the electricity sector, but coal is also used in other industrial sectors, including industrial processes to make heat and steam, for coking metallurgical silicon and steel, and to a lesser extent for home and commercial heating applications. Sometimes, this coal is distinguished as metallurgical coal (Box 3.1). The United States exports a large portion of this kind of coal to other countries.

In China, a majority of coal production comes from regions in the north of the country. Most of these coal-fired power plants are located in the coastal regions; the development of coal as a cheap energy source required the development of railway infrastructure to move coal at low cost. Much like the rest of the world, China's coal industry is undergoing rapid mechanization and using few workers per unit of energy mined. By the time coal ends its reign as a major source of energy, China will have consumed about 50% of all global coal ever produced (Wang and Li 2016).

Coalbed methane or coal seam gas is an unconventional method of extracting natural gas or methane, from coal deposits by fracturing the coalbed and extracting the gas. In the 1970s coal-to-liquids became important to many countries reliant on unstable sources for their supplies of petroleum for gasoline, diesel, and heating fuels. The extraction methods for coalbed methane development require removing groundwater to allow the gas to be extracted easier, so the process creates large amounts of effluent during production. Wells are drilled, groundwater withdrawn, and the gas then begins to flow.

Terraforming Landscapes

Mining for coal transforms landscapes. From MTR to surface mining, these transformations often result in wholesale trans terraforming of landscaping to new shapes and contours. The primary means that coal is taken from the ground is by developing underground tunnels, mountain top removal, or surface or pit mining. In the United States, the major coal mining region of the world, the Powder River Basin, most of the coal is taken from surface mining. These projects look like stadiums set into the landscape, with their cutways that allow vehicles to move into the pit.

MTR is a coal mining practice using large equipment to remove topsoil, vegetation, and rock—so-called over-burden—to expose large seams of coal, made popular in West Virginia and Kentucky. Draglines are massive cranes that take the material left over from MTR, the rocks, tree debris, and other overburden. The trucks used to move coal and overburden have wheels over 10 feet in diameter, some a lot larger. Mechanized production processes are better for workers than the small underground mines. Among the more concerning impacts of MTR are severe water quality and riparian impacts. MTR begins with benches cut into the hillsides to allow access to places to set off explosives. In the Appalachia region of Eastern North America, MTR mainly occurs in mixed mesophytic forests, important seedbeds for the continental forests. Once the coal seams are revealed, they are removed to the next stage of processing. Most overburden is moved by dragline into low-lying areas or used to reconstruct contours.

According to the US EPA, a majority of industrial particulate matter and mercury are from coal. US EPA 2023. Progress Report - Emissions Reductions https://www.epa.gov/power-sector/progress-report-emissions-reductions. Lesser amounts of cadmium, antimony, dioxins, furans, and lead emissions can also occur from coal-fired generators. These power plants contribute to local air pollution known to contribute to adverse health outcomes. They also cause regional pollution. When sulfur dioxide and other sulfur and nitrogen compounds (SO_x and NO_x) emissions from burning coal are mixed with water in the hydrosphere, acid rain is formed from nitric and sulfuric acids. Weather systems can deposit acid rain far away and harm aquatic life, soils, and vegetation sensitive to pH.

Mercury pollution hot spots are usually found near either mining activities or coal-fired power plants. Mercury is a contaminant found in coal, and after combustion, it is deposited in the environment as methylmercury emitted from coal-fired generation (see Box 3.5). Heavy metals biomagnify up the food chain biogeochemically, moving from krill to krill-eating fish through piscivorous fish and ultimately the human food supply and other large mammals or living organisms. Other air pollutants from burning coal include dioxins, furans, arsenic, selenium, cadmium, and molybdenum. Pollution control equipment can significantly reduce these emissions. The best available technologies used in air pollution control for fossil fuel combustion include electrostatic precipitators, filters, scrubbers, and fluidized bed absorbers.

Fly ash is a solid waste product from burning coal that contains highly toxic heavy metals and furans. Fly ash made into a slurry is moved into ponds located on site with

Box 3.5 Mercury Methylation and Bioaccumulation

Coal impacts the carbon cycle but also plays a major role in the cycling of mercury in addition to legacy mining activities. Coal is responsible for up to 90% of global mercury emissions today. The methylation of mercury occurs when mercury changes oxidation states. These compounds can be neurotoxins and cause numerous developmental issues, as well as various cancers. Exposures to mercury can biomagnify or bioaccumulate in ecosystems through food webs. Fish eating fish, for example, have much higher levels of mercury as they eat fish who feed at lower trophic cascades. Eliminating coal power will significantly reduce these pollution exposures, although we will be living with centuries of mercury pollution into the future.

the power plant and is often held back from other water bodies by earthen dams. These ponds are linked to dredge cells where the materials are evaporated and solids settle to the bottom before being sent off for permanent disposal at a waste treatment facility or landfill. These dams have been known to break open and spill liquid fly ash into water bodies and neighborhoods. One well-known case came near the Kingston coal-fired generator in the Tennessee Valley Authority's (TVA) service area, when millions of gallons of toxic sludge—residual waste from burning coal—poured into a neighborhood and the Emory River, a tributary of the Chattanooga River (Sullivan 2024). The power plant was the largest in the world when completed in 1954, and fly ash was stored in holding cells until it broke and flooded a community on December 22, 2008. It was the largest industrial waste spill in U.S. history (Sullivan 2024). Some other uses for fly ash include bricks and wallboard, so long as the metals can be sealed or removed.

Some environmental impacts are related to inputs to the coal supply chain. 10,000 gal of a coal-cleaning chemical called 4-methylcyclohexane methanol or "crude MCHM" was spilled by Freedom Industries into West Virginia's Elk River, which led to over 300,000 people losing drinking water in the state for many weeks (Turley and Caretta 2023). Making these invisible parts of the coal production system more visible can help us see otherwise abstract ideas about impact as suggested by activist-scholars interested in understanding coal frontiers from a critical perspective (Fabricant 2021).

Most mining developments, including coal mines, require reclamation after the project is over. This does not mean full restoration of the landscape, but instead is more ensuring that the landscape does not slump or slide and ensuring that water runoff can be retained onsite, usually with a retention basin. Research by a group of political ecologists found that sites of former coal mines are important places for activists to hold mining companies accountable for cleaning up and rehabilitating old abandoned mines, using legal strategies drawing on a case study from coal fields in Kalimantan, Indonesia (Toumbourou et al. 2020). The U.S. Surface Mining Control

Figure 3.3 Mountain top removal reclamation in eastern Kentucky. In summer 2022, eastern Kentucky was hit by a historic storm with the flood stage of the Kentucky River exceeding its record height by over 7 feet.

and Reclamation Act of 1977 requires coal mines on federal land to be returned to the "approximate to the original contour" (Udall 1978). Many states in the United States require that companies make some concerted effort to return the site to its prior condition—often referred to as reclamation. Reclamation remains challenging as, often, the soils and seedbeds have been removed (Figure 3.3).

Developing Economies, Developing Coal, Just Transitions

Coal remains important to developing economies today in the form of employment and revenue generation. There is a long history of dependency on these natural resources dating back to the time of company towns. St. Claire, a small mining town in Pennsylvania, illustrates resource dependency in the nineteenth century (Wallace 1988). In some ways, taking a critical approach to coal means interrogating our ideas about coal and the communities it supports. This brings to light the tensions between phasing out coal because of its pollution while understanding the vital role that it still plays in many lives today. It also helps us understand reactions to coal transitions, including an understanding of the emergence of state-sanctioned political violence associated with these social processes (Figure 3.4) (Kumar 2024).

Another aspect of justice in global energy transition is ensuring the transitions benefit typically marginalized communities. The normative idea is that policymakers

Figure 3.4 Open-pit coal mining. Source: Beyond Coal & Gas Image Library / Flickr.

Box 3.6 Just Transitions from India to Montana

Coal-fired electricity is coming under increasing competition from natural gas, solar power, wind power, and other now cheaper sources of electricity. The question facing developing countries and rural regions generally is how to adequately supply their residents with power and industries with the energy needed to grow their economies and bring their poorest residents out of poverty. Built into international climate agreements is the notion that it is important that all countries act on climate change, but that some part of the world—namely the advanced developed economies—should play an outsized role in the advancement of clean energy and other climate solutions. The notion of "common but differentiated responsibilities" implies that India and other countries that have not yet developed as many of their resources have the opportunity to do that.

and communities should embrace a just transition, a more equitable sharing of the benefits and burdens of energy and natural resource use (see Box 3.6). Sandeep Pai and colleagues have investigated the labor situation in the coal commodity supply chain reaching back into rural India (Cha 2020; Pai et al. 2020). This research shows the importance of coal mining as a livelihood strategy and how international policies will need to take great care in ensuring that the people most affected by the loss of these mining and power generation activities are not left behind in the transition. The solution resides in developing a plan to ensure that communities facing significant disruption due to changes in energy supply chains have access to new opportunities and protection from increasing vulnerabilities. At the end of 2021, India announced it would no longer invest in coal-fired power plants. This helps explain why so much just transitions research is written about the coal sector.

Colstrip, Montana is a remote and rural community that for centuries has been dependent on coal mining and coal-fired electricity generation and at the center of just transitions conversations (Haggerty et al. 2018). Located on the northern edge of the Powder River Basin, the community is the largest single supply of coal in the United States (Roemer et al. 2021). The community depends on coal because it provides employment, associated economic activity, and tax revenues. Colstrip and other mining and coal-fired power plant communities across the Organization for Economic Cooperation and Development countries are planning for a future beyond coal in the near term. In some places public policies will encourage public investments in reskilling and the next generation of employment opportunities.

Where the coal industry goes from here is an interesting question with several important themes, such as its possible use for other materials via coal-to-chemicals, as described is happening in China. Even with combustion slated to phase out in the coming decades, the projections for other uses of coal vary, as do emissions and land use intensities (see Box 3.7). Climate change and global warming projections require projections of economic activity to inform the climate models. The IPCC is a nongovernmental body that aims to synthesize and propose actions on climate science including energy policies. Representative Concentration Pathways (RCPs) are projections of economic activity including resource development patterns. The most damaging of these climate scenarios is termed RCP 8.5. The pathway down RCP 8.5 depends on a significant growth in the amount of coal produced. In fact, researchers that have investigated these models suggest that development would need to expand many times over, something that is not borne out by projections of resource use. There

Box 3.7 Emissions and Land Use Intensity from Coal Power

There are different emission factors for various energy resources, including several types of coal. But knowing a little about the coal in question, we can estimate the greenhouse gas pollution from a power plant, and with some basic information about the resource, we can start to approximate the land use intensity of coal.

Land Use = Land use footprint of coal resource + land use footprint of infrastructure
Land use footprint of coal resource = specific density × volume of coal × recovery rate
Energy generation = power x time x capacity factor
Carbon intensity = emissions factor / energy generation

are still questions about climate feedbacks, how energy use patterns evolve, tipping points, etc., but it is generally a good thing that civilization is no longer on a realistic pattern with this scenario of coal combustion for electricity.

Takeaways/Highlights

- Coal is a key part of today's energy system for electricity and heat, and it is also a feedstock for producing steel and solar photovoltaics.
- Emissions from coal power are the most carbon-intensive of electricity sources.
- Most countries have goals to phase out coal for electricity generation.
- A critical approach to energy transitions in coal communities look at how access to the benefits and exposure to the vulnerabilities and burdens play out.
- Moving away from coal will have tremendous benefits to air and water quality, but comes with challenges as places move away from the sector as a source of economic revenue and employment.

Questions for Review

1. Why does coal have higher greenhouse gas emissions per unit energy than other energy resources including other fossil fuels?
2. Where does the global supply of coal come from? What are the leading producers and users of coal? What does the future of coal look like?
3. What kind of complications might arise from a fossil fuel non-proliferation treaty? What are the strengths and weaknesses of such a proposal?

Exercise 3.1

Estimate the lifetime greenhouse gas emissions and land use intensity from a coal-fired power plant. The plant operates 24 hours a day for 340 days of the year, for 50 years, with the occasional downtime being for maintenance or upgrades. Here are some parameters for the quality of the coal used and the operations of the coal-fire power plant.

Energy density = 25 GJ/metric ton
Conversion efficiency = 55%
Capacity factor = 340/365 days
Specific density of coal seam: 1.5 tons per m^3
Coal seam thickness = 20 m
Coal recovery rate = 97%
Area of coal mine and power plant: 1,000,000 m^2

Exercise 3.2

Explain the source and fate of pollution from coal in the environment. How do patterns of air pollution from coal-fired generation vary between carbon dioxide, sulfur dioxide, and mercury emissions? How do these scales of environmental impact speak to the types of approaches to reducing pollution that might work better for carbon dioxide and greenhouse gases than air emissions of mercury? What are the similarities and differences between the materiality of the emissions and emissions reductions, and who and how do these emissions impact which communities and environments?

Exercise 3.3

Imagine you are a policymaker in a developing country with high rates of poverty and unemployment. What strategies would you use to transition your community away from coal? What kinds of opportunities might benefit this community? How could you raise the revenues to pay for it? How would you measure its success?

References

Cardoso, A. and Turhan, E. (2018). Examining new geographies of coal: dissenting energyscapes in Colombia and Turkey. *Applied Energy* 224: 398–408.

Cha, J.M. (2020). A just transition for whom? Politics, contestation, and social identity in the disruption of coal in the Powder River Basin. *Energy Research & Social Science* 69: 101657.

Chengcheng, Q. and Schäpe, S. (2024). Analysis: China's coal-to-chemicals growth risks climate goals. *Dialogue Earth*. December 13, 2024. https://www.eco-business.com/news/analysis-chinas-coal-to-chemicals-growth-risks-climate-goals/ (accessed 8 November 2025).

Denchak, M. (2018). *Fossil Fuels: The Dirty Facts*. Natural Resources Defense Council https://www.nrdc.org/stories/fossil-fuels-dirty-facts.

Fabricant, N. (2021). Learning about coal frontiers: from the mountains of Appalachia to the streets of South Baltimore. In: *The Routledge Handbook of Critical Resource Geography* (eds. M. Himley, E. Havice, G. Valdivia), 309–318. Routledge.

Haggerty, J.H., Haggerty, M.N., Roemer, K., and Rose, J. (2018). Planning for the local impacts of coal facility closure: emerging strategies in the US West. *Resources Policy* 57: 69–80.

International Energy Agency (2024a). World Energy Outlook 2024. https://www.iea.org/reports/world-energy-outlook-2024 (accessed 8 November 2025).

International Energy Agency (2024b). Coal 2024: analysis and forecasts to 2027. https://www.iea.org/reports/coal-2024 (accessed 8 November 2025).

Kumar, M. (2024). Violent transitions: towards a political ecology of coal and hydropower in India. *Climate and Development* 16 (9): 751–761.

Lempriere, M. (2024). China responsible for 95% of new coal power construction in 2023, report says. *Carbon Brief*, April 11, 2024. https://www.carbonbrief.org/china-responsible-for-95-of-new-coal-power-construction-in-2023-report-says/#:~:text=China%20accounted%20for%2095%25%20of,the%20global%20coal%20power%20industry (accessed 8 November 2025).

Lewis, M.C., Voisin, S., Hazra, S. et al. (2014). Stranded assets, fossilised revenues. *Energy Transition & Climate Change*. Kepler Cheuvreux. April, 24.

Newell, P. (2021). The road to a fossil fuel non-proliferation treaty. *Green European Journal*. November 29, 2021. https://www.greeneuropeanjournal.eu/the-road-to-a-fossil-fuel-non-proliferation-treaty/ (accessed 8 November 2025).

Pai, S., Harrison, K., and Zerriffi, H. (2020). *A Systematic Review of the Key Elements of a Just Transition for Fossil Fuel Workers*, 196–220. Ottawa, ON: Smart Prosperity Institute.

Roemer, K., Raimi, D., and Glaser, R. (2021). Coal communities in transition: a case study of Colstrip, Montana. https://media.rff.org/documents/RFF_Report_21-01_Colstrip_Case_Study.pdf (accessed 8 November 2025).

Scott, R.R. (2009). In Appalachia and the construction of whiteness in the United States. *Sociology Compass* 3 (5): 803–810.

Semieniuk, G., Holden, P.B., Mercure, J.F. et al. (2022). Stranded fossil-fuel assets translate to major losses for investors in advanced economies. *Nature Climate Change* 12: 532–538. https://doi.org/10.1038/s41558-022-01356-y.

Smil, V. (1991). *General Energetics: Energy in the Biosphere and Civilization*. New York: Wiley.

Smil, V. (2016). Examining energy transitions: a dozen insights based on performance. *Energy Research & Social Science* 22: 194–197.

Sullivan, J. (2024). *Valley So Low. One Lawyer's Fight for Justice in the Wake of America's Great Coal Catastrophe*. Penguin/Random House.

Thurber, M.C. (2019). *Coal*. Wiley.

Toumbourou, T., Muhdar, M., Werner, T., and Bebbington, A. (2020). Political ecologies of the post-mining landscape: activism, resistance, and legal struggles over Kalimantan's coal mines. *Energy Research & Social Science* 65: 101476.

Turley, B. and Caretta, M.A. (2023). Narrative and folklore as methodologies for studying emotions, embodiment, and water during the 2014 Elk River chemical spill in West Virginia. *Social & Cultural Geography* 24 (10): 1771–1789.

Udall, S. (1978). H.R.2 - 95th Congress (1977–1978): Surface Mining Control and Reclamation Act of 1977. (1977, August 3). https://www.congress.gov/bill/95th-congress/house-bill/2.

Wallace, A.F. (1988). *St. Clair, a Nineteenth-Century Coal Town's Experience with a Disaster-Prone Industry*. Ithica: Cornell University Press.

Wang, Q. and Li, R. (2016). Journey to burning half of global coal: Trajectory and drivers of China's coal use. *Renewable and Sustainable Energy Reviews 58*: 341–346.

Further Reading

Bridge, G., Barr, S., Bouzarovski, S. et al. (2018). *Energy and Society: A Critical Perspective*. London: Routledge https://doi.org/10.4324/9781351019026.

Della Bosca, H. and Gillespie, J. (2018). The coal story: generational coal mining communities and strategies of energy transition in Australia. *Energy Policy* 120: 734–740.

Eldridge, E.R. (2015). The continuum of coal violence and post-coal possibilities in the Appalachian South. *Journal of Political Ecology* 22 (1): 279–298.

Goodell, J. (2006). Big Coal: the dirty secret behind America's energy future. New York, Houghton Mifflin.

Lewis, M.C., Voisin, S., Hazra, S. et al. (2014). Stranded assets, fossilised revenues. *Energy Transition & Climate Change*. Kepler Cheuvreux. April, 24.

Marsden, B. (2002). *Watt's Perfect Engine: Steam and the Age of Invention*. Columbia University Press.

Pai, S. and Carr-Wilson, S. (2018). *Total Transition: The Human Side of the Renewable Energy Revolution*. Rocky Mountain Books Ltd.

Pai, S., Harrison, K., and Zerriffi, H. (2020). *A Systematic Review of the Key Elements of a Just Transition for Fossil Fuel Workers*. Ottawa, ON: Smart Prosperity Institute.

Rhodes, R. (2018). *Energy: A Human History*. Simon and Schuster.

Roemer, K.F. and Haggerty, J.H. (2021). Coal communities and the US energy transition: a policy corridors assessment. *Energy Policy* 151: 112112.

Wallace, A.F. (1988). *St. Clair, a Nineteenth-Century Coal Town's Experience with a Disaster-Prone Industry*. Cornell University Press.

Yeh, E.T. (2015). *Political Ecology in and of China. The International Handbook of Political Ecology*. Edward Elgar.

Chapter 4

Petroleum and Natural Gas

With improvement of the internal combustion engine a vast new source of power was opened up, fully equal to the old coal beds in importance, even if doomed to be consumed at a possibly more rapid rate.

—Lewis Mumford, 1934, Technics and Civilization

Fossil fuels are energy-dense resources. Petroleum is used globally for transportation in part because it is easier to move liquid or gaseous fossil fuels than solid ones. Large oil tankers and pipelines have moved oil around the world for a century. Natural gas is one of the world's fastest-growing energy resources used in a large number of applications from heat to electric power. Developments nearly 50 years ago in hydraulic fracturing, pipelines, and liquified natural gas transportation have seen the growth in natural gas displace much of the coal-fired electricity. This chapter will describe the life cycle of natural gas production and the various end-use sectors. Hydraulic fracturing practices have raised questions about potential impacts to freshwater, specifically groundwater and drinking water. Water quality can be affected by methane migration into drinking water or heavy metal leaching. Groundwater risks are posed by damaged well casings or by damaged or unlined wastewater pits. Where fracking is for oil, methane is often vented or flared, but is increasingly captured for energy.

Necessary ingredients for fracking include sand, which have led to mining impacts in areas with these resources. Other industrial materials that are made from natural gas include ammonia-based fertilizers and plastics. The social challenges with the natural gas system start at the fenceline communities, exposed to water and air pollution risks, and the boom-and-bust cycle, under-investments in the community that go with it. Natural gas geopolitics is also a growing area of international dispute and rivalry.

Global Transportation Runs on Petroleum

Despite the excitement surrounding emerging trends in electric vehicles, especially in places with high rates of adoption, most global transportation is still fueled by petroleum products. Petroleum—or crude oil—is the raw material used to produce a variety of fuel products that keep the world moving. The fuels include gasoline, diesel, bunker fuels for shipping, jet and aviation fuels, as well as widespread materials like plastics and asphalt—all are derived from petroleum or natural gas.

This chapter describes the physical properties and geographies of petroleum and its many products. Readers from Europe may refer to some of these final products as "petro"—but petroleum refers specifically to the primary energy resource extracted from the Earth. In addition to motor gasoline and diesel, petroleum is refined into distillate fuel oil, kerosene, liquefied petroleum gas (LPG), petroleum coke, still gas, and residual fuel oil. Each of these end products has a distinct chemical formulation and end uses. Consequently, each fuel and its associated supply chains have unique geographies, social dynamics, emissions profiles, and uses and demands, refining processes, and amounts of primary energy used. Oil and gas, since their inception, have been tuned into feedstock for energy sources and fuels such as kerosene, gasoline, diesel, and jet fuel, as well as essential industrial products like chemical feedstocks, plastics, solvents, and lubricants. The variety of petroleum grades found on the Earth, along with the wide array of products derived from them, underscores the complexity and versatility of this global energy resource.

One characteristic that made petroleum easier to transport compared relative to gaseous and solid fuels is its liquidity, allowing it to be pumped through pipelines or carried over long distances by tanker ships. But this property contributed to some of the largest and costliest environmental disasters in history. Oil spills can be very difficult to stop and contain once they start, and they are notoriously difficult to clean up. Some ecosystems affected by major oil spills have not fully recovered.

Petroleum pollution from oil spills was a catalyst in the rise of the environmental movement in North America and Europe. Major oil spills include the UK, Santa Barbara in 1969, Mexico, the 1989 Exxon Valdez spill, and the 2011 BP Deepwater Horizon spill in the Gulf of Mexico. These events caused severe and long-lasting damage to wildlife and water quality in sensitive marine and coastal habitats. Major oil spills, especially in deep water or remote places, can be difficult to manage, clean up, and remediate.

Petroleum fuel combustion takes place in billions of small engines around the world. These smaller sources are a major contributor to air pollution, as they are often subject to less stringent emissions standards than, for example, new automobiles manufactured in the same year. In some power plants and generators, oil is burned directly to generate electricity. Diesel fuel is also commonly used to "prime" or start natural gas power plants or cogeneration facilities—especially in regions such as the Middle East, Caribbean, and Central America.

For instance, Jamaica relies on oil for over 90% of its electricity, and the US state of Hawai'i generates more than 75% of its electricity by burning oil. These figures are likely to change rapidly, particularly in these relatively sunny regions, as renewable energy sources like solar become more viable. Diesel is also widely used in backup generators for hospitals, large buildings, corporate campuses, and data centers. Even in tech-forward regions like Silicon Valley, California, diesel generators remain prevalent, with several gigawatts (GWs) of capacity approved by the state public utilities commission in 2020. Environmental justice is a key concern, as pollution from refineries often disproportionately affects fenceline communities. Moreover, communities rich in petroleum resources do not always benefit from them—a phenomenon known as the "resource curse," and in the context of oil, often referred to as the "curse of the black gold" (Watts 2008) (Figure 4.1).

Many consider the first major oil strike to be "Drake's Folly" in 1859, in Titusville, Pennsylvania. This first successful commercial oil production operation set into motion an oil rush in the region, which soon ushered in a new era of fuels. In the early years of the petroleum industry, the heavier components of crude oil were widely used as lubricants for machinery, while the lighter fractions were used for lighting and heating. The development and perfection of the combustion and diesel engines

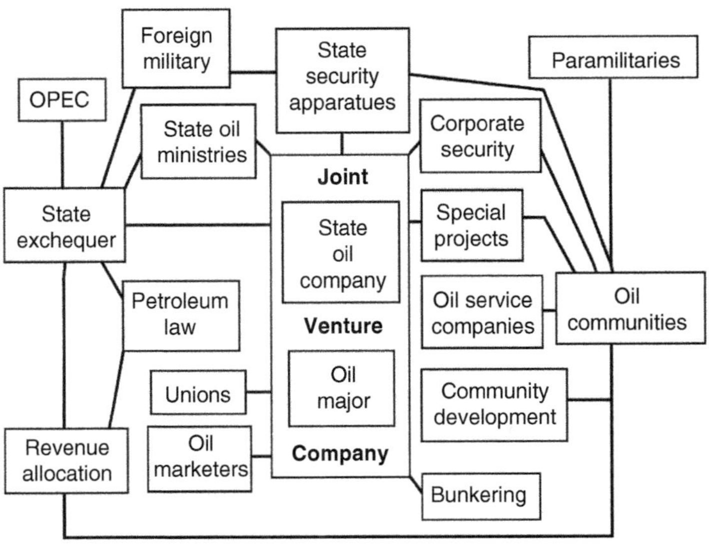

Figure 4.1 The political economy of oil.

opened up new markets for these lighter fractions of crude oil (Mumford 1934). By 1892, gasoline was in widespread use, especially with the rise of the burgeoning automobile industry.

During both twentieth-century World Wars, fossil fuels—particularly petroleum-derived fuels—became essential to military success, powering aviation, mechanized transport, and the synthesis of explosives. In fact, the primary product extracted and sold from petroleum in the early 1900s was kerosene, often called "coal oil." Much of the early Rockefeller fortune was built on Standard Oil's kerosene profits, which became a popular fuel for lamps, replacing whale oil and ethanol (see Box 4.3).

By 1970, crude oil supplied nearly half of the world's total energy: 43% in the United States and 50% in the European Union (Smil 2017). That same decade witnessed a series of oil crises, characterized by price spikes, shortages, and heightened geopolitical and military tensions around the protection of petroleum resources and infrastructure. These events triggered calls to transition away from oil, driven largely by concerns over energy security. Oil production and pricing were strongly influenced by decisions made by the Organization of the Petroleum Exporting Countries (OPEC), prompting growing concern that national foreign policies were increasingly shaped by the need to secure energy supplies, particularly petroleum. Some scholars have argued that national dependence on petroleum has led to military conflict and, in some cases, served to uphold global power structures or maintain empire (Harvey 2003; Klare 2003).

A shift from petroleum fuels to electrification could reshape global demand for metals and minerals, as discussed in Chapter 8. The US Department of Energy's Energy Information Administration (EIA) projects that the use of petroleum in transportation will decline by 2050. However, petroleum—along with natural gas—is still projected to remain one of the most widely consumed energy resources. Climate modelers and energy transition analysts argue that much steeper reductions in both petroleum and natural gas production are necessary unless significant decreases in the carbon intensity of these fuels can be achieved. Even scientists at prominent fossil fuel companies called attention to the issue (see Box 4.2).

While oil industry leaders frequently reference the "energy transition," there tends to be less attention paid to oil and gas workers compared to those in coal or natural gas power plant sectors. Yet, oil and gas production shares a number of skillsets that could be transferable to renewable energy industries. For instance, geothermal energy development relies on expertise in underground geology, drilling, well casing construction, turbine operation, and wastewater management—all of which overlap with skills found in the oil and gas workforce. With modest retraining, it is possible that many oil and gas workers could transition into roles in geothermal, wind, or solar industries. These workforce transitions could and should be pursued in parallel with broader decarbonization efforts.

The geopolitical dimensions of oil and gas have been well documented for over a century (see Box 4.1 for the latest). In the early twentieth century, Great Britain transitioned its naval fleet from coal to oil, seeking greater speed and tactical agility.

Box 4.1 War in Ukraine and Energy Security for Europe

If you were following the news in the winter of 2022, you were likely struck by the shocking war in Ukraine. But what are the connections between energy supplies and the crisis in Ukraine? What are the potential spillover effects and broader implications for energy development, supply chains, and global energy security?

Russia is widely considered an energy superpower. It is a major exporter of natural gas and oil, as well as numerous other key commodities such as nickel, cobalt, aluminum, gold, and copper. These materials are critical not only for fossil fuel-based systems but also for the production and deployment of renewable energy technologies.

In early 2022, following decades of warnings and discussions about the strategic energy corridors through Ukraine and their implications for European energy security, Russia invaded Ukraine. For many years, Russia has been a primary energy supplier to Europe, a region with limited domestic fossil fuel resources—particularly petroleum and natural gas. In the 1990s, around 90% of Russian natural gas exports to Europe transited through Ukraine. Today, the European Union imports nearly 95% of its energy resources, and countries like Germany and its immediate neighbors are especially dependent on Russian energy for heating, electricity, and transportation.

Some argue that these international exchanges—where energy flows in one direction and money flows in the other—can help preserve peace by creating economic interdependence and stability during times of political or social turmoil. However, this relationship is complex. In resource-based economies, especially in the petroleum sector, rising prices tend to benefit oil companies and petro-states disproportionately. Natural gas, due to its more limited and inflexible distribution networks, can be more easily weaponized. It is essential for heating homes and buildings in Europe, producing fertilizers, and generating electricity.

Energy scholar Cullen Hendrix (2017) argues that violent conflict initiation is more common among oil-exporting states and that there is a correlation between oil wealth and armed conflict. High revenues from petroleum exports may embolden states to pursue aggressive international agendas. In thin markets—where buyers have few supplier alternatives, exporting nations are somewhat shielded from the economic consequences of conflict. As Rupert Russell discusses in *Price Wars*, volatility in energy prices is not just an economic concern—it can trigger instability and upheaval on a global scale.

This shift, however, significantly increased the country's dependence on petroleum imports—much of which had to be sourced from beyond its own borders, despite later discoveries in the North Sea. In contrast, when the navy was powered by coal, such geopolitical concerns were minimal, as coal was abundant and domestically accessible.

Oil is also a political actor in its own right. The prevailing wisdom holds that economic dependence on cheap oil arises from cabals of powerful political elites, the collective imperatives of imperialism, or even rampant conspicuous consumption. In these dominant narratives, oil takes on agency—it becomes a force that drives wealth and poverty, boom and bust cycles, and shapes both freedom and authoritarianism. Geographer Matthew Huber (2013) offers a compelling framework for understanding the deep cultural entrenchment of oil in American life. He challenges the notion that US oil addiction is simply about the substance itself, arguing instead that oil should be seen as a socio-ecological relation. Drawing on Marx's concept of productive forces, Gramsci's theory of hegemony, and Foucault's analysis of micro-politics and subject formation, Huber shows how oil permeates daily life through embodied practices that are central to social reproduction—the activities and relations necessary to sustain life and return workers to work.

"Without oil, life, family, and everything else could not be reproduced," Huber writes. This individualistic reliance on oil ties energy prices to a form of cheap energy populism, one that often harbors authoritarian tendencies. When energy prices rise, they disrupt the foundations of social reproduction, threatening the perceived freedom and personal autonomy essential to daily life. These tensions are visible in today's cultural and political conflicts. By the 1980s and 1990s, concerns about running out of oil had largely faded. There was plenty of petroleum—it had simply become more expensive to extract, refine, and deliver. However, by the 2000s, energy prices had re-emerged as a major political issue in the United States. This time, public outrage over high energy costs increasingly clashed with rising concerns about climate change and the moral implications of foreign conflicts—often framed as "blood for oil."

A World of Petroleum Pollution

Powering the world's transportation systems with petroleum has resulted in extensive air, water, and soil pollution. Our transportation-related investment decisions have also significantly impacted the climate system. Every stage of petroleum's life cycle—development, extraction, transportation, storage, and use—contributes to environmental degradation, with air pollution often being the most visible and measurable impact on the landscape. At each step in the production and use of petroleum, noxious emissions of primary air pollutants are common. However, the most harmful emissions typically occur at the point of combustion—what's often referred to as "end-of-pipe" pollution.

Environmental damage is also sometimes a direct consequence of war. During the Iraq War, for example, over 700 oil wells and storage facilities were set on fire as the Iraqi army retreated from the Rumaila oil field—a contested area that contributed to the escalation of the conflict. The Iraqi military also dug trenches, filled them with oil, and set them ablaze. The resulting black carbon plumes darkened the skies over Kuwait and accounted for nearly 50% of all oil fires in the history of the petroleum industry. These fires, which began in February 1991 and were not fully extinguished until November of that year, consumed an estimated one billion barrels of oil—making

Table 4.1 Notable global historic oil spills and fires.

Oil spill	Year	Volume—barrels	Tons
Lakeview Gusher	1910–1911	9,000,000	1,200,000
Union Oil, Santa Barbara	1969	80,000–100,000	
Amoco Cadiz, France, Brittany	1978	1,635,000	223,000
Ixtik (Ixtoc), Gulf of Mexico	1979	1,381,000–3,300,000	
South Africa, Saldana Bay	1983	1,848,000	252,000
Exxon Valdez	1989	275,000–750,000	35,065–103,896
Persian Gulf, Kuwait fires	1991	1,000,000,000	
Angola, offshore	1991	1,907,000	260,000
Uzbekistan	1992	2,090,000	285,000
Kolva River Spill, Russia	1994	2,000,000	
Deepwater Horizon	2010	4,300,000	

it one of the most catastrophic examples of war-induced environmental destruction (Table 4.1).

The 1969 Union Oil spill, caused by a blowout on a platform near Santa Barbara, California, brought national attention to the offshore oil and gas industry and to environmental policy more broadly (Clarke and Hemphill 2002). The spill remains the most extensive and damaging in California's history and ranks third in the US by volume and mass loss. Union Oil Company's platform, located six miles off the Southern California coastline, suffered a massive blowout that released approximately 220,000 gal of oil into the Pacific Ocean, contaminating over 800 mile^2 of ocean. The oil slick spread from the platform to the northern Channel Islands.

The disaster dominated headlines for weeks and received widespread coverage across television and radio, both nationally and internationally. Many suggest that the spill helped catalyze the creation of the US Environmental Protection Agency (EPA) as part of a broader push to prevent future environmental disasters. Around the same time, Congress passed major environmental legislation, including the Clean Air Act, Clean Water Act, Endangered Species Act, and Marine Mammal Protection Act—laws that have since shaped how oil and gas damage assessments and regulatory requirements are handled. States also took action where possible. In California, some oil production takes place in federal waters, while other operations fall within the state's coastal zone. Areas under state control have seen efforts by local governments to reclaim regulatory authority. After years of legal disputes, the city of Santa Barbara reached a compromise with oil companies to ban drilling within a certain distance of the city.

As new petroleum resources are located in deeper waters and more remote areas, new risks emerge. The Deepwater Horizon disaster, for example, released over four million barrels of oil into the Gulf of Mexico, severely impacting tourism, fisheries, and other coastal livelihoods (Figure 4.2). Oil cleanup is a complex, multistep process that typically begins with wildlife rescue and containment. Oil is often removed by skimming the contaminated water surface. In some cases, more extreme methods like "in situ burning"—burning the oil on the water's surface—are used.

Figure 4.2 Deepwater Horizon offshore drilling platform on fire. Source: Ideum - ideas + media / Flickr.

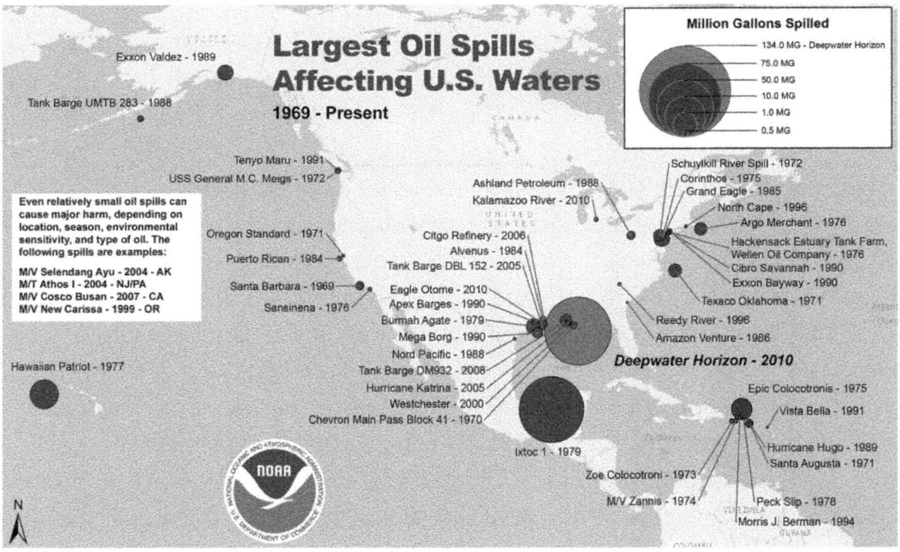

Figure 4.3 Major oil spill affecting United States represented by volume. Source: NOAA 2024. https://www.noaa.gov/education/resource-collections/ocean-coasts/oil-spills.

Chemical dispersants, such as those employed during the BP spill, may also be applied to break down oil, but studies (e.g., Beyer et al. 2016) suggest that these chemicals can create new problems deeper in the water column.

In US coastal states like California and Florida, new drilling is prohibited within three miles of the shoreline. Figure 4.3 shows some of the major oil spills in the United States. In California, the State Lands Commission has jurisdiction over tidelands and waters extending 3 miles offshore. Currently, there are 23 oil platforms in federal waters off California and 4 in state waters, meaning that the majority fall under federal oversight.

California has a diverse geological landscape and a long history of oil extraction, both offshore and onshore. The state's first known use of hydraulic fracturing (fracking) dates back to 1953. The Santa Barbara Channel is considered one of the most geologically favorable regions for offshore fracking, containing abundant fossil fuel reserves alongside exceptionally rich marine biodiversity. The southern California coast also hosted one of the world's first offshore platforms at the Summerland Oil Field. Even small, persistent leaks—from pipelines, transfer points, or transport vehicles—contribute to cumulative pollution in coastal waters. These diffuse sources add up over time, reinforcing the need for robust monitoring and regulation (Box 4.2).

Criteria air pollutants are one of the most significant impacts today from combustion emissions from coal, oil, natural gas, and even many types of biomass, as we will discuss in the next chapter. The public health costs of ground-level ozone pollution are enormous, estimated in the tens of billions of dollars. Particulate matter from everything from fossil fuels to biomass to wildfires is a leading cause of heart disease and pulmonary disease. Leaded gasoline was used as an anti-knocking agent, as it was an important air pollutant to track, especially in urban areas in the twentieth century. Exposure to lead can cause developmental problems, especially in children, still a persistent problem, for example, with drinking water in many communities. Lead

Box 4.2 Deny, Deceive, Delay, Exxon Lied. Did the Oil and Gas Industry Know About the Potential Impacts to Climate from Fossil Fuels?

"Exxon Lied." That's the claim made by advocates pushing for a transition away from fossil fuels, who argue that the oil supermajor engaged in a decades-long campaign of denial, deception, and delay regarding climate action. The allegation is that Exxon's own scientists had a clear understanding of the planetary consequences of burning large quantities of carbon-based fossil fuels—but that these findings were deliberately withheld from the public. By preventing the disclosure of this critical information, the company is accused of delaying meaningful responses to climate change, ultimately benefiting executives and shareholders at the expense of global environmental well-being. In 2021, executives from ExxonMobil and other major oil companies were called to testify before the US Congress about whether their internal scientific research had revealed the impact of fossil fuel emissions on the atmosphere. In a striking example of irony, Humble Oil—a predecessor of Exxon—once ran an advertisement boasting that its fuel had enough energy to melt a glacier. That marketing message, once meant to celebrate energy power, is now a reminder in light of contemporary climate challenges.

exposures were more widespread and severe when lead was in more commonly used products from fuel to paint to lawn fertilizer.

As oil and gas development expands into more remote and sensitive areas—both on land and offshore—it increasingly comes into conflict with ecologically important habitats. A prominent example is the ongoing debate over the Arctic National Wildlife Refuge (ANWR), a 19.3-million-acre protected area in northeastern Alaska, located above the Arctic Circle. Established in 1960 and managed by the US Fish and Wildlife Service, ANWR was created to preserve its unique ecosystems, which include tundra, mountains, and coastal plains. The region is renowned for its thousand-mile caribou migrations, particularly the Porcupine caribou herd, which relies on the coastal plain—known as the 1,002 Area—for calving and post-calving habitat. Several rivers within ANWR are also designated as Wild and Scenic, reflecting their ecological and recreational value. The central controversy concerns whether oil and gas development should be permitted in the 1,002 area, a 1.5-million-acre section of the coastal plain believed to contain significant petroleum reserves. Advocates argue that tapping this resource could enhance US energy security, while opponents warn of severe consequences for local wildlife, including caribou, polar bears, and migratory birds, as well as for Indigenous communities that rely on these ecosystems for cultural and subsistence practices (Schlosser 2006) (Figure 4.4).

Figure 4.4 The Arctic National Wildlife Refuge contains the largest calving grounds for the Porcupine caribou herd, the longest-migrating terrestrial mammal in North America. It regularly figures in US political debates over whether to open the area to oil and gas development. Source: U.S. Fish & Wildlife Service / Flickr.

Oil and Gas Booms and Busts

Throughout its history, petroleum extraction has often been associated with boom-and-bust cycles in towns and regions. A community discovers a valuable natural resource, and its development brings an influx of workers, revenue, and investment to support public services, infrastructure, and economic activity. However, over time, something typically disrupts this growth: a new technology, the emergence of a cheaper or more competitive resource, or the depletion of the original resource as it passes peak production and enters permanent decline. In many cases, the rapid pace of development leads to unsustainable growth, straining public services, housing availability, infrastructure, and social cohesion.

Uneasy but not unexpected alliances between the oil industry and labor unions have also shaped the political landscape around energy transitions. Historically, some labor unions have opposed environmental regulations out of concern for job security and economic stability. However, many scholars argue that these alliances could be crucial in ensuring that workers are fairly compensated and supported through future transitions. Just Transitions scholar Mijin Cha (2024) emphasizes the significance of class and racial politics, the role of labor unions, and the need to center the interests of working-class and marginalized communities. She advocates for a movement rooted in global solidarity, combining climate action with environmental justice, and addressing the potential loss of public revenues from fossil fuel-dependent industries that fund social services at the city and state levels.

This requires starting with a focus on labor organizing, empowering workers more broadly, pursuing community development, promoting economic diversification, and exploring public ownership models to confront the structural tendencies of capitalism. Geographer Matthew Huber (2022) usefully distinguishes among different forms of capital—extractive, industrial, electricity, financial, and green capital—each with distinct interests and influences. He highlights the growing power of the rentier class (those who own and extract income from capital) in shaping political-economic outcomes. Because much of this capital operates on a global scale, Huber argues that international solidarity among working-class movements is essential to confront climate change and push for a more just energy transition (Box 4.3).

Ultimately, oil and gas are finite natural resources, which leads many to ask whether global supplies are running out. In the early 1970s, oil production in Texas peaked and began a steady decline. Around the same time, overall US oil production also began to fall. It was not until the advent of horizontal drilling and hydraulic fracturing (fracking) in the early twenty-first century that production in these areas rebounded dramatically. As of 2025, the United States is considered a global fossil fuel superpower, leading the world in both oil and natural gas production.

The concept of "peak oil," often associated with Hubbert's Peak theory, posited that after a certain point in time, oil production would enter an irreversible decline due to geological limits, with costs rising as easier-to-access reserves were depleted (Inman 2016). According to this view, oil would become increasingly expensive and

Box 4.3 The Seven Sisters

The "Seven Sisters" were the seven major Western (primarily Anglo-American) oil companies that dominated global oil production and distribution from the 1920s through the 1970s. Their dominance was largely a legacy of the breakup of Standard Oil and the consolidation of a few other powerful multinational oil firms. The term *Seven Sisters* was popularized by Enrico Mattei, head of Italy's state oil company ENI, to describe these companies' oligopolistic control over the world's petroleum reserves during the early to mid-twentieth century. These companies formed a powerful consortium that coordinated pricing and access to global oil resources, particularly in the Middle East, and they worked together to maintain control over oil supplies in the face of rising challenges, such as the nationalization of energy resources by host countries. Over time, their dominance declined due to the rise of state-owned oil companies, the formation of OPEC in 1960, and growing national control over oil reserves in countries such as: Brazil (Petrobras), Malaysia (Petronas), Russia (Gazprom), Iran (National Iranian Oil Company, NIOC), and especially Saudi Arabia (Saudi Aramco).

The original Seven Sisters were Anglo-Iranian Oil Company (now BP), Royal Dutch Shell, Standard Oil of New Jersey (which became Exxon, later ExxonMobil), Standard Oil of New York (Mobil, merged with Exxon in 1999), Standard Oil of California (Chevron), Gulf Oil (merged with Chevron in 1984), and Texaco (merged with Chevron in 2001). By the 1980s, the landscape had shifted toward state-owned oil companies and new oil majors, as well as an increase in commodity trading firms that reshaped oil markets.

In recent years, the term "New Seven Sisters" has been used to describe the most powerful state-owned oil and gas companies—mostly outside the Organization for Economic Cooperation and Development nations—that now control a significant share of global energy resources. These include Saudi Aramco (Saudi Arabia), Gazprom (Russia), CNPC (China National Petroleum Corporation—China), NIOC (Iran), PDVSA (Venezuela), Petrobras (Brazil), and Petronas (Malaysia). Collectively, these companies are estimated to control about one-third of global oil production and nearly one-third of the world's oil and gas reserves. Unlike their private-sector predecessors, these firms are closely tied to national governments and often play key roles in geopolitics, energy security, and state finance.

scarce, triggering a long-term decline in supply. Today, however, most geologists and energy analysts agree that a physical shortage of petroleum is unlikely in the near term. Instead, the challenge lies in the economic and environmental costs of extracting oil from more difficult and expensive sources—such as deepwater offshore fields, Arctic reserves, and unconventional shale formations. It is also notable that the rate of new oil and gas field discoveries peaked in the 1960s and has generally declined since. Recent increases in US oil production have largely resulted from the commercial viability of shale and tight rock formation resources, made possible by technological advancements and favorable market conditions beginning in the mid-2000s.

Although often treated as a homogeneous commodity, petroleum exists in a variety of types, each with distinct physical and chemical properties. Two key characteristics that determine how crude oil is classified and processed are viscosity and sulfur content. These properties are typically described using the terms light versus heavy (viscosity) and sweet versus sour (sulfur content).

Light crude oil is less viscous, flows more easily, and is generally easier and less expensive to refine into high-value products like gasoline and diesel. Whereas sweet crude oil has a low sulfur content, making it less corrosive and less environmentally damaging when burned. Conversely, heavy and sour crudes are thicker and/or contain more sulfur, which makes them more challenging and energy-intensive to process. Sour crude, in particular, can corrode refinery equipment if not properly treated, requiring specialized facilities or upgraded infrastructure.

Some refineries are equipped to handle sour or heavy crude, but the light, sweet combination remains the most desirable, especially for gasoline production. Petroleum is also traded through global benchmark markets, which serve as price references for different grades of crude oil. The three most widely used benchmarks are as follows:

- West Texas Intermediate (WTI)—A light, sweet crude primarily produced in the US and delivered to Cushing, Oklahoma.
- Brent Crude—Sourced from the North Sea; slightly heavier and less sweet than WTI, but still considered light and sweet.
- Dubai Crude—A heavier, sour crude used as a pricing benchmark for Middle Eastern exports to Asia.
- Urals Crude (Russia) is another benchmark for sour, heavier oil, mainly exported to Europe.

These benchmarks represent crude with access to key infrastructure, such as pipelines, storage terminals, and shipping ports, and their prices are widely quoted in financial news and commodity indexes. Light, sweet crude is very valuable, as are natural gas condensates. If production of light, sweet crude is disrupted—due to repairs, accidents, or geopolitical conflicts—gasoline prices often rise. This is because light, sweet crude is more efficiently refined into gasoline, and its absence increases the energy and cost required to meet demand using heavier, more sulfur-rich alternatives.

Conventional petroleum production involves drilling a well, constructing a well pad, and eventually installing auxiliary equipment such as pumps, tanks, and separators. Natural gas—particularly methane—is often co-produced with crude oil, along with natural gas liquids (NGLs) like ethane, propane, and butane. When this associated gas cannot be captured or utilized economically, it is often flared—burned off at the wellhead. Flaring contributes to localized air pollution, particularly in large oil fields with multiple flares or in areas prone to temperature inversions, which can trap pollutants near the surface. There is considerable global disparity in flaring practices. The United States, for instance, has made progress in reducing flaring intensity in some oil-producing regions, though significant flaring still occurs in areas like the Permian Basin. By contrast, countries such as Russia and Nigeria have historically exhibited higher levels of flaring, though improvements have been made. In Nigeria, for example, flaring declined from approximately 25 billion cubic meters (bcm) in 2000 to 7 bcm in 2020, according to the World Bank (2021).

Oil fields naturally decline in output by about 3–6% per year. As pressure drops and easily recoverable oil is depleted, production becomes increasingly expensive unless secondary or tertiary recovery methods—such as water injection or carbon dioxide (CO_2) flooding—are used. These are forms of enhanced oil recovery (EOR), which, along with hydraulic fracturing (fracking), are classified as unconventional techniques. One notable case is the Bakken Shale in North Dakota (not South Dakota), where fracking enabled previously uneconomic shale oil to be extracted, triggering a dramatic boom-and-bust cycle in oil production. Originally, shale oil was considered unrecoverable due to its tight geological formations, but advances in drilling and completion technologies shifted this calculus. These cycles raise ongoing questions: What happens if oil prices fall during a period of scarcity? Will it accelerate the transition to alternative energy, or make future oil development less attractive due to high upfront costs that may not be recovered?

Moving away from petroleum will be challenging—but more so in some countries than others. Why are certain nations especially dependent on petroleum? In the case of the United States, high oil consumption is largely due to a transportation system dominated by light-duty vehicles and freight trucks. The country lacks the extensive public transit systems and high-speed rail networks found in regions like Europe and Japan, which use significantly less energy per capita to move people and goods regionally. In Europe, the widespread use of diesel-powered vehicles—which have higher thermal efficiency than gasoline-powered internal combustion engines—has historically contributed to lower fuel consumption. Diesel engines typically convert around 35–40% of the fuel's energy into motion, compared to about 20–25% for gasoline engines, meaning less energy is wasted as heat. However, this advantage has come under increased scrutiny due to air quality concerns and nitrogen oxide (NO_X) emissions, especially in urban areas.

Electrification is accelerating in many European countries. Norway, for example, is leading the transition: by 2025, it plans to end the sale of new internal combustion engine vehicles, aiming for 100% zero-emission vehicle sales. That transition has been made possible through a combination of strong policy incentives, public investment in charging infrastructure, and high fuel taxes. In contrast, the slow pace of reducing petroleum dependence in US transportation has been linked to decades of weak or stagnant fuel economy regulations. In the US, these are known as Corporate Average Fuel Economy (CAFE) standards. Periods of regulatory inaction—along with lobbying by automakers—have led to long stretches in which vehicle fleet efficiency barely improved, resulting in minimal gains in miles per gallon. While vehicle electrification is now gaining momentum in the US, progress may still be gradual, due in part to the slow turnover of the global vehicle fleet. It can take 15–20 years or more for a vehicle fleet to fully transition, meaning that petroleum will remain a major transportation fuel for the foreseeable future.

Human Rights and Petrochemicals Industries

For as profitable as the natural gas, petroleum, and petrochemicals industries are, these industries have long been connected to human rights abuses. Geographer Michael Watts (2001) has examined questions around the involvement of Royal Dutch Shell in the 1995 execution of poet laureate Ken Saro-Wiwa and other activists working on behalf of the Ogoni people in Nigeria against oil development in Ogoniland. The story here is more complicated than the simple resource curse narrative. The Royal Dutch Shell had been operating in Nigeria since the 1950s, but the country saw coups and regular changes in military control. The resulting governance resulted in large revenues from oil production, but a limited return of benefits to the Ogoni people. Shell distanced itself by saying the responsibility of distributing the benefits rests with the Nigerian government (Holzer 2007). Similar poor human rights relations have been uncovered with Chevron (at the time Texaco) and their treatment of lands and Indigenous peoples in Ecuador. Notoriously, a pipeline spill at the Lago Agrio oil field resulted in contaminated drinking water with polycyclic aromatic hydrocarbons from nearly 900 unlined open wastewater pits impacting Indigenous peoples and wildlife (Kimerling 1990). These problems are not just confined to publicly traded companies. Indigenous groups in Ecuador have asserted that their own government has auctioned off the mineral rights and land access to state-owned companies in China. Complicating these tensions are questions about fuel subsidies and energy access. See Boxes 4.4 and 4.5 for examples of how costs of fuel can make or break accessibily, but also how cheap energy is entrenched in the economy and society, making it harder to transition.

The dynamics of global oil production are heavily influenced the OPEC, which includes member countries Saudi Arabia, Kuwait, Algeria, Angola, Congo, Ecuador, Equatorial Guinea, Gabon, Iran, Iraq, Libya, Nigeria, Qatar, Saudi Arabia, United Arab Emirates, and Venezuela. The cartel operates by ensuring that prices and production outputs are set to achieve favorable outcomes for members. Saudi Arabia is often referred to by oil economy experts as a swing producer because they can move production up or down on short notice because of a significant amount of oil production capacity, giving it great control over the marginal price of a barrel of crude oil. But recent years have seen some diminishment in the power of OPEC and the rise of the US and natural gas industries over petroleum ones. The Permian basin, which crosses from west Texas to southeastern New Mexico, is considered the second most productive oil field in the world. In 2022, Saudi Arabia and OPEC's disagreement with

Box 4.4 Kazakhstan and the Elimination of Fuel Subsidies

The nation of Kazakhstan covers a vast area, nearly the size of Canada. It has long been championed as the most successful post-Soviet countries in Central Asia. Kazakhstan has vast oil, gas, coal, and supplies 40% of the supply of uranium, so it is an important primary energy supplier to the world. Yet, only 155 people account for 55% of the country's total wealth in a country where people make less than $100 per month (Kantchev 2022). The country cut subsidies for liquified petroleum fuel in early 2022, resulting in escalating energy prices and soon after violent clashes in the streets of the nation's largest city. The nation's natural resource wealth does not trickle down to its citizens, and the results have been disastrous. Some protesters pointed out that global energy prices were up, meaning the government and elites should have had more revenues to work with. The example shows that the idea of eliminating subsidies can face a backlash, especially if high energy prices undermine people's livelihood, well-being, or sense of fairness.

Box 4.5 The "Iron Law" of Climate Policy

Political scientist Roger Pielke Jr. (2010) suggests that there is an "iron law" with climate policy that says, "When policies on emissions reductions collide with policies focused on economic growth, **economic growth will win out every time**." There is no more important area of energy policy to watch the iron law's effect than oil and natural gas prices, particularly as oil use for gasoline and home heating fuels, and natural gas for heating and electricity, remains. These are both critically important to modern economies. We have seen time and again social disruption caused by a lack of access to energy, including the overthrow of entire governments. The yellow vest movement in France was a protest movement that attempted to shut down roads carrying fuels and other products with rising prices in response to increased diesel, gasoline, and home heating fuel costs. Chants of "drill, baby, drill ..." in the United States began to take off in the 2010s, also a time of rising fuel prices and costs of living. There is a presumed right to access to energy that can become a flashpoint of social protest when the prices of energy are too high, which should give pause to climate policy makers to ensure that customers and consumers are not underexposed to price shocks, which could lead to political backlash that undermines climate efforts.

the US over the production quotas showed possible hints of a new configuration of oil production and consumption with OPEC, Russia, and China more deeply connected through the oil trade. In 2016, Russia and OPEC agreed to production quotas, with some referring to the group as OPEC+. The implications of this include moving some global trade in oil done outside the dollar, long a force stabilizing the value of US currency. But Russian oil assets are also not as flexible to operate as Saudi Arabia, for example, Russian oil pipelines would freeze with production stoppages that are used in other places with warmer climates, where oil can be bottled up in pipelines temporarily. Russia also has influence over Iran, an OPEC member that is at odds with Saudi Arabia (Laron 2023).

With Texas the leading oil producer in the US, California ranks third after North Dakota in 2022. In addition to opening up new resources, the introduction of hydraulic fracturing has helped restore hundreds of old oil and natural gas wells, adding to the tens of thousands of active oil and natural gas wells today. California oil is viscous and can be produced by steam injection, a particularly energy- and carbon-intensive extraction technique. Oil and gas production contributes to California's economic activity and employs hundreds of thousands. The Monterey Formation, located below much of the southern half of California, contains nearly 24 billion barrels of oil. Technological advancement with hydraulic fracturing and drilling may make these resources more easily extracted. California's contradiction between oil and gas production, while being a major renewables producer, offers a microcosm of the influence of economic factors shaping climate policy. Similar contradictions can be seen in countries like Norway.

Globally, there are a handful of major players in offshore oil production. Norway successfully utilizes its oil wealth to support social services. Not all governance is so well managed; West African countries like Angola and Nigeria are examples where the benefits are less evenly distributed with many getting few benefits despite living with the public health risks of living among oil infrastructures. As pressure to develop fossil fuel and deep-sea resources in international waters grows, it will be critical that these pursuits abide within the framework set out by the United Nations framework called the Law of the Sea. This international treaty establishes where economic zones extend into the ocean for countries to make sovereign decisions about whether to preserve or develop the natural resources of the sea floor and seabed. Exclusion economic zones usually extend sovereignty 200 miles in the sea.

Natural Gas

Natural gas remains a critical energy resource in many parts of the world, and will continue to do so for the foreseeable future, although there are strong tides pushing back to keep natural gas energy in the ground. Natural gas, the resource taken from the Earth, is called wet natural gas. It is called wet natural gas in part because it contains not only water but also other very important hydrocarbons that are used throughout society as feedstocks for chemical industries, fertilizers, etc. Wet natural gas also contains sulfur and other contaminants that need to be removed from the methane, ethane, butane, propane, etc. Natural gas, the fuel sold to customers, is also called dry natural gas. Dry natural gas has the other hydrocarbons separated and any impurities removed. When it's sold, it also contains a small amount of mercaptan, which is an additive that gives natural gas odor, which lets us know if there are any leaks.

Natural gas has also been central to the boom-and-bust cycles discussed earlier in this chapter. The widespread development of hydraulic fracturing (fracking) in the United States has generated significant tax revenues and created new jobs, both directly in shale gas extraction and in related support industries. As a result of this surge in production, natural gas prices in the US have fallen dramatically, making shale gas a dominant fuel for electricity generation. In addition to its domestic use, US shale gas is now poised to play a major role in liquefied natural gas (LNG) exports, particularly to markets in Europe and Asia. However, while shale gas development continues to yield substantial volumes, these resources are not unlimited. Over time, production will become more expensive, as drillers must access deeper formations and lower-quality reservoirs, increasing the cost and complexity of extraction.

The major areas for growth in the oil and gas industries are petrochemical production, led by ethane, LPG, and naphtha (IEA 2021). The manufacture of plastics of one of the more profitable enterprises of the oil and gas industries, so in addition to new sources of electricity and fuels, there is an effort to replace petrochemical feedstocks. We will explore these sectors in more detail in a later chapter. The issue of petrochemicals is also an end-of-life waste issue. Over eight million tons of plastic waste enter our oceans every year, much of which is single-use plastics. Plastics are seen as a major growth area for the oil and gas industries in the coming decades.

Natural gas liquids are the hydrocarbons contained in wet natural gas besides methane, including ethane, propane, butane, and natural gasoline; ethane and propane are about 80% of overall production. These are made into both final fuels and feedstocks for plastics and fertilizers. NGLs are also a byproduct of oil production and processing. The US is the number one producer of NGLs, while leading importers in 2022 include Canada, China, and Japan.

Natural gas liquids are not to be confused with LPG, but instead are a mixture of butane, propane, and isobutane—this is what folks might be most familiar with as it is sold commercially as "propane." Comprised of several gases, propane is an important product found in wet natural gas. It is widely used as a heating and cooking source of energy. Most commonly, smaller tanks are bought over the counter, while larger tanks might be refilled by a delivery truck. Propane is widely used in warehouses to power folklifts and other warehouse equipment. Propane is also widely used in agricultural farm and food processing equipment. Propane is used to dry crops like rice, a critical step in the processing of foodstuffs for storage of something eaten by billions of people on the Earth.

Tar Sands, Oil Sands, and Oil Shale

There are a few very unconventional fossil fuel resources that also provide or may provide some of our energy supplies. Oil sands are a combination of sand, clay, water, and bitumen. Bitumen is a kind of hydrocarbon and can be used for any products of oil and for gasoline. Nearly 75% of the tar sands on Earth are in Alberta, Canada. Producing and processing tar sands oil creates roughly 14% more greenhouse gas emissions than regular conventional oil. There are many controversies over using tar sands as a main energy source since it creates high amounts of carbon dioxide emissions. Some refer to oil sands as "tar sands." The oil and gas industry prefers oil sands to emphasize the resource, whereas most environmental organizations refer to it as tar sands because it conveys some of the properties of the viscous material that can be more challenging to clean up.

Tar sands are at the center of the controversy over the Keystone XL pipeline. For every gallon of gasoline produced by tar sands, about 5 gal of freshwater are consumed during the extracting, upgrading, and refining process. When tar sands are mined on the surface, wastewater is stored in evaporation ponds. Tar sands are only viable as a resource for making gasoline at relatively high prices. The production of fuels from tar sands will always be a water- and greenhouse gas–intensive even compared to other fossil fuel resources. Tar sands are also geographically limited by challenges associated with moving them around. Whether by pipeline or train, moving tar sands is a hazardous proposition, and few will welcome infrastructures that activists have nicknamed "bomb trains."

The Keystone Pipelines takes oil from Canada to the US in four phases. The first phase takes petroleum from Alberta, Canada, through the Dakotas, Nebraska, and ends in Illinois. The second phase of the pipeline starts at the connection in Nebraska and then ends in Cushing, Oklahoma, where large volumes of oil are stored before moving them to refineries in Texas along the third phase. First proposed in 2008, the fourth phase—The Keystone XL Pipeline—is more controversial. This phase connects Hardisty, Alberta, Canada, through Montana, South Dakota, and eventually the rest of the system in Steele City, Nebraska. The purpose of the pipeline, according to project developer TransCanada, is to provide additional oil supply and create much-needed jobs for Americans, and increase tax revenues. One area of contestation is that the pipeline would run over sandhills in Nebraska, which are interconnected by the Ogallala Aquifer. Many farmers do not want this pipeline because they worry about how a tar sands spill might impact water quality, which is critical to the farm economy there. A leak could contaminate the aquifer that so many depend on. In summer 2020, Keystone XL was dealt a major legal setback.

Another controversial pipeline in recent years is the approximately 1,200-mile pipeline from North Dakota to Illinois refineries called the Dakota Access line. In this case, in addition to environmental and climate action groups, many indigenous persons opposed the Dakota Access pipeline (DAPL) for reasons related to not just water quality but cultural resources, prior informed consent, and injustices related to a settler-colonial type (Whyte 2017). The pipeline needed to cross under the tribes water resource, Lake Oahe. Seeing this conflict in a settler-colonial frame allows for a decentering of procedural injustices as they are reflected in environmental review or formal consultations, where disagreements span only the minutiae of specific project elements or design features, allowing for broader introspection to zoom out and see situations as part of a larger colonial history of broken treaties, land grabs, and systematic racism. Much like the tar sands pipeline, DAPL was ordered to shut down, dealing a hard-fought victory to indigenous activists and their climate justice collaborators. But the pause was only temporary. The key permit to cross the water body is still lacking and will likely be decided on by 2025.

Oil shale is a viscous energetic fluid sometimes called kerogen oil. It is a different natural resource than the previously described shale oil, like that obtained from the Bakken or Permian Basin shale oils. Oil *shale* is mostly a distinct material niche or pilot or experimental extraction technology in a few places on the planet. The US supply of oil shale is mostly found in Wyoming, Utah, and Montana, and this represents most of the oil shale on the planet, although some oil shale is found in Russia, Brazil, and the Democratic Republic of Congo.

The process most commonly used to extract oil shale is called retorting, which essentially boils water to extract the fuels in the mixtures of water, kerogen, clay, and sand. This means the extraction process requires large amounts of water, which is especially problematic in the dry areas where it occurs. Oil shale attracted a lot of research and development support during the energy crises of the 1970s. It is very unlikely this technology will be developed for gasoline, given the direction of climate policy and the high cost of development. Yet, shale gas was once thought to be uneconomical. Now, it's a leading primary source of energy.

New Horizons for Oil and Gas: Hydrogen, Carbon Capture, Net Zero

So what do critical analyses of oil and gas industries have to offer compared to *acritical* accounts of these industries? These are the studies that focus on the power relations and inequities produced through the development of these resources. Geographer Anna Zalik (2009) shows through case studies of oil production off the coast of West African and also on the Mexico side of the Gulf of Mexico about how zones of exclusion are created to favor corporate extractive industries over other users. But how some users were able to recoup remunerations from the state in the case of Mexico.

Anthropologist James Ferguson (2005) commented on the extensive amount of foreign direct investment in mineral extraction, notably oil in Africa, and riffing off James Scott (2020), said that to understand patterns of underdevelopment, it is important to "see like an oil company." That is the skepticism toward the oil and gas industries that critical studies invite (Battersby 2024). On the horizon remain questions about the oil and gas industries' roles in schemes that perpetuate operations or defer mitigations or claim to aspire to "net zero" such as direct air capture, carbon capture and sequestration (Buck et al. 2024), hydrogen (Vezzoni 2024), or carbon offsets and biofuels, as we turn to next.

One important product derived from oil and gas production is hydrogen, which plays a growing role in energy and industrial systems. There are several methods for producing hydrogen, but as of 2022, the dominant process globally is steam methane reforming (SMR), which accounts for approximately 95% of hydrogen production—particularly in countries with abundant natural gas. SMR is a chemical process in which methane (CH_4) from natural gas reacts with steam (H_2O) at high temperatures (typically 700–1,000 °C) in the presence of a catalyst, most commonly nickel. The reaction produces hydrogen gas (H_2) and carbon monoxide (CO), followed by a water-gas shift reaction that converts CO and steam into additional hydrogen and carbon dioxide (CO_2):

$$CH_4 + H_2O \rightarrow CO + 3H_2$$

Water gas shift reaction

$$CO + H_2O \rightarrow CO_2 + H_2$$

$$CH_4 + 2H_2O \rightarrow CO_2 + 4H_2$$

Precious metals like palladium or platinum may be used in alternative catalytic processes or in hydrogen purification steps, but nickel is the standard catalyst for SMR due to its cost-effectiveness. Despite its dominance, SMR is carbon-intensive, which is why interest is growing in low-carbon hydrogen production, such as electrolysis (powered by renewable electricity) or SMR with carbon capture and storage (CCS)—often referred to as blue hydrogen.

Hydrogen can also be made via electricity from water in a process called electrolysis. In fact, if hydrogen can be affordably produced from seawater, the supply of hydrogen would be unending, given the volume of the ocean on the planet. Imagine a world where offshore wind turbines make this so-called "green hydrogen" reaction, which not only requires some kind of catalyst (often platinum), but it could also utilize new catalyst materials such as nickel silicide. As platinum can be extremely expensive, finding cheaper replacement materials or ways to use less can go a long way.

Hydrogen is also an important feedstock for some renewable energy technologies, such as photovoltaics, where hydrogen is used in the manufacturing and distillation of polysilicon. Box 4.6 describes the coloring scheme for different hydrogen sources and processing.

There is a bit of a chick-and-egg problem with hydrogen in that the infrastructure and devices need hydrogen fuel, but investments in greening that fuel are waiting for industry to scale more. Development of green hydrogen may be hastened by the development of blue hydrogen, and some would even include grey hydrogen here. The German government has developed a policy to eschew blue hydrogen, while the Canadian government prefers to include fossil fuel sources that have carbon capture to prevent most of the greenhouse gas emissions. The irony here is once again natural gas is being used alongside the bridge fuel narrative. As we will see in Chapter 9, producing green hydrogen via electrolysis requires pipeline infrastructures and

Box 4.6 Hydrogen Color Schemes

Hydrogen production is often classified using a color-coded scheme that distinguishes different production methods and associated emissions. While these color labels are not standardized globally and come with many caveats, they offer a useful way to introduce the origins and environmental impacts of various hydrogen supply chains.

Gray hydrogen—This is the most common form of hydrogen production today. It is made primarily through SMR of natural gas, though oil and coal can also be used. In this process, methane (CH_4) reacts with steam (H_2O) to produce hydrogen (H_2) and carbon dioxide (CO_2). The CO_2 is released into the atmosphere, contributing to greenhouse gas emissions.

Blue hydrogen—Like gray hydrogen, blue hydrogen is also produced via SMR, but with an important difference: CCS technology is used to capture and store the CO_2 emissions. The captured carbon is typically transported via pipeline and stored underground or used in industrial applications. While cleaner than gray hydrogen, blue hydrogen still relies on fossil fuels, and CCS efficiency varies.

Production involves methane derived from oil, coal, or gas and the capture of the carbon emissions.

Green hydrogen—This is produced through electrolysis, a process that splits water (H_2O) into hydrogen and oxygen using electricity from renewable energy sources, such as wind, solar, or hydropower. Green hydrogen is carbon-free if both the electricity and water inputs are sustainably sourced. It is often considered the gold standard for clean hydrogen.

Red or pink hydrogen—Produced via electrolysis powered by nuclear energy. The label varies: *pink hydrogen* is the more commonly used term in international discourse, while *red hydrogen* is sometimes used interchangeably or regionally. This method produces zero emissions during production and can offer a base-load hydrogen supply, depending on the nuclear infrastructure.

Turquoise hydrogen—Made through methane pyrolysis, which splits methane into hydrogen and solid carbon using thermal energy. If the heat is sourced renewably and the solid carbon is sequestered or used, this method could offer a low-emissions alternative.

White hydrogen—Hydrogen naturally found in underground deposits.

substantial quantities of electricity. Germany began operation of a hydrogen-powered train in the summer of 2022, while Santa Cruz, California, has the largest order of hydrogen buses in the US. At the start of 2025, 98% of the global hydrogen supply was made from fossil fuels. For the time being, these fuels will not have much emissions reduction, if any, compared to natural gas or electricity.

Environmental justice communities have strong reservations about the use of hydrogen from fossil fuels because it continues the operations that cast heavy burdens on fenceline communities. Hydrogen production, including that which may eventually include CCS as discussed below, perpetuates the operations of refineries and oil and gas wells, which disproportionately harm those living around these facilities and infrastructures.

Carbon capture and storage/sequestration

Addressing the climate change problem while still using fossil fuels will require advancements in CCS technology. For fossil fuel production, this often involves the conversion of coal to syngas. After impurities are removed, the syngas is burned in a gas turbine, and waste heat is harvested to turn a second steam turbine.

$$\text{Coal} + O_2 + H_2O \rightarrow H_2 + CO$$

There are several strategies for capturing carbon, including from the combustion stream or pulling it directly out of the air. The latter, so-called direct air carbon capture (DACC), draws carbon out of the ambient air streams. Pulling CO_2 out of waste streams or flue gas from a combustion process can result in avoiding over 90% of the emissions in the most optimistic scenarios. A newly proposed CCS facility in Texas will capture CO_2 from German gas producer Linde's hydrogen operations in a joint project with oil super giant BP. Several other organizations have emerged to help different regions of the world develop. The Africa Green Hydrogen Alliance aims to develop green hydrogen through a collaboration with Kenya, South Africa, Namibia, Egypt, Morocco, and Mauritania. A group calling itself the Copenhagen Infrastructure Partners proposed building out hydrogen electrolysis and offshore wind on a small island off Denmark in the North Sea.

In the US, an important provision of the 2022 Inflation Reduction Act included important tax credits to encourage the use of CCS technologies and processes. Some groups suggested these were giveaways to the oil and gas industry. But CCS could also benefit other industries, including fertilizer, cement, bioenergy, paper, and steel manufacturing, as we discuss in later chapters.

Takeaways/Highlights

- Petroleum fuels power global transportation.
- Natural gas energy resources are used in numerous applications in a diverse array of sectors.
- Oil and gas are natural resources that are more important to other sectors of the global economy beyond the energy sector.
- The transport and processing of oil and gas resources are some of the most disproportionate impacts in human history.

Questions for Review

1. Why does oil and gas development benefit some communities while burdening others? Consider factors such as geography, regulatory frameworks, environmental risk, and ownership of resources. The oil and gas sector produces not only fuels but also a wide range of petrochemical products, including plastics, fertilizers, lubricants, and synthetic materials.
2. What are the potential alternatives for replacing oil- and gas-derived products? Which sectors might face the greatest challenges in transitioning away from fossil-based inputs?
3. The "iron law" suggests that when climate policies conflict with economic growth, growth wins. How can energy and climate policy be designed to avoid triggering this trade-off? What strategies might align economic interests with emissions reduction goals? Can you think of a real-world policy example that attempts to defy or reinforce the "iron law"?
4. For the oil and gas industry to persist in a carbon-constrained world, CCS may play a key role. What are the advantages of large-scale carbon storage for emissions mitigation?
5. What are the potential risks or unintended consequences, such as technological lock-in or continued fossil dependence? Should CCS be prioritized over renewable deployment, or only applied to hard-to-abate sectors?

Exercise 4.1

Discuss the trade-offs associated with including gray and blue hydrogen in energy policy incentives. Compare the life cycle emissions, water use, and energy inputs of green, blue, and gray hydrogen. How are communities impacted at each stage of hydrogen production—from extraction to processing to use? What kinds of jobs, pollution risks, or infrastructure needs are involved in each method? What policies or practices might mitigate the environmental and social harms of non-green hydrogen pathways? Who should be included in these decisions?

Exercise 4.2

Much of political ecology focuses on who has access to natural resources and the resulting inequities in development outcomes.

- Identify a community—domestic or international—that has been positively or negatively impacted by oil and gas development.

- What are the short- and long-term outcomes of this development for that community?
- Which institutions (local, regional, or international) could help facilitate a just transition away from fossil fuel dependence?
- What role can community engagement, labor organizations, or environmental NGOs play?

Exercise 4.3

Oil and gas development often follows a boom/bust cycle that can erode the social and ecological fabric of rural and resource-dependent communities. Describe the key economic, environmental, and social challenges of this cycle. How does the experience of boom/bust dynamics differ across regions (e.g., North America versus Sub-Saharan Africa versus the Middle East)? What makes some communities more vulnerable to busts than others (e.g., lack of diversification, weak institutions)? What role can policy, planning, and economic diversification play in preparing communities for post-extraction futures?

References

Battersby, S. (2024). "Net zero" may need a rethink to keep climate targets within reach. *Proceedings of the National Academy of Sciences of the United States of America* 121 (18): e2407160121.

Beyer, J., Trannum, H.C., Bakke, T. et al. (2016). Environmental effects of the Deepwater Horizon oil spill: a review. *Marine Pollution Bulletin* 110 (1): 28–51.

Buck, H.J., Markusson, N., and Carton, W. (2024). Racial capitalism's role in mitigation deterrence from carbon removal. *Environmental Science & Policy* 160: 103865.

Cha, M. (2024). *A Just Transition For All*. Cambridge: MIT Press.

Clarke, K.C. and Hemphill, J.J. (2002). The Santa Barbara oil spill: A retrospective. *Yearbook of the Association of Pacific Coast Geographers* 64: 157–162.

Ferguson, J. (2005). Seeing like an oil company: space, security, and global capital in neoliberal Africa. *American Anthropologist* 107: 377–382.

Harvey, D. (2003). *The New Imperialism*. Oxford University Press.

Hendrix, C.S. (2017). Oil prices and interstate conflict. *Conflict Management and Peace Science* 34 (6): 575–596.

Holzer, B. (2007). Framing the corporation: royal Dutch/Shell and human rights woes in Nigeria. *Journal of Consumer Policy* 30 (3): 281–301.

Huber, M.T. (2013). *Lifeblood: Oil, Freedom, and the Forces of Capital.* University of Minnesota Press.

Huber, M.T. (2022). *Climate Change as Class War: Building Socialism on a Warming Planet.* Verso Books.

IEA (2021). Analysis and forecast to 2026. https://www.iea.org/reports/oil-2021 (accessed 10 November 2025).

Inman, M. (2016). *The Oracle of Oil: A Maverick Geologists Quest for a Sustainable Future.* New York: W.W. Norton.

Kantchev, G. (2022). Kazakhstan's elite got richer on Natural Resources. Then Came the Unrest. Wall Street Journal. January 7, 2022. https://www.wsj.com/articles/kazakhstans-elite-got-richer-on-natural-resources-then-came-the-unrest-11641572839 (accessed 10 November 2025).

Kimerling, J. (1990). Disregarding environmental law: petroleum development in protected natural areas and indigenous homelands in the Ecuadorian Amazon. *Hastings International and Comparative Law Review* 14: 849.

Klare, M.T. (2003). For oil and empire? Rethinking war with Iraq. *Current History* 102 (662): 129–135.

Laron, G. (2023). The OPEC+ Puzzle: why Russian-Saudi Cooperation starts – and Stops – with oil prices. Wilson Center. January 19, 2023. https://www.wilsoncenter.org/article/opec-puzzle-why-russian-saudi-cooperation-starts-and-stops-oil-prices (accessed 10 November 2025).

Mumford, L. (1934). *Technics and Civilization.* University of Chicago Press.

Pielke, R. Jr. (2010). *The Climate Fix: What Scientists and Politicians Won't Tell You About Global Warming.* Basic Books.

Schlosser, K.L. (2006). US national security discourse and the political construction of the Arctic National Wildlife Refuge. *Society and Natural Resources* 19 (1): 3–18.

Scott, J.C. (2020). *Seeing Like a State: How Certain Schemes to Improve the Human Condition Have Failed.* Yale University Press.

Smil, V. (2017). *Oil: A Beginner's Guide.* New York: Simon and Schuster.

Vezzoni, R. (2024). How "clean" is the hydrogen economy? Tracing the connections between hydrogen and fossil fuels. *Environmental Innovation and Societal Transitions* 50: 100817.

Watts, M. (2001). Petro-violence: community, extraction, and political ecology of a mythic commodity. In: *Violent Environments* (ed. N.L. Peluso and M. Watts), 189–212. Cornell University Press.

Watts, M. (2008). Blood oil: The anatomy of a petro-insurgency in the Niger delta. *Focaal 2008* (52): 18–38.

Whyte, K. (2017). The Dakota access pipeline, environmental injustice, and US colonialism. *Red Ink: An International Journal of Indigenous Literature, Arts, & Humanities* 19 (1): 154–169.

World Bank (2021). Global gas flaring tracker report. April 2021, Global gas flaring reduction partnership. https://thedocs.worldbank.org/en/doc/1f7221545bf1b7c89b850dd85cb409b0-0400072021/original/WB-GGFR-Report-Design-05a.pdf (accessed 10 November 2025).

Zalik, A. (2009). Zones of exclusion: offshore extraction, the contestation of space and physical displacement in the Nigerian Delta and the Mexican Gulf. *Antipode* 41 (3): 557–582.

Further Reading

Inman, M. (2016). *The Oracle of Oil: A Maverick Geologists Quest for a Sustainable Future.* New York: W.W. Norton.

Kashi, E. and Watts, M. (2008). *Curse of the Black Gold: 50 Years of Oil in the Niger Delta*, 44. Brooklyn, NY: Powerhouse Books.

Smil, V. (2015). *Power Density: A Key to Understanding Energy Sources and Uses.* MIT Press.

Yergin, D. (2011). *The Prize: The Epic Quest for Oil, Money & Power.* Simon and Schuster.

Chapter 5
Bioenergy

Bioenergy is human civilization's oldest form of energy. Bioenergy derived from carbon-based materials comprises a large set of energy types, from combustion or pyrolysis of wood and plants, to liquid fuels such as ethanol and biodiesel, to biogases of biogenic methane or hydrogen. This chapter is organized around the bioenergy pathways to energy supply and situates these fuels in their social and environmental context. As we will discover, there are impacts of each of the potential feedstocks to make them. Commodity chains for charcoal, palm biodiesel, corn ethanol, and sugarcane ethanol all intersect with socio-ecological processes in different ways, and this chapter will draw upon a wide range of research studies to describe these systems. Bioenergy systems like agriculture are the cause of land use change, such as with palm oil in Malaysia and Indonesia, where biodiverse tropical rainforest and orangutan habitat are converted to palm monocultures. Advanced biofuels offer opportunities to grow more biomass with fewer energy inputs, but they also require innovations in the lab and changes to agricultural practice. Since land use and nitrogen pollution are key themes in biofuels, the nitrogen cycle and tools to estimate land use change are introduced. Bioenergy is a very land-intensive energy system, raising many concerns and debates about food versus fuel and land grabs in developing countries.

Bioenergy From the Sun

Renewable energy refers to energy sources that replenish naturally on human-relevant timescales. This replenishment can occur daily, such as the Earth's rotation driving solar energy; seasonally, such as the wind and precipitation patterns powering wind and hydropower; or annually, as with bioenergy crops like corn, sugarcane, or soybeans grown for fuels.

For millennia, humans have relied on biomass—wood, dried plants, dung, and later charcoal—for heating, cooking, and metallurgy. Often, this dependence led to localized deforestation, particularly as societies intensified their demand for fuel and materials (Mwampamba et al. 2013). Bioenergy, derived from biological materials, is humanity's oldest energy source—and, more broadly, it is the foundation of energy for most life on Earth, since photosynthesis stores solar energy in plant matter.

Bioenergy or biofuels are forms of organic matter that can be used for energy and are available on recurring timescales, whether seasonally or annually. Through photosynthesis, plants and other organisms store solar energy in chemical bonds. When bioenergy is burned, these bonds are broken, releasing heat and energy. Traditional forms of bioenergy—cordwood, charcoal, and plant oils—were widely used long before fossil fuels. Bioenergy feedstocks can be grouped into sugar crops, such as sugarcane, grains, such as corn and wheat, and oilseed crops, such as soy, canola, and camelina. Biowastes—organic waste from agriculture, forestry, or municipal sources—also contribute to bioenergy. While often lower in energy density, they are abundant and can make meaningful contributions to energy supplies. These include crop residues, manure, forest slash, municipal solid waste, and even aquatic biomass like algae and seaweed.

Even if not directly combusted, most biomass eventually releases energy through decomposition, a slower form of oxidation. Under specific geological conditions, decomposed biomass may be prevented from oxidizing and instead become fossil fuels over millions of years.

Bioenergy production is often land-intensive and closely tied to agricultural systems. This creates overlaps with issues in agroecosystems, such as land use and land tenure, food versus fuel trade-offs, pollution and nutrient runoff, and access and equity in land distribution. Many bioenergy crops require significant inputs—water, fertilizer, and labor—and are often subsidized. For example, corn, a common feedstock for ethanol in the US, is high-yielding in part because it responds strongly to nitrogen fertilizer (Box 5.1).

Biofuels are frequently promoted for rural economic development. Financial incentives and investments in biofuel infrastructure can lead to job creation, increased tax revenues, and/or expanded public services. This is where the critical approach can be helpful, because these can be taken-for-granted assumptions.

Bioenergy's supporters also emphasize the role of domestically produced fuels in reducing reliance on foreign oil, aligning biofuels with national energy security goals.

Box 5.1 How Bioenergy Affects the Water and Nitrogen Cycles

Agricultural and agroforestry systems have significant implications for both the water and nitrogen cycles. The energy use and pollution associated with the use and production of nitrogen fertilizers can be significant. Nonpoint runoff from agricultural production can substantially impair rivers, streams, groundwater, and where wells are impacted, drinking water. Without thinking through the multiple dimensions of bioenergy's impacts, some aspects could turn out very bad. For example, it might be low-carbon biogas collected from an animal feedlot, but it could result in worse water pollution and even impact people's drinking water to let such problems persist.

However, this can lead to political clashes between the energy lobby and agricultural or food producer lobbies. Nevertheless, biofuel production contributes to several Sustainable Development Goals (SDGs), including responsible consumption and production and decent work (IEA 2024).

Is bioenergy a climate solution? Bioenergy is sometimes framed as a tool or resource for decarbonizing energy systems. Living organisms fix carbon from the atmosphere, and energy is released when the chemical bonds in plant material are broken—primarily through combustion. Photosynthesis is an extraordinarily powerful natural process. On land and in oceans, it captures energy at a rate of roughly 120,000 terawatts (TW)—about 10 times global human power consumption. At any given time, roughly 1.5 zettajoules (ZJ) (1 ZJ = 10^{21} j) of solar energy is stored in global biomass—about 30 times the annual energy produced by human activities. However, unless sequestered, this carbon is eventually re-released as carbon dioxide or methane during decay or combustion. In some rare cases, sedimentation or anoxic conditions may prevent oxidation and allow the biomass to form fossil fuels, but most of it is returned to the atmosphere on relatively short timescales.

According to the International Energy Agency (IEA 2020), land-use change, largely driven by deforestation for food and energy crop expansion, contributes roughly 15% of global greenhouse gas emissions. Most biofuels today are derived from crops that are also grown for food, such as corn, sugarcane, and soybeans. These "energy crops" can be highly input-intensive and subject to the same environmental pressures as conventional agriculture.

Biofuels now displace roughly 10% of gasoline (ethanol) and 10% of diesel (biodiesel) in regions like the US and Europe. The supply chains for these fuels often resemble those of industrial agriculture or agroforestry, meaning the social and environmental impacts of producing biofuels—such as methane from manure or pesticide runoff from corn fields—are not unique to the fuel itself, but tied to the broader production system.

At its core, bioenergy is a form of solar energy. Only about 1% of incoming solar radiation is converted by plants into usable biomass. Due to this relatively low conversion efficiency, bioenergy is land-intensive. Historically, the use of wood and charcoal as primary energy sources led to widespread deforestation. For example, sixteenth-century Great Britain experienced a severe wood shortage due to growing demands from papermaking (after the invention of the printing press), shipbuilding (naval and commercial fleets), and metallurgy (silver, copper, and iron smelting using charcoal). This wood crisis eventually spurred the transition to coal, helping launch the Industrial Revolution (Nef 1977). The invention of the steam engine, fueled by coal, followed soon after.

Humans have used oilseed crops for millennia to produce liquid fuels for lighting and heating. During the Bronze Age, crops like camelina were used for oil lamps. The earliest automobiles ran on blends of camphor, turpentine, and ethanol (Rhodes 2018). Later, the lighting industry relied heavily on whale oil, particularly from sperm whales. This led to a boom in whaling, which peaked around 1,846, with about 735 whaling ships in operation (Coleman 1995). Just 50 years later, only about 200 remained, and the industry collapsed soon after due to overharvesting. This is a classic case of over-extraction, with severe consequences for whale populations, especially one of the most intelligent and socially complex species in the ocean.

Energy Crops for Ethanol, Biodiesel, Renewable Diesel, Methanol, Charcoal, and Biochar

Bioenergy's nomenclature—how it is classified into categories—is a little bit confusing. But one attempt puts them into the broad categories of sugar crops, grain crops, oilseed crops, and biowastes. Most bioenergy sources are agricultural crops, while biowastes refer to fuels that are the waste products from other activities, including human and animal waste that can be turned into methane. These are typically lower in energy content but can be plentiful in supply where landfills and wastewater treatment plants are underdeveloped. Some bioenergy sources include agricultural residues and other plants, ranging from trees removed from orchards, to shells of nuts, to seaweed to grasses. Sometimes, bioenergy and biofuels as terms are used interchangeably, colloquially. Biofuel usually refers to a solid, liquid, or gaseous fuel from organic material. Bioenergy is generally a broader term. Energy crops refer to high-yield species with higher energy content. These are more expensive to grow because they require more inputs like fertilizers and fuels, but are also highly productive for the same reasons. Most forms of biopower, bioenergy, biomass, biogas, and biofuels have supply chains that are barely distinguishable from production processes for other agricultural commodities like corn, soy, wheat, sugarcane, and palm oil, or sometimes waste from animal agriculture or municipalities. Some biofuels are produced solely from agricultural surplus or waste.

Ethanol is sometimes classified as part of the sugar platform for biofuels. Fermentable sugars or starches capable of saccharification can all be readily converted to ethanol. Corn ethanol is a major liquid fuel and supplies about 10% of US transportation fuels as of 2020. Ethanol is also made from sugarcane and wheat, as they power vehicles that drive gasoline-ethanol blends or flex-fuel vehicles. These production systems are linked to water pollution from agricultural runoff in the regions where they are grown. The delta of the Mississippi River in Louisiana, USA, is often an anoxic zone depleted of oxygen from fertilizers, leading to a lack of life in some parts of the water column. The agricultural commodity aspect means it becomes difficult to trace these crops sometimes back to the farms that produce them, making accountability up the supply chain more challenging in this space. Globally, sugarcane contributes a significant portion of ethanol production, and those supply chains have been linked to accusations of child or even slave labor (Carrillo 2017). The key difference between corn and sugarcane ethanol is that corn requires an extra processing step called saccharification, where enzymes are introduced to convert starches to fermentable sugars. Sugarcane already has fermented sugars present.
Photosynthesis, carbon dioxide to sugars

$$6CO_2 + 6H_2O + \text{light} \rightarrow C_6H_{12}O_6 + 6O_2$$

Fermentation to alcohols

$$C_6H_{12}O_6 \rightarrow 2C_2H_5OH + 2CO_2$$

Combustion or decay to carbon dioxide

$$C_2H_5OH + 3O_2 \rightarrow 2CO_2 + 3H_2O + \text{heat}$$

There have been many debates and claims made about ethanol's carbon intensity compared to gasoline (Van der Horst and Evans 2010). Some point to its modest reductions but also the development of more innovative fuels with much lower carbon intensity (Table 5.1). But in many cases, the greenhouse gas emissions from some kinds of ethanol production result in a carbon intensity that is higher than gasoline. Some ethanol processes can be 50% or even 15% of the emissions of gasoline (see Table 5.2), which is very low. Gasoline needs an oxygenate to optimize combustion. Prior to ethanol, gasoline contained methyl tertiary butyl ether (MTBE), a fuel additive used until it was discovered to be a major groundwater contaminant (Table 5.1 and Figure 5.1).

Ideas to power transportation by ethanol and biodiesel date back to the earliest days of the automobile, when a vision advanced by Henry Ford argued that farmers should grow the fuel for automobiles. Reformulated gasoline is the major reason we blend ethanol into gasoline today, as policy enacted with the 1990 Clean Air Act amendments. Originally used to stop engine "knocking"—the oxygenate helps eliminate volatile organic compounds, precursors to smog, and air toxins like benzene and 1,3-butadiene. A report by the USEPA (EPA 1999) suggested there would be significant impacts to water from widespread and continued use of MTBE that warranted its elimination in favor of ethanol (which has its own water quality impacts) as a fuel

Table 5.1 Carbon intensity of selected liquid biofuel feedstocks and processes.

Fuel	Feedstock/energy/process	Place	Cl (gCO_2e/MJ)
Gasoline	Petroleum/natural gas	California	**98**
Ethanol	Corn/natural gas dry mill	Nebraska	66.52
Ethanol, cellulosic	Corn fiber/n. gas dry mill	Texas	23.13
Ethanol	sugarcane juice, molasses	Brazil	45.06
Ethanol	Sorghum	Kansas	65.06
Diesel	Petroleum/natural gas	California	**102**
Biodiesel	Soybean oil	Texas	53.43
Biodiesel	Canola oil	Illinois	53.00
Renewable diesel	Tallow, renewable Naphtha	California	33.42
Renewable diesel	Used cooking oil	Singapore	28.64
Liquefied natural gas	Wet natural gas	California	**84.55**
Liquefied natural gas	Landfill gas	California	84.55
Compressed natural gas	Wet natural gas	California	**78.36**
Compressed natural gas	Landfill gas	Ohio	41.61
Compressed natural gas	Food scraps	California	−79.91

Source: Adapted from CARB (2022).

Table 5.2 Bioenergy types and selected feedstocks.

Liquid fuels	Feedstocks
Ethanol	Corn, sugarcane, wheat, barley, and sorghum
Biomethanol	Wood waste, black liquor (paper mills), biomethane, and municipal waste
Biobutanol	Corn grain, seaweed/microalgae, and agricultural residues
Biodiesel	Soybeans, corn oil, palm oil, algae, and used cooking oil
Renewable diesel	Soybeans, vegetable oils, algae, palm oil, and animal fats
Hydrotreated oils	Animal tallow, used cooking oil, and fish oil
Solid fuels	
Woody biomass	Trees, municipal green waste, forestry residues, and logging slash
Wood pellets	Trees and forest thinnings, post-industrial scrap wood, and sawdust
Biochar	Carbon-based materials, wood chips, and nutshells
Residue briquettes	Corn stover, wheat straw, and rice husks
Gaseous fuels	
Biogas	Wastewater solids, animal feedlots, landfills, and food scraps
Biomethane	Wastewater solids, animal feedlots, landfills, and food scraps
Bio-Hydrogen	Algae, gasification processes, and fark fermentation

oxygenate. The problem emerged when it was learned that MTBE was contaminating water through leaking underground storage tanks at gasoline distribution stations.

As a result of that effort to protect drinking water and soils, ethanol is the primary oxygenate in the US today, with very few states allowing MTBE (Oklahoma is an example). In fact, the Energy Policy Act of 2005 repealed the fuel oxygenate requirement and replaced it with the so-called ethanol mandate requiring 36 billion gallons of ethanol production by 2022. Ethanol is also a source of octane in gasoline. Higher octane gas has higher performance and burns hotter, so it can be more desirable. But it is also more expensive, so when the overall picture of fuel sales is in view, there is a need to supply each of the range of octane performances as allowed. In Europe, Asia, Africa, and South America, MTBE is still used widely as a fuel additive. The US still produces MTBE mostly for fuel exports. The third most significant octane booster and oxygenate is methanol, which is widely used in China but is soon to be phased out in favor of ethanol.

But as critics suggested nearly 20 years ago, the greenhouse gas benefits of corn ethanol range from limited to dubious. A research paper in the *Proceedings of the National Academy of Sciences* in 2022 suggested blending 10% ethanol into gasoline "increased annual nationwide fertilizer use by 3 to 8%, increased water quality degradants by 3 to 5%, and caused enough domestic land use change emissions such that the carbon intensity of corn ethanol … is no less than gasoline." There may be lower carbon intensity than gasoline in most cases, but mostly marginal reductions on the order of 20% lower. The best-performing ethanol from a greenhouse gas perspective is probably on the order of 10–15% of gasoline. But these are much less common and more expensive to make than the corn or sugarcane that partly fills fuel tanks in the US and Brazil, respectively (Figure 5.1).

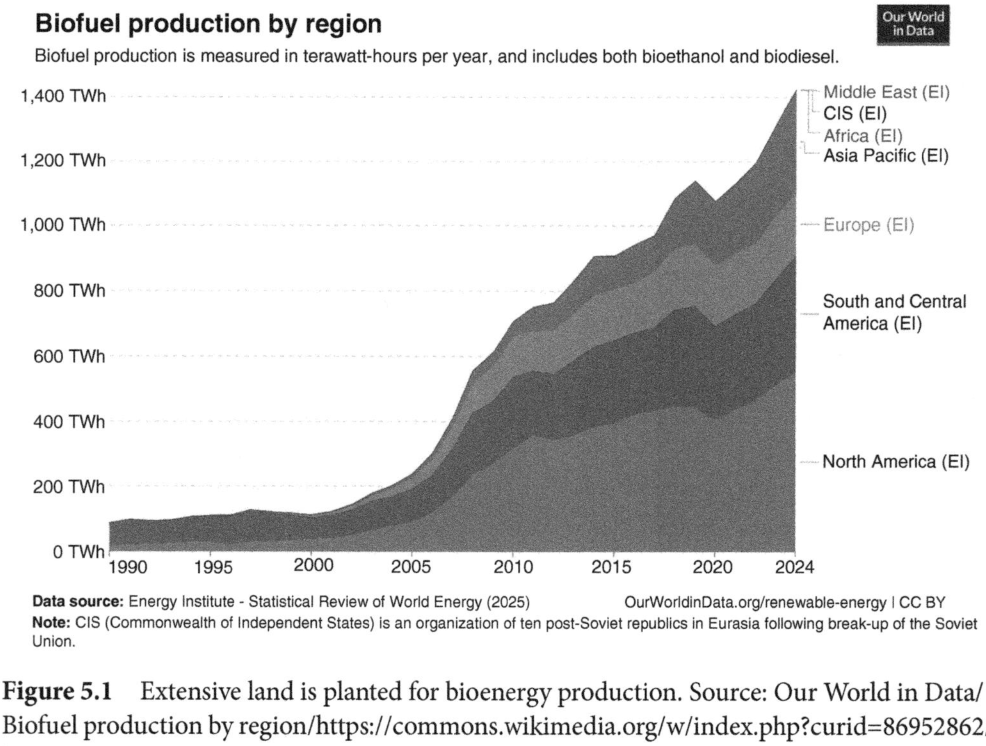

Figure 5.1 Extensive land is planted for bioenergy production. Source: Our World in Data/Biofuel production by region/https://commons.wikimedia.org/w/index.php?curid=86952862/last accessed on September 18, 2025.

The use of mostly corn for ethanol production has led many to question the implications for food security. Ethanol adoption in the 2000s led to price spikes that caused food security issues throughout the world, including Mexico. Loss of rural employment owing to such shifts could cause major disruptions in some places, as evident in information about uneven impacts from changes in Brazil's sugarcane industry (Hodbod et al. 2015; Machado et al. 2017). Sugarcane provides about 50% of transportation liquid fuels in Brazil, whereas this is about 10% in the US. Even in developed countries, agriculture relies on farmworker labor, where some environmental justice problems can be most acute (Carrillo 2017). Ethanol plays a larger role in Brazil, where it is made from sugarcane on farms in the eastern-most seaboard of the country near the megacity Sao Paulo. Numerous non-food alternative feedstocks, such as various types of algae, waste-to-ethanol, and synthetic biofuels, show better environmental performance than corn and sugarcane.

Biogas and Biomethane

Bioenergy can also be produced in gaseous form and used to generate electricity or can be made into heating or transportation fuels. Electric utilities buy from electricity suppliers that produce biomethane or biogas from anaerobic digestion at landfills, dairies, confined feeding operations, and wastewater treatment facilities. There are slight differences in the energy content of methane (natural gas or biomethane) and biogas. There are small amounts of carbon dioxide and other gaseous impurities that make the energy value of biogas slightly lower. The methane content of biogas can vary from 45% to 75%, which translates to 16 to 28 MJ/m^3 (IEA 2020). Biomethane is upgraded to be closer to 99% methane. Syngas—a mixture of carbon monoxide, hydrogen, and methane is another biogas produced by pyrolysis, where biomass is heated in the absence of oxygen under high pressure. Hydrogen can also be produced biologically.

Most biogas in the world is made from manure, with the US the leading producer. The US also uses more biogas in the transportation sector than any other place in the world. Wastewater treatment facilities are common sites of biogas production, as are landfills and food scrap recovery facilities from homeowners, restaurants, or food processors. Thailand produces a significant amount of biogas from its cassava starch industry's wastewater treatment plants (Papong et al. 2014; Suwanasri et al. 2015). Efforts to collect food scraps and convert them to biogas for heat and/or electricity, or upgrade to biomethane, are intended to avoid landfill greenhouse gas emissions.

Biogas that is processed and upgraded to biomethane is of a quality that can be injected directly into pipelines without concern for mixes of gases, moisture, impurities, etc., entering the pipeline system. Biomethane is "pipeline-ready," compared to biogas that is produced from similar types of facilities—namely, dairies, landfills, fresh food processors, wastewater treatment plants, and food waste processing facilities that

Figure 5.2 Dairies and confined feeding operations for cattle are major sites of biogas and biomethane production. Source: Martina Nolte / Wikimedia Commons / CC BY-SA 3.0.

use biogas usually onsite for heat, hot water, or electricity (Figure 5.2). Some biogas facilities convert their infrastructures to produce biomethane because there are incentives to do so. A critical question for just transitions is whether efforts to use biomethane are just a way to let combustion fuels continue further into the future, or whether incentives would be better spent elsewhere, such as on electrification.

To decarbonize natural gas delivered through pipelines, some propose blending in lower-carbon biomethane or even hydrogen. This could work, especially if the materials help avoid methane venting or flaring. A critical assessment of biomethane would look at how the markets are structured and whether this really delivers the reductions as promised. For example, some utilities buy biomethane carbon offsets or renewable electricity certificates and call fossil natural gas biomethane by virtue of this purchase of carbon credit, without physically delivering the gas to the customer who wants to offset their emissions. This policy incentive in spirit works with the idea that it encourages more producers to deliver biomethane to the pipeline system more broadly. But should a buyer be able to claim that their natural gas is low-carbon biomethane if it is not physically delivered? That might be more complicated. What are these kinds of financial arrangements? They just incentivize more dairy cows, What would be the implications of these offsets on the continuation of uneven impacts to a community living near a dairy on air or water quality? These are the types of critical analyses that are needed compared to the more acritical approaches that would simply see these schemes as incentives to grow green industries.

Bioenergy is also an important resource for improving access to better-quality energy in many developing countries. For example, it is common to use charcoal, biomass, or dung indoors for cooking, which is a leading cause of indoor air pollution exposures in some parts of the world. Cleaner-burning or all-electric cookstoves and other cleaner-burning fuels can help alleviate some of the worst injustices, but even still, do so unevenly (Damgaard et al. 2017). Exposures to indoor air pollution disproportionately burden the poor, making the clean cookstove sector another area where it will be important to watch both the fuel and the distribution of devices (Simon et al. 2014). Cookstove supply chains can help meet the objectives of just transitions, especially where communities get both the benefits of cleaner stoves and local jobs in assembling or distributing stoves. The transition to cleaner cooking fuels is also a complex cultural question as well, one that may take innovative ideas to move people away from cooking things in a flame (Venkataraman et al. 2010).

Combustion emissions from biomethane are generally similar to fossil fuels for indoor air quality (Kashtan et al. 2023). Carbon monoxide, particulate matter, SO_x, NO_x, and formaldehyde are all common indoor air pollutants that are widely recognized to cause significant respiratory and other public health problems. Researchers have found benzene and other very hazardous air pollutants causing indoor air quality problems with open flames.

Land Grabs for Bioenergy

Dispossession is a social process that takes people away from the lands and resources they once used without their consent and long predates the fossil fuel era. Large-scale acquisition of lands of agriculture—sometimes today referred to as "landgrabs" (Box 5.2)—was a key force of colonialism. Yet the development of supply chains for biofuels can have implications for communities dependent on agriculture or forestry, from land use changes, increased economic activity, patterns of ownership, public services and investment, and a myriad of complex outcomes. Understanding these histories is critical to understanding power dynamics among stakeholders, as settler-colonial relations role in dispossessions across the Earth is still contested and ongoing. What, on the one hand, might seem like an innovative low-carbon fuel company might be seen as part of a larger historical process where land is taken from locals or public spaces are privatized.

The land-use intensity of biofuels means they require a large amounts of land, which increases the likelihood of land use conversion and large-scale aquisition. Some crops require extensive land holdings to make production economically viable, resulting in land-grabbing or the more anesthetized "large scale land transactions." By some account, the amount of land that has been transferred is 45 million ha (170,000 $mile^2$) in the past decade (Liao and Agrawal 2024). Land-grabbing occurs when organizations might use their "green" credentials to appropriate or alienate local farmers from their land (Fairhead et al. 2012). Lands have been sold all over the

Box 5.2 Land Grabs for Biofuels

Sometimes new public policies for national security that poise an industry for growth can set off patterns of investment that lead to land grabs, especially for timber, minerals, and biofuels (Borras and Franco 2012). Since 2007 or so, there has been a land rush to acquire land across the world from Africa to Asia to South America, although scholars debate what the ultimate drivers are. Changing property relations are often indications of agrarian change, but drivers can also include labor dynamics or intertwined histories of colonialism (Edelman et al. 2013). The United Nations (2014) recognizes an acute need to ensure that women and Indigenous communities are adequately represented on questions about their use and rights in the context of land grabs. Vulnerable populations should have mechanisms for formal grievances or to appeal land grabs that impact them via some arbitration or dispute resolution. The UN Declaration on the Rights of Indigenous Peoples and International Labor Organization Convention No. 169 protects the rights of indigenous peoples and aims to empower them through everything from better participation and consultations to new treaties and land back.

world to large companies that operate agricultural production systems. Sometimes these shifts lead to rapid mechanization or loss of labor activities more generally, so fewer jobs. Such changes in land access can lead to further marginalization of peoples who have already been dispossessed of their land (Li 2018). Loss of land access can have implications for food security because production networks can be important sources of livelihoods.

The United Nations Inter-Agency Support Group on Indigenous Peoples' Issues (2014) has recognized land grabs, sometimes simply described as large-scale land purchases, as an important issue to address, especially where land use change is happening under other initiatives like clean energy and energy access. One way that land can be appropriated is by classifying or signifying that land otherwise not put to biofuels production is a wasteland, as has been described in India (Baka 2017; Baka and Bailis 2014). Often, these lands turn out to be important spaces of use to rural and urban households. The classification of wastelands upon critical interrogation turns out to be a misunderstanding.

Large-scale land purchases are also often for plantation-style agribusiness interests. Important biodiverse ecosystems that have been developed for biofuels, including ones that have been linked to tropical deforestation. Palm oil is an important feedstock for biodiesel production. Palm oil development can lead to deforestation, which then causes loss of biodiversity, habitat degradation, and loss of endangered species. West Kalimantan has lost nearly 30% of its tropical forest cover to palm oil production for energy and food, mostly for global exports. Furthermore, in places like Borneo, one of

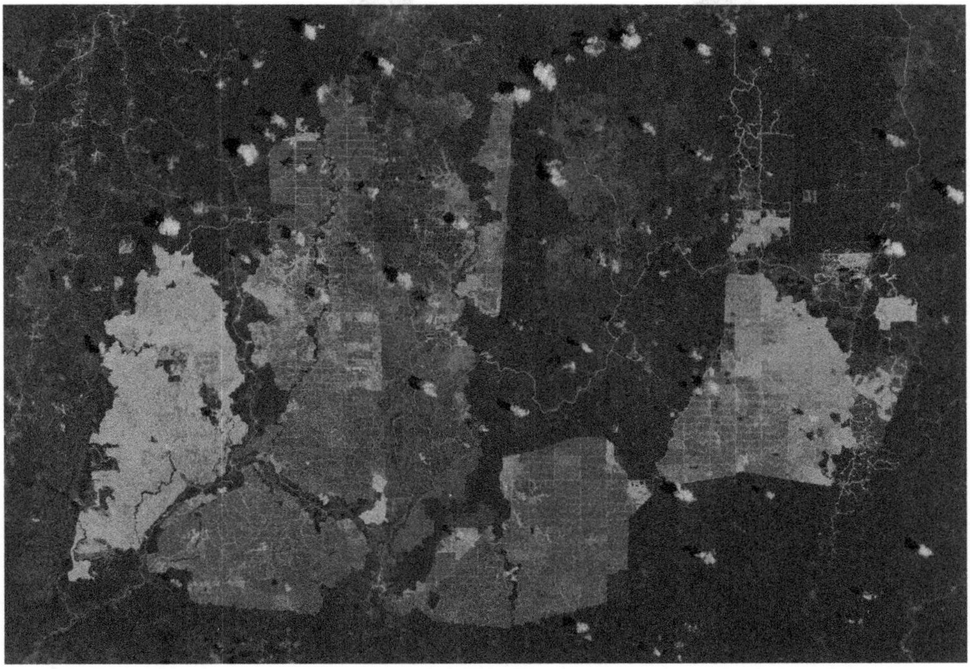

Figure 5.3 Land clearing for palm oil production. Source: European Space Agency / Wikimedia Commons / CC BY-SA 3.0.

the Earth's biodiversity hotspots, Indigenous groups are displaced from traditional lands, especially where tribes do not have formal land rights by treaty or otherwise (Figure 5.3).

Sustainable Bioenergy Metrics

There have been many that have contended from its earliest suggestion as a climate solution that sustainable bioenergy is a misnomer or oxymoron. There are a number of important ways we might quantify some of the biggest and most important impacts from bioenergy systems, covering the air pollution where fuels are combusted to the water pollution near where bioenergy crops are grown. Bioenergy needs also to be understood in the context of gasoline use; it may be a major to slight improvement in some things (carbon), but not much in others (smog). Combustion still results in some unavoidable pollution.

Land use change is probably one of the most important measures of bioenergy because it means either consumptive land use or the displacement of other people or economic activities (McDonald et al. 2009). Land use change is more complicated than it sounds. For example, direct land use change usually refers to the direct impact of developing a parcel of land, whereas indirect land use change might be the effect of cascading land use decisions. For example, corn may replace soybeans, and it may appear that there is not any deforestation land use change, but what if the displaced soybeans are then grown in an area that was once forest? This is a difficult topic to model as decisions about crop choice are often mediated by prices, which are determined by multiple things on a year-to-year basis. Certainly, there is some impact from these indirect land use changes related to bioenergy production, especially if the crop is receiving premium prices so is the crop of choice to plant for farmers and growers.

Energy Return on Investment (EROI) is a metric used to represent the energy returns from different sources, processes, and technologies. A useful analogy is a financial investment where one dollar is invested yielding let's say, US$5 in return, you have a return on investment of 5. EROI would be similar except for energy in comparison to the energy yielded from the energy investment. EROI metrics are important for energy systems because they can tell us about the bang for the buck, so to speak. If the same energy investment could be used to yield more energy, why not?

A meta-analysis of ethanol energy production finds that corn-to-ethanol EROI ranges between 0.75 and 1.7. This is largely due to the fertilizer used to grow the primary feedstock, corn, and the use of coal for heating in some facilities. Other crops could show higher EROIs on the order of ten. Beyond that, the limits of solar insolation make it challenging to get much more energy out. If land is a constraint, a photovoltaic system can generate more energy per acre.

$$\text{Carbon Payback Time} = \text{Carbon}_{\text{landsource}} - \text{Carbon}_{\text{biofuelcrop}} / \text{Biofuel}_{\text{carbonsavings}} \text{per hectare} / \text{yr}$$

Energy and carbon payback times are similar metrics. The energy payback time (EPBT) is meant to illustrate how long the energy investments in a renewable energy source take to pay off the initial or life-cycle energy investments. Bioenergy sources have an energy payback, a time period of production to recover energy used in production and transport. One study found the EPBT for biofuels can range from 1 to 162 years (Elshout et al. 2015). Where the carbon payback time similarly represents a payback only for carbon emission invested. This research is especially important to understand in the context of expanding production of biofuels in the tropics, where with high-yielding crops, deforestation for production could take from 40 to 120 years to pay back, much longer for low-yielding biofuels (Gibbs et al. 2008). This suggests that biofuel production in areas where deforestation is linked to production through direct or indirect uses could have severely negative implications for greenhouse gas savings from biofuels.

The opportunities and challenges facing bioenergy are multifaceted and require careful scrutiny, especially as they pertain to land use, water, and air pollution, where there might be tradeoffs. There are likely short-term opportunities in both the biomethane space and renewable diesel to displace some greenhouse gas emissions. The more long-term solutions that bioenergy could be part of could be drop-in sustainable aviation fuels, where it is unlikely there will be major shifts towards electrification in the near term. Bioenergy will continue to be an important source of energy for heating, cooking, and transportation. Blending bioenergy with fossil fuels into the near future seems like its most likely fate for now, and the challenges to sustainability and environmental justice from land use to combustion will remain important questions of critical energy scholarship.

Takeaways/Highlights

- Bioenergy resources can be very land-use intensive owing to the low conversion efficiency of photosynthesis into energy stored in plants and recoverable as energy fuels.
- The land uses for biofuels often make them subject to green grabbing or land grabs, particularly in the global south, where large agribusiness can buy large parcels of land for agricultural development.
- Bioenergy from waste resources can be helpful to avoid more potent greenhouse gas emissions like methane.
- Biogas and biomethane in the future could help reduce the carbon intensity of natural gas through blending opportunities, but still present health impacts due to indoor combustion.

Questions for Review

1. Biofuels are a renewable energy resource but often require extensive fossil fuel inputs or result in very inefficient land use conversions to energy. There are several metrics commonly used to measure their environmental performance, but these should be considered in the context of nonquantifiable impacts like dispossession, marginalization, and land degradation.
2. The greenhouse gas benefits of biofuels vary considerably, but, in general, they show modest improvements over the carbon intensity of gasoline. The best-performing biofuels are on the order of 1/10th the greenhouse gas intensity of gasoline but make up only a small portion of the market.
3. Bioenergy systems sit at the interface of agriculture and energy. Bioenergy production systems and feedstock most often have undifferentiated agricultural impacts and socio-ecological phenomena, but sometimes there are key differences to tease out. Bringing an interdisciplinary lens to the pursuit of sustainable and just biofuel production requires becoming versed in various aspects of energy and agricultural production, from both technological and human dimensions.
4. Some argue that companies that participate in Corporate Social Responsibility (CSR) initiatives also have better tools for worker empowerment and to drive better environmental outcomes in corporations. Are there certain features of some kinds of bioenergy commodities that are not appropriate sustainability certifications? Who would be important to include in the design of a sustainability certification for a biofuel system?

Exercise 5.1

The amount of energy it takes to produce 1,000 acres of corn and process it into ethanol is 31,000 MJ, while the amount of energy that can be extracted is 600 gal of ethanol, or 80,100 MJ. What are the EROI and EPT of this system and feedstock?

Exercise 5.2

If the fuel is used in an average fuel economy internal combustion vehicle, and is 90% gasoline at a carbon intensity from the table and 10% ethanol. What is the carbon intensity of the fuel blend when varying the carbon intensity of various feedstocks for ethanol and gasoline? Find examples from the literature or lookup tables and illustrate a range of possible outcomes. The units should be reported in grams of CO_2e per megajoule. Compare with gasoline at 100 g of CO_2e per megajoule.

Exercise 5.3

Explain the challenge of land grabs in a developing country. Why do these projects often get championed by elites in some parts of the world? How do they impact communities? What are the risks and benefits to communities? What do new energy landscapes promise, and what experiences might give pause to uncritical unacceptance of these claims? Search for a case study using search terms combining bioenergy, biomass, or biofuel with land grabs, landscapes, and commodity chains. Link your explanation to one policy or effort to correct the challenges with land grabs and land scale large transfers for agricultural production.

References

Baka, J. (2017). Making space for energy: wasteland development, enclosures, and energy dispossessions. *Antipode* 49 (4): 977–996.

Baka, J. and Bailis, R. (2014). Wasteland energy-scapes: a comparative energy flow analysis of India's biofuel and biomass economies. *Ecological Economics* 108: 8–17.

Borras, S.M. Jr. and Franco, J.C. (2012). Global land grabbing and trajectories of agrarian change: a preliminary analysis. *Journal of Agrarian Change* 12 (1): 34–59.

California Air Resources Board (2022). LCFS Pathway Certified Carbon Intensities. https://ww2.arb.ca.gov/resources/documents/lcfs-pathway-certified-carbon-intensities (accessed 10 November 2025).

Carrillo, I.R. (2017). When farm work disappears: labor and environmental change in the Brazilian sugar-ethanol industry. *Environmental Sociology* 3 (1): 42–53.

Coleman, J.L. (1995). The American whale oil industry: a look back to the future of the American petroleum industry? *Nonrenewable Resources* 4 (3): 273–288.

Damgaard, C., McCauley, D., and Long, J. (2017). Assessing the energy justice implications of bioenergy development in Nepal. *Energy, Sustainability and Society* 7 (1): 8.

Edelman, M., Oya, C., and Borras, S.M. Jr. (2013). Global Land Grabs: historical processes, theoretical and methodological implications and current trajectories. *Third World Quarterly* 34 (9): 1517–1531.

Elshout, P.M.F., Van Zelm, R., Balkovic, J. et al. (2015). Greenhouse-gas payback times for crop-based biofuels. *Nature Climate Change* 5 (6): 604–610.

Fairhead, J., Leach, M., and Scoones, I. (2012). Green grabbing: a new appropriation of nature? *Journal of Peasant Studies* 39 (2): 237–261.

Gibbs, H.K., Johnston, M., Foley, J.A. et al. (2008). Carbon payback times for crop-based biofuel expansion in the tropics: the effects of changing yield and technology. *Environmental Research Letters* 3 (3): 034001.

Hodbod, J., Tomei, J., and Blaber-Wegg, T. (2015). A comparative analysis of the equity outcomes in three sugarcane–ethanol systems. *The Journal of Environment & Development* 24 (2): 211–236.

International Energy Agency (2020). Outlook for biogas and prospects for organic growth World Energy Outlook Special Report biomethane. https://www.iea.org/reports/outlook-for-biogas-and-biomethane-prospects-for-organic-growth/an-introduction-to-biogas-and-biomethane (accessed 10 November 2025).

International Energy Agency (2024). Bioenergy Review 2023. https://www.ieabioenergyreview.org/ (accessed 10 November 2025).

Kashtan, Y.S., Nicholson, M., Finnegan, C. et al. (2023). Gas and propane combustion from stoves emits benzene and increases indoor air pollution. *Environmental Science & Technology* 57 (26): 9653–9663.

Li, T.M. (2018). After the land grab: infrastructural violence and the "Mafia System" in Indonesia's oil palm plantation zones. *Geoforum* 96: 328–337.

Liao, C. and Agrawal, A. (2024). Towards a science of 'land grabbing'. *Land Use Policy* 137: 107002.

Machado, P.G., Walter, A., Picoli, M.C., and João, C.G. (2017). Potential impacts on local quality of life due to sugarcane expansion: a case study based on panel data analysis. *Environment, Development and Sustainability* 19 (5): 2069–2092. https://doi.org/10.1007/s10668-016-9823-6.

McDonald, R.I., Fargione, J., Kiesecker, J. et al. (2009). Energy sprawl or energy efficiency: climate policy impacts on natural habitat for the United States of America. *PLoS One* 4 (8): e6802. https://doi.org/10.1371/journal.pone.0006802.

Mwampamba, T.H., Ghilardi, A., Sander, K., and Chaix, K.J. (2013). Dispelling common misconceptions to improve attitudes and policy outlook on charcoal in developing countries. *Energy for Sustainable Development* 17 (2): 75–85.

Nef, J.U. (1977). An early energy crisis and its consequences. *Scientific American* 237 (5): 140–151.

Papong, S., Rotwiroon, P., Chatchupong, T., and Malakul, P. (2014). Life cycle energy and environmental assessment of bio-CNG utilization from cassava starch wastewater treatment plants in Thailand. *Renewable Energy* 65: 64–69.

Rhodes, R. (2018). *Energy: A Human History*. New York: Simon and Schuster.

Simon, G.L., Bailis, R., Baumgartner, J. et al. (2014). Current debates and future research needs in the clean cookstove sector. *Energy for Sustainable Development* 20: 49–57.

Suwanasri, K., Trakulvichean, S., Grudloyma, U. et al. (2015). Biogas–key success factors for promotion in Thailand. *Journal of Sustainable Energy & Environment Special Issue* 25: 30.

United Nations (2014). Lands territories, and resources. Inter-Agency Support Group on indigenous peoples' issues. Thematic Paper.

USEPA (1999). Achieving clean air and clean water: The report of the blue ribbon panel on oxygenates in gasoline. USEPA. https://archive.epa.gov/oust/mtbe-a/web/pdf/r99021.pdf (accessed 10 November 2025).

Van der Horst, D. and Evans, J. (2010). Carbon claims and energy landscapes: exploring the political ecology of biomass. *Landscape Research* 35 (2): 173–193.

Venkataraman, C., Sagar, A.D., Habib, G. et al. (2010). The Indian national initiative for advanced biomass cookstoves: the benefits of clean combustion. *Energy for Sustainable Development* 14 (2): 63–72.

Further Reading

Arshad, M. (ed.) (2021). *Sustainable Ethanol and Climate Change: Sustainability Assessment for Ethanol Distilleries*. Springer International Publishing. https://doi.org/10.1007/978-3-030-59280-6.

Bhardwaj, A.K., Zenone, T., and Chen, J. (2015). *Sustainable Biofuels: An Ecological Assessment of the Future Energy*. Higher Education Press, De Gruyter, Inc.

International Energy Agency (2020). Outlook for biogas and prospects for organic growth World Energy Outlook Special Report biomethane. https://www.iea.org/reports/outlook-for-biogas-and-biomethane-prospects-for-organic-growth/an-introduction-to-biogas-and-biomethane (accessed 10 November 2025).

Chapter 6
Hydropower

By erecting thirty thousand dams of significant size across the American West, they dewatered countless rivers, wiped out millions of acres of riparian habitat, shut off many thousands of river miles of salmon habitat, silted over spawning beds, poisoned return flows with agricultural chemicals, set the plague of livestock loose on the arid land—in a nutshell they made it close to impossible for numerous native species to survive.

—Mark Reisner, Cadillac Desert (Reisner 1993)

Hydropower will remain the world's largest source of renewable electricity through 2028 (International Energy Agency 2024). From small diversions to mega-dams to pumped-hydro energy storage projects, hydropower systems are connected to electricity grids around the globe. In some regions, hydroelectric dams form the backbone of national energy supply, while in others, hydropower availability is highly seasonal, depending on rainfall and snowmelt patterns. Historically, hydropower, even before electricity, was harnessed by industries and cities for mechanical tasks. With the advent of electricity, hydropower was a favored development tool promoted by institutions such as the World Bank and other international development agencies. Projects like the Three Gorges Dam in China (see Box 6.3 and Figure 6.1)—which displaced over one million people—highlight the social and environmental costs associated with large-scale hydropower. Hydropower has long been an object of protest and resistance, with significant opposition movements in countries like India and Brazil, where communities have mobilized against major dam projects. In the realm of environmental literature, Edward Abbey's classic *The Monkey Wrench Gang* features a protagonist driven to sabotage the Glen Canyon Dam, a fictional response to the loss of his childhood home beneath its reservoir—symbolizing the deep cultural

Figure 6.1 Three Gorges Dam on the Yangtze River in China from space. Source: NASA / Wikimedia Commons / Public domain.

and ecological disruptions caused by dam building. Today, some of the largest dams in the western United States are contributing to the decline of wild salmon populations, underscoring the ecological trade-offs of hydropower. Hydropower facilities are also multipurpose—supporting irrigation, transportation, flood control, and recreation—making them critical to more than just electricity production. This chapter will introduce students to the hydrological cycle, explore the engineering and ecological dimensions of hydropower, and examine the major transformations of riparian ecosystems, particularly those on the world's major river systems. Special attention will be given to the social, environmental, and political challenges of managing hydropower in an era of climate change and rising demand for clean energy.

Harnessing the Power of the Water Cycle

It should be no surprise that a planet covered by water is also shaped by the water cycle, and the power of water that carved the world's canyons and floodplains, and posed risks to human settlements would be developed as a resource for humans. Weather, ecologies, and weathering are all driven in part by the power of water as it moves through the hydrosphere. While water shaped the world, the harnessing of water to do work for societies and economies came much later. Other species like beavers reshaped rivers to improve the habitat around them. But human transformations of riparian areas are some of the most substantial and dramatic terra-transformations on Earth.

Hydropower was likely first used in China over 2000 years ago to crush grains, break ore, and other simple mechanical processes. In the nineteenth century, hydropower was used to generate electricity. From there, the use of hydropower quickly spread across nations, with the United States and Canada leading the way. Today, global hydropower production is led by China, Canada, and Brazil, with China nearly producing three times as much energy as the next highest producing country. It is said that the Three Gorges Dam and reservoir are so large that they shifted how weight rotates around the planet, slowing Earth's rotation by 0.06 ms, changing local temperatures and rainfall patterns as well. Dammed rivers literally reshape the Earth.

Hydropower development is found on every continent on Earth, except Antarctica. Early American and European hydropower dams harnessed power for mechanical power. The massive dams built in Egypt, China, Uruguay, Brazil, India, and so many parts of the world generate electricity, as well as provide other services, including flood control and navigation. But they are a key infrastructure of development over the past century or more. They are also featured in decarbonization strategies because they do not directly generate greenhouse gas emissions and are more or less renewables. Geographer Tyler Harlan describes the extensive development of hydropower in western China rather than a decarbonization strategy as often assumed by observers; it is better described as part of a socio-spatial process linked to codependency with interlinked external economies that undermines the benefits that these regions can accrue (Harlan 2023). These processes also reallocated property rights and prioritized infrastructure investments, so help explain the uneven patterns of benefits.

Hydropower is a long-standing major renewable energy workhorse of industrial civilization, from eleventh-century watermills to twenty-first-century mega-dams and electric power generators. Environmental historian Donald Worster describes water development for irrigation across the American West as a foundation to the American empire (Worster 1992). Hydropower is renewable in that the water cycle is generally continuous, with snowfall and rainfall translating to gravitational potential energy that can be harnessed as kinetic energy as the water flows downhill.

Early uses of hydropower were primarily focused on mechanical applications, such as driving textile production and grain "grist" mills. Early settlers often established sawmills near gulches and rivers, taking advantage of flowing water to power lumber mills, paper mills, and other forms of early industry. In arid regions, however, the use of hydropower was typically seasonal, dependent on rainy seasons and vulnerable to periodic and episodic droughts. To manage water flow and extend operability, mills commonly included retention ponds upstream, allowing operators to regulate the flow of water and ensure a more consistent power supply.

Hydropower was an energy source powering the early days of electricity. In 1880, the Grand Rapids Electric Light and Power Co. began to generate direct current electricity on a belt-dynamo that lit 16 arc street lamps. The first large direct current hydroelectric station came into operation two years later with a hydropower plant that powered a paper mill in Appleton, Wisconsin. The first alternating current hydroelectric power plant came online near the Willamette Falls in Oregon in 1889. By 1890, there were 200 hydroelectric power plants in the United States. The first pumped hydropower project was built in New Milford, Connecticut, in 1930 by the Connecticut Electric Light and Power Co. A powerhouse on the American River was one of the first power plants in the country, powering homes and businesses in Sacramento. When it opened in 1895, it operated the longest transmission line in US history, and one of the first to use the three-phase alternating current used today.

Dams for industrial and later electricity transformed ecologies of places in significant ways. In New England, fisheries collapsed, and many historic flows remained impounded. China's Three Gorges Dam is believed to have resulted in the extirpation of a freshwater dolphin (Fearnside 1988). The Rhine River in Europe transformed the riparian ecosystem there but served a similar early role in energy development on that continent (Lagendijk 2016). The river was the backbone of an electric generation system in Western Europe. Hydropower development in the Rhine is a case history in cooperation among competing nations at a time of nationalist impulses.

The power of a river is analogous to the power of electricity in some ways. Power is flow as regards both water and electricity; the power of electricity and water flow are both carried by current. Voltage is the "push" behind electric current and would be analogous to the steepness of the river. In both cases, to get a lot of power the current needs to have a lot of flow or be very steep. There is an equation that can be used to estimate the power potential of any water flow down any steepness (Box 6.1). There also needs to be many assumptions made about the efficiencies of various parts,

Box 6.1 Estimate the Power Potential of a Dam

The proposed dam would have a height of 400 ft and a flow of 2,000 gal per second. How much sustained power could it produce?

The efficiency is the product of transferring energy from turbine to gear (88%), from gear to gear (90%), from gear to generator (92%), and then generator to electricity (90%).

$$\text{Efficiency} = 0.88 \times 0.9 \times 0.92 \times 0.90 = 0.66$$

The power of a river equation is as follows:

$$P = \frac{0.58 \times 2{,}000 \times 400}{5.3} = 87{,}547\,\text{W} = 87.5\,\text{kW}$$

The amount of power is 87.5 kW.

inter-seasonal or inter-annual rainfall changes, and the like. But a rough estimate of power potential can go a long way to estimating the power of a river.

Hydropower is power that comes from the falling or flow of water. Most hydroelectric power plants work by transferring potential energy to kinetic energy to electricity or simply by just transferring kinetic energy to electricity. Hydroelectric energy comes from capturing the power of moving water; as the water is released from the reservoir, water flows through a turbine, exerting pressure on the blades, making the blades spin around a center point. The spinning action generates kinetic energy, which is converted to electricity; the motion activates the generator, by producing electricity for the power grid (Figure 6.2).

The amount of power generated by a specific dam and river system depends on several factors. The key variables are the flow rate, the height (often referred to as the head), and the efficiency of the system. The efficiency of the system can be reduced to four dimensionless factors—the efficiency of the turbines, the flow through the penstock, the gears, and the turbine. The efficiency factors are multiplied together to get an overall flow rate.

Hydropower is renewable insofar as the flow of water is dependent on the sun and the water cycle. Hydropower is also versatile, as dams can control the flow of water and electricity generation. Dams can go from producing no electricity to giving maximum output in an instant, making energy available on demand. Farmers can also use the stored water to irrigate crops. The reservoirs of hydroelectric plants can be locations for recreational activities such as boating and fishing, potentially stimulating the economy in the area. While hydropower has many benefits, it also has many

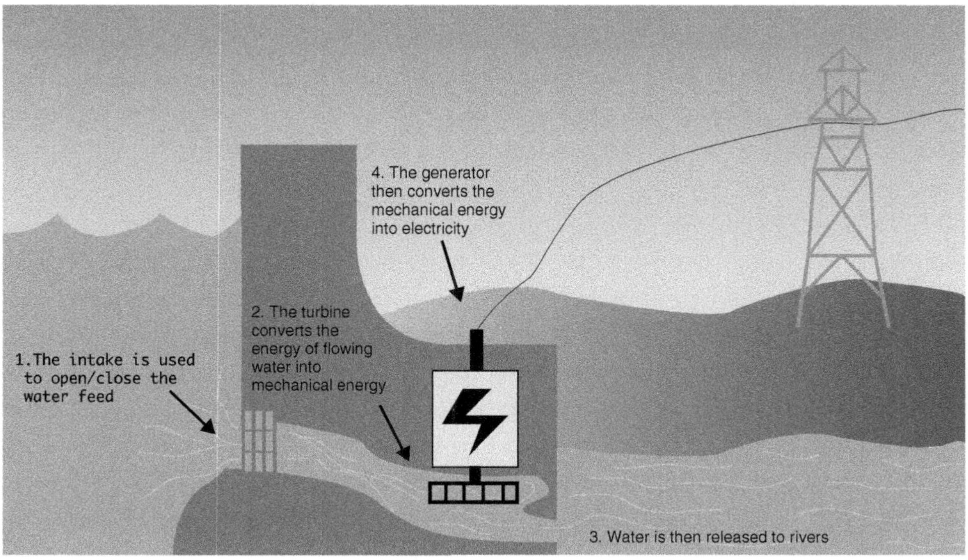

Figure 6.2 How a dam and reservoir are used to generate electricity. Source: JensenMCD/Hydropower Dam/https://commons.wikimedia.org/wiki/File:Hydropower_Dam_Diagram.svg/last accessed on September 18, 2025.

environmental costs. Over 84,000 dams were built in the United States in the twentieth century (Reisner 1993). In the United States today, there are 19,700 large dams (over 26.3 ft tall) and storage capacity of 20,000 m^3 (Figure 6.2) (East and Grant 2023).

Legendary North American folk musician Woodie Guthrie was hired by the Bonneville Power Administration (BPA) in the 1930s to travel and write about the people, communities, and landscapes of the Columbia River Valley. Many of the dams that were built before the 1960s are without hydroelectric power, an estimated 80,000 unpowered dams. Research often finds trade-offs between ecological values and energy. Dams in streams may cause less harm to migratory fish, but dams in larger waterways have a greater value to fish habitat because of the larger flow and storage capacity. Design principles are needed for dams aimed to increase ecological sustainability, including adding freshwater reserves around the remaining watershed catchments, building dams in areas of tributary watersheds and not in the main downstream areas of rivers, making sure that habitats are provided between dams that will encourage and support species reproduction and preserve offspring, avoid using groundwater in arid areas, and finally, making decisions at the scale of river basins. Some of the benefits attributed to large hydropower dams are displaced fossil fuels, storage for urban water supplies, flood control, irrigation, and even the creation of jobs. Hydroelectric power generation has some major impacts, including loss of habitats, forests, streams, restriction of river flow, and negative effects on aquatic species like salmon.

Hydropower can change stream flows, increase turbidity, displace or lock-in habitats, change fish migration patterns, degrade wilderness resources, and increase erosion and landslides. Large mega-dams cause ecological, environmental, and social impacts. Like many large infrastructures with a lot of rebar and cement, dams have a habit of going over budget during construction. New dam projects are subject to social opposition or may require ecological mitigations that can delay projects or lead to their outright cancellation. Here, the social frictions that emerge in different visions of the river as a socio-ecological force are fruitful areas for critical energy scholarship.

Developing countries or poorer regions of wealthier countries are often at the frontlines of new dam development. An important area undergoing energy transitions toward increased hydropower is in eastern India, in the Himalayas where monsoons can bring substantial precipitation but may impact local people and ecologies (Rampini 2022). These projects may harm indigenous or landless people's livelihoods, but are also controversial because of their strategic geopolitical location near China. In Brazil's River Xingu basin in Amazonia, dams are promoted as clean energy sources while they transform ecosystems and livelihoods (Hall and Branford 2012). Yet, there are accusations of human rights violations and severe impacts to rain forests and wetlands for reservoir construction. Research on dams has identified different development practices and has suggestions for indigenous consultation, consent, and incorporation, if not leadership, in planning (Kelly et al. 2024).

In the United States, the Federal Energy Regulatory Commission (FERC) is the primary authority responsible for licensing and regulating nonfederal hydroelectric power facilities. Initially, FERC was not required to consider environmental protection in its project licensing decisions. This changed with the passage of key environmental laws: the National Environmental Policy Act (NEPA) in 1970, the Clean Water Act in 1972, and the Endangered Species Act in 1973 (Schramm et al. 2016). Further regulatory obligations were added with the Electric Consumers Protection Act (ECPA) of 1986, which mandated that FERC consider the protection, mitigation of damage to, and enhancement of fish and wildlife, environmental quality, and recreational opportunities during licensing decisions (Schramm et al. 2016). In addition to these statutory requirements, FERC is obligated to incorporate fishway prescriptions issued by the US Fish and Wildlife Service or the National Marine Fisheries Service. It must also establish minimum streamflow levels to comply with state water quality certifications, typically under Section 401 of the Clean Water Act. While these changes significantly strengthened environmental oversight, they have also increased the complexity and duration of the licensing process. Meeting mitigation and environmental compliance requirements has become more challenging for hydroelectric project developers.

Dams are a relatively cost-effective source of low-carbon electricity, particularly when compared to other renewable energy technologies such as solar, wind, and geothermal. In addition to generating power, dams often serve multiple functions simultaneously. They create water reservoirs that provide water storage, flood control, and opportunities for recreational activities such as boating, fishing, and rafting.

There are several types of hydropower systems. Run-of-river facilities divert water from a river or stream into a turbine system, where electricity is generated before the water is returned to its natural channel. These systems typically have minimal water storage and are dependent on seasonal flow patterns. Pumped storage facilities, on the other hand, operate as energy storage systems (Box 6.5). During periods of low electricity demand, water is pumped from a lower-elevation reservoir to a higher-elevation reservoir. During peak demand, the stored water is released back down to drive turbines and generate electricity. This system acts like a giant battery, helping to balance supply and demand on the grid.

Dam failures throughout history have caused some of the most devastating human disasters. In 1975, the Banqiao Dam in Henan Province, China, collapsed after record-breaking rainfall from Typhoon Nina overwhelmed the structure. The disaster caused a cascade of dam failures and resulted in the deaths of an estimated 85,000 to over 200,000 people, with as many as 11 million left homeless. It remains one of the deadliest dam failures in history. Even ancient civilizations experienced the risks of dam collapse. The Great Dam of Marib in present-day Yemen, one of the most sophisticated engineering feats of the ancient world, controlled seasonal floodwaters from around 800 BCE until its final breach in the sixth century CE. The collapse contributed to the decline of the Sabaean civilization, triggering migrations and social disruption. In the United States, the Johnstown Flood of 1889 was one of the nation's worst disasters. The South Fork Dam, located upstream from Johnstown, Pennsylvania, collapsed after days of heavy rain, unleashing a wall of water that killed 2,209 people. It remains the deadliest dam failure in US history. Another major disaster occurred in 1928, when the St. Francis Dam, located near Los Angeles, collapsed in the middle of the night, killing over 450 people. The flood swept through downstream communities and migrant labor camps, exposing flaws in engineering oversight and contributing to major changes in dam safety regulation.

By the twentieth century, dams were being built larger and taller, especially in the western United States. Major rivers—including the Colorado, Snake, Columbia, Rio Grande, Virgin, and Sacramento Rivers—were dammed to support hydropower, irrigation, flood control, and urban growth. One of the most significant structures, the Glen Canyon Dam, opened in 1963. Standing at 216 m (710 ft), it is one of the tallest dams in the United States and forms Lake Powell, the country's second-largest reservoir by capacity. By the 1970s, with financial support from institutions such as the World Bank, large-scale hydropower projects expanded into developing countries. These dams became central to national development strategies, offering the promise of low-cost electricity due to free and abundant water resources. However, their construction often brought significant social, ecological, and political challenges, particularly in regions with weak governance, vulnerable communities, or ecologically sensitive watersheds.

Dams are considered low-carbon energy sources, but they are not zero-carbon. Reservoirs can produce significant amounts of greenhouse gas emissions—particularly methane—as a result of decaying plant material in the flooded areas surrounding the reservoir. These areas often become anaerobic (oxygen-free) environments, which facilitate methanogenesis, a microbial process that releases methane. This effect is especially pronounced in tropical regions, where higher temperatures and dense vegetation increase biological activity and decomposition rates. In addition to methane emissions, reservoirs tend to trap sediments, which can concentrate nutrients and pollutants over time. These accumulated substances can lead to eutrophication, harming aquatic ecosystems, killing wildlife, and degrading water quality. Sediment buildup also reduces reservoir capacity and may eventually require costly dredging or remediation, while further altering downstream flow patterns and compact waterways.

Hydraulic Societies and the Age of Mega-dams

Hydraulic societies are those that depend heavily on complex water management infrastructures, such as dams, canals, and irrigation systems, to support urban water supply, agriculture, and flood control. These societies often emerge in arid or semi-arid environments, where centralized control of water resources is essential for sustaining large populations and productive landscapes. While early thinkers like Karl Marx and Friedrich Engels noted how control over water and labor shaped the organization of certain agrarian states, the most influential formulation of the concept came from anthropologist Karl A. Wittfogel. Writing in the 1950s and 1960s, Wittfogel developed the theory of "hydraulic societies"—also referred to as hydraulic empires or oriental despotisms—to describe ancient civilizations that constructed large-scale irrigation systems and, in doing so, developed centralized, bureaucratic, and often authoritarian governance structures. In his landmark book *Oriental Despotism* (1957), Wittfogel argued that the technical demands of water control—such as organizing labor, maintaining infrastructure, and regulating flow—led to the rise of state power and social stratification. He created a typology of hydraulic societies, showing how they varied across regions, especially those where agriculture was dominant and climatic conditions made artificial irrigation essential. However, Wittfogel's thesis has also been widely debated and critiqued. Critics argue that his framework risks oversimplifying diverse historical trajectories, overlooks local agency, and overstates the link between hydraulic engineering and political authoritarianism. Nonetheless, his work continues to influence studies of infrastructure, state formation, environmental history, and political ecology. Modern scholars often revisit the concept of hydraulic societies to explore how water infrastructures shape power relations, not only in ancient civilizations like Mesopotamia, Egypt, and China, but also in contemporary water-stressed regions, where access to water remains a source of political control, inequality, and conflict (Wittfogel 1957) (see Box 6.2).

Box 6.2 Hydrosocial Cycles of Water Infrastructure-Dependent Communities

The hydrological cycle depicts flows of water in Earth's biosphere. Water evaporates, makes its way into storm systems, and is delivered to land masses, sometimes to high terrain, where it slowly makes its way back to groundwater or the ocean. The hydrosocial cycle aims to put the water cycle into social context by highlighting the political and social dimensions of water (Linton and Budds 2014). By highlighting these aspects, one can more easily attend to the governance challenges and priorities for water systems from a socio-ecological perspective rather than focus on water flows absent from human context. The "hydrosocial" is a term that implies a deep connection between nature and society, one that might be called dependent. Hydrosocial imaginaries, for example, is a term that implies that people bring with them different views of water as a resource, commodity, cultural resource, etc.

Marc Reisner's landmark book *Cadillac Desert* (1993) offers a sweeping account of the conflicts and ambitions surrounding water resources in the American West, particularly focusing on the transformation of the Colorado River and the water engineering projects that enabled the growth of Southern California. Reisner documents the profound ecological, political, and financial consequences of massive water development schemes, including reservoirs, aqueducts, and irrigation systems. In the spirit of critical energy studies introduced in Chapter 2, his work critiques the logic of "reclamation" and the dream of turning deserts into farmland through federally subsidized megaprojects, exposing the tensions between environmental sustainability and expansionist infrastructure dreams.

The themes Reisner explores build on a longer tradition of critical writing about water and dams in the American West. John Muir, the founder of the Sierra Club, opposed the damming of Hetch Hetchy Valley, a battle that became one of the first major national debates over conservation. Edward Abbey's *The Monkey Wrench Gang*, a foundational text for environmental activism, targeted the Glen Canyon Dam as a symbol of ecological and spiritual loss. Meanwhile, Stuart Udall, a former Secretary of the Interior and environmental thinker, also expressed concern over the unchecked expansion of water infrastructure and the environmental costs it imposed.

In Richard White's *The Organic Machine*, the Columbia River is examined not just as a body of water but as a shaped and reshaped energy system. Once known for its mighty rapids and remarkable anadromous fish runs, especially salmon that sustained Indigenous communities for millennia, the Columbia was transformed by a series of dams into what White calls a "machine"—a system engineered for power production, industrial growth, and national defense. Many of these dams were constructed to serve the demands of World War II-era manufacturing, including powering aluminum plants and nuclear weapons production at the Hanford Site. Today, if you include the capacity of its key tributary, the Snake River, the Columbia River hydropower system generates over 36 GW of electricity—roughly equivalent to the total electricity demand of the state of California. This scale of energy production makes it one of the most significant hydroelectric systems in North America, but it has also come at great ecological and cultural cost, particularly for Tribal nations whose fishing rights and cultural ties to the river have been deeply affected.

The Columbia River is also entangled in the history of progressive-era infrastructure development. During the New Deal, the BPA led efforts to promote federal investment in dams as tools for rural electrification and economic uplift. In a fascinating intersection of culture and policy, folk singer Woody Guthrie was hired by the BPA in 1941 to write songs and perform concerts advocating for public support of the Bonneville and Grand Coulee dams. Guthrie's music celebrated the democratic promise of publicly owned power and helped craft a narrative of progress and national unity tied to river development (Figure 6.1 and Box 6.3).

Box 6.3 Three Gorges Dam and the Yangtze River

Visible from space, the Three Gorges Dam is the world's largest hydroelectric power facility, with an installed capacity of 23 GW. It is located on the Yangtze River—the third-longest river in the world—which stretches approximately 3,900 miles and drains into the Pacific Ocean near Shanghai. The dam itself spans 1.5 miles across the river, while its reservoir extends over 400 miles upstream. In addition to its energy-generating capabilities, the dam was also constructed for flood control, designed to protect an estimated 15 million people living downstream on the Yangtze floodplain, as well as approximately 3.7 million acres of agricultural land (NASA 2007). Planning for the Three Gorges Dam began in the 1980s, with construction starting in the 1990s. The dam's power plant began full operations in 2009.

According to scientists at NASA's Jet Propulsion Laboratory in Pasadena, California, the dam's massive reservoir may have regional climatic effects, including a "lake effect" that contributes to increased rainfall and cooler temperatures in the surrounding area due to the large, cooler body of water. Despite its engineering significance, the project has been highly controversial. It led to the forced resettlement of approximately 1.3 million people and has had significant ecological consequences, including the likely extirpation of the Yangtze River dolphin (baiji)—a freshwater species once native to the river system.

Dams and Displacement

Dams and reservoirs are some of the most transformative effects on riparian and arid landscapes. They also have tremendous impacts on the livelihoods of people who live there or who are involuntarily resettled (Roy 2008). Hydropower has historically led to displacement and resettlement of people or major disruptions in ecosystems, with important examples of hundreds of thousands resettled from China, India, Laos, Brazil, Nigeria, and Malaysia (Aiken and Leigh 2015). But even further back, Native American Tribes, residents also of the Colorado and Tennessee River valleys, everywhere dams were built people are displaced, especially in arid areas because rivers can be the only source of water. There are varied experiences with dams and development around the world, which are reshaped by hydrosocial relations of power, knowledge, and expertise (Boxes 6.2 and 6.4).

Mega-dam projects in the twentieth century were very much about the development of "cheap" sources of electricity. The fact that hydropower is a more or less renewable energy resource, means that countries and regions can diversify their fuel supplies in ways that decrease reliance of fuel imports, which can be complicated when countries have unstable currencies or are in situations where they can be shut out of energy supplies altogether. So places around the world aiming to develop cheaply have substantially invested in hydropower capacity.

Geographers Atkins and Hope use hydropower case studies in Bolivia to show renewed interest in their development because of their green attributes, but it looks more like an extension of extractivism (Atkins and Hope 2021). Building off prior work by geographers Lefebvre and Ahlers et al., they describe different domains perceived, conceived, and lived, in cases of uneven development.

The disruption of livelihood strategies is a key impact in developed countries with high levels of poverty. The World Commission on Dams has long sought to forcibly resettle agrarian/agriculturalists who occupy lands they need for the reservoir and water storage. Despite some attention to the challenges of equitably resettling agriculturalists, many key recommendations have not yet been implemented by the organizations that hope for equitable outcomes.

One complication in some hydropower development schemes is that the resulting partnerships between a regional utility and the dam developer result in the privatization or partial privatization of the electric power supply. In places without adequate governance, there would be concern that price spikes could result in fleecing of customers through overcharging.

The twenty-first century has seen quite an explosion in the growth of massive-scale mega-dams in regions and countries such as Laos, Brazil, Uruguay, and India. Geographer Costanza Rampini writes about how the development of the Himalayas in eastern India has resulted in the restructuring of socio-ecological relations. Costa Rica is a country that developed significant energy resources by way of dams, but seems generally to have had favorable ecological management outcomes and the country is often seen as a model for sharing in the benefits of infrastructure developments.

Resettlement programs are among the complicated social interventions in the development of any energy, flood control, or irrigation project. Resettlement programs often result in rural land users moving to urban areas, and the programs are intended to provide a safety net from poverty in addition to any compensation for land. Research by geographers shows how dam resettlements programs that intend to improve people's livelihoods, but end up commodifying other aspects of life transformed by hydropower resettlement (Green and Baird 2016).

Dams can be sites of social resistance. The São Luiz do Tapajós Dam would have been the second largest in Brazil, but was postponed indefinitely when it was determined that it would submerge land of the Munduruku, an Indigenous people of the Amazon Basin (Hess and Fenrich 2017). The impact on Indigenous peoples is one reason that free and prior informed consent is often related to hydropower (and mining) infrastructures.

The Mekong Delta is a tremendously important agricultural area of Southeast Asia. The area is one of the largest rice growing regions in the world, which is depended upon by hundreds of millions of people for the staple food. In recent years, it has become the site of conflicts of mega-dam construction in the region, including in the headwater around Laos. Some of this is a proxy conflict between the United States,

Japan, and China according to scholars of the region who explain development patterns on shifting regimes of finance (Souvannaseng 2022).

South America's steep terrain also makes it a very suitable resource for hydropower. Peru is favored by a stable and growing economy and the availability of indigenous sources of energy for electricity generation, including hydro and natural gas. The Peruvian electricity sector is among the few in Latin America and the Caribbean that have not confronted an energy crisis in recent years. The power sector in Peru was reformed and restructured between 1991 and 1993, followed by a privatization and concession process. A modern legal and regulatory framework was established in the electricity concessions law of 1992–1993 (see Box 6.4).

Box 6.4 Trust in Numbers: Should We Build the Dam? How Does Numeracy Work in Public Policy?

Science and Technology Studies (STS) scholar Theodore Porter has argued that the bureaucratization of expertise and the growing reliance on quantitative metrics have shaped the development of public infrastructure systems for decades. In his influential work *Trust in Numbers*, Porter examines how quantification becomes a strategy for making decisions appear objective and apolitical, especially in large, bureaucratic institutions. His research focuses on the decision-making practices of the United States Army Corps of Engineers, particularly their use of cost-benefit analysis (CBA) as a central tool for evaluating infrastructure projects. In one case study of a proposed dam and reservoir, Porter shows how the Corps' reliance on quantifiable indicators—such as projected flood damage reduction, land value enhancement, and power generation—was constrained by the types of information they chose to include and the assumptions built into their models.

The case highlights how infrastructures are not purely technical systems, but rather contested socio-political objects shaped by local politics, institutional cultures, and epistemic commitments. Porter's analysis also reveals a broader pattern: societies increasingly rely on numbers as a form of authority, not necessarily because they are more accurate, but because they carry an aura of neutrality, standardization, and impersonality. Quantification, in this sense, serves as a mechanism for creating social distance—allowing decision-makers to present their choices as rational, impersonal, and universal, while often obscuring the specific, local, and subjective dimensions of the issues at hand. The dam project thus becomes not just a story about engineering, but a case study in how numbers can depoliticize contested decisions and erase local knowledge in favor of abstract, aggregated models.

Source: Adapted Porter (1996).

Impacts to Ecosystems and Fisheries

The interventions to the flow of water through ecosystems can have tremendous impacts on local ecologies, plant communities, and the types of animals present in the waterways. Anadromous fish are one of the spectacular outcomes of the evolution of organisms on planet Earth. Fish born in gravel in freshwater pools, migrate to the sea, and spend one or more years in the ocean, before returning to the river they spawned to lay eggs in a gravelly freshwater pool again. The prolific runs of salmonids along the Pacific Coast fed grizzly bears and no doubt fertilized the redwood forests of California, adding not only blue carbon to the sea but also to carbon on land (Figure 6.3).

Concerns about salmon runs on the Columbia River and Snake Rivers have led to several lawsuits, which could result in the dismantling of several dams. The National Marine Fisheries Service, which is in charge of offering expert opinion on infrastructure projects that could put a species into jeopardy under the Endangered Species Act, prepared an evaluation that said the management protections were insufficient to ensure the species survival (Box 6.5).

After many years of building dams, we are entering an era of dam removal. Four key dams on the Lower Klamath River, California's second longest river, were removed, and in 2024 the first Chinook salmon in over a century swam upstream. These dams were the Copco 1 Dam in 1922, Copco 2 Dam in 1925, J.C. Boyle in 1958, and Iron Gate Dam in 1964, with production capacities of 20, 27, 98, and

Figure 6.3 The fish ladder at the Bonneville Lock and Dam on the Columbia River allows various anadromous fish to move upstream. Source: Steven Baltakatei Sandoval / Wikimedia Commons / CC BY-SA 3.0.

Box 6.5 Pumped Hydropower in Low Carbon Energy Systems

Pumped hydropower energy storage systems will get increasing attention because at a very basic level it is much easier to store water than it is to store energy. Pumped hydropower works by pumping water uphill when power might be plentiful and having it available when needed, for example, on peak loads or when renewable intermittencies occur. Many pumped hydropower projects contain enough energy to operate at full power for several days. A proposed pumped hydropower on the Hudson River atop Storm King Mountain was at the center of one of the first major tests of environmental laws and wilderness areas in the United States (Schuyler 2018).

An acritical approach to pumped hydropower storage sees the technology from a technocratic perspective, perhaps only looking at impacts through the lens of greenhouse gases or renewable energy integration. Geographer Alida Cantor and colleagues have conducted a critical examination of pumped hydropower in Goldendale, Washington, contested on the grounds of environmental justice and reproducing colonial power asymmetries with Tribes.

Source: Adapted from Cantor et al. (2023).

18 MW and their removal is an important case study in Tribal leadership and cooperation and good example of eco-cultural revitalization of the local Karuk Tribes cultural practices (Diver et al. 2024). These kinds of understandings are important for energy transitions, as many would view the removal of hydropower as contradictory to the goals of climate action. But there is a bigger picture that puts the questions of greenhouse gas emissions and renewable energy in the context of other important socio-ecological processes, such as the Tribe's relation to salmon. The return of management of the Klamath to the Karuk is an example of restorative justice, not just as in riparian restoration, but the restoration of relations shattered by broken promises, stolen land, and genocide. The full restoration of the Klamath River will end when the diversions of water for agriculture in the upper Klamath Basin are restored to the river. Over 20 rivers in the United States have involved Tribes in dam removal, including the Ottaway, Penobscot, and Elwha Rivers involving the Saint Regis Mohawk Tribe, Penobscot Indian Nation, and Lower Elwha Klallam Tribe (Figure 6.4) (Fox et al. 2022).

One of the major changes facing hydropower in the face of climate change is snowpack and the availability of water. Some climate models project snow levels rising to higher levels in some parts of the world. This could result in less available water during the driest parts of the year and directly translate to less hydropower available.

Figure 6.4 After many years of protest, the Yurok Tribe helped negotiate the removal of several dams on California's second longest river. Source: Patrick McCully / Flickr.

Mini and Micro-Hydro-Political Ecologies

Mini, micro, and pico-hydropower—sometimes grouped as small hydropower—projects are, as they sound, smaller versions of power plants that take water to generate electric power. What are the opportunities for sustainable hydroelectric power production at smaller scales? Can hydropower be used as a domestic renewable energy source to run appliances, for heating, or cooling a home? These developments are usually described as pico-hydropower or micro-hydropower. The sources can be easily obtained as one of the simplest and most consistent forms of renewable energy, depending on the local climate and competing water demands. Micro-hydropower is best where water flowing across a landscape is present. These systems can produce between 5 and 100 kW, depending on the resource and design. A micro-hydro- power system can easily generate power for smaller farm, resorts, or large homes.

The critical energy studies of small-scale hydropower start by interrogating its core assumptions, namely its tendency towards more democratic governance and providing more community benefits. Many small micro hydro projects are related to the Clean Development Mechanism set up under the Kyoto Protocol, where carbon offset proceeds are used to make renewable energy infrastructure investments (Bin 2021).

Small hydropower projects have also proliferated in many developing countries and critical examinations of these trends find a few surprising outcomes. Geographers studying patterns of small hydropower development have found that despite claims of a global boom in small hydropower development, a closer inspection finds a longer trend of small development rather than the rapid growth implied in a boom (Ptak et al. 2022).

With pico-hydropower, the turbines are even smaller and produce around 1–5 kW to similarly generate enough energy to power the same living standards as a micro-hydropower system. Pico-hydropower systems are rarely fed into the power grid, but instead use electricity on-site or store it. Attaining the water source for pico-hydropower usually begins with identifying where water enters and exits from a creek, whether any permit is required or there are water rights issues, and where to locate how the water enters and exits the household. Other potential sites of pico-power include rain gutters, sprinklers, hoses, dishwashers, sinks, showers, toilets, and washing machines. New technologies can provide power with in-pipe turbines near water facilities. Spherical turbines can be installed inside gravity-fed water pipelines. A system like this is installed for the Portland Water Bureau in Portland, Oregon. Using four turbines, the micro-hydro system produces 900 MWh per year for the electricity grid.

Takeaways/Highlights

In this chapter, we have learned that

- Hydropower transformations are some of the most significant landscape transformations on Earth displacing ecosystems and wildlife, people, and livelihoods.
- The development of hydropower is a widely used development strategy, especially in developing countries. But for these reasons they often result in dispossession as the populations moved can be some of the less well off or more vulnerable.
- Knowing the flows by volume and height of a dam can inform us about the approximate amount of power it can deliver.
- Smaller hydropower systems can add up to a lot of power collectively, but pose their own challenges, even social ones.

Questions for Review

1. Why are mega-dams proposed and built? What are the motivations for their construction? What has been the pattern of construction for mega-dams? Where were they built first, and where are they being built today? Do mega-dams make for reasonable development strategies?
2. What might remedy conflicts over hydropower projects? What are the interventions that might help bring benefits to communities, with fewer impacts to ecologies, if not enhance local communities?
3. What are the key differences between a mega-dam hydroelectric project and micro-hydropower one, and how might these differences "structure" social relations in each differently?

Exercise 6.1

Hydropower projects are often framed as clean energy solutions, but they raise important social and ecological questions.

a. Create a list of key socio-ecological considerations involved in dam development. These may include biodiversity loss, displacement of communities, sediment flow disruption, Indigenous rights, or water quality.
b. Choose a case study of a hydropower project from anywhere in the world. Summarize the ecological impacts on the river ecosystem, such as changes in species populations, water temperature, or migratory patterns (e.g., salmon, dolphins, amphibians).
c. Reflect: How were the environmental trade-offs justified or contested? Were there mitigation strategies in place, and were they effective?

Exercise 6.2

Controversies over dams are common and often politically charged.

a. Find a hydropower project that has faced public opposition—either historical or contemporary. Examples include the Belo Monte Dam (Brazil), Glen Canyon Dam (U.S.), or Myitsone Dam (Myanmar).
b. Identify the key stakeholders: governments, international lenders, corporations, Indigenous communities, environmental nongovernmental organizations, etc.
c. Describe each group's perspectives and concerns. What trade-offs are they weighing?
d. Consider the counterfactual: What are the consequences of *not* developing the dam (e.g., missed economic growth, lack of electrification, ongoing fossil fuel dependence)?
e. Analyze what types of development or resistance followed the controversy. Did it reshape policy, environmental law, or regional planning?

Instructor note: Assign students different dam projects from various continents to explore how geography, economy, and culture shape hydropower debates.

Exercise 6.3

Drawing on insights from STS scholars such as Theodore Porter in the Box 6.4 in this chapter, consider how numbers and metrics shape hydropower planning.

a. Research the use of CBA or environmental impact assessments (EIA) in a dam approval process (e.g., by the U.S. Army Corps of Engineers or a World Bank-funded project).
b. What kinds of data and assumptions are included in the analysis? Which social or ecological impacts are difficult to quantify and might be left out or downplayed?
c. Reflect: Do numbers help create a sense of objectivity, or do they mask power dynamics and value choices? Provide specific examples.
d. Propose how more inclusive decision-making tools could be integrated into infrastructure planning.

Exercise 6.4

Hydropower development often erases long-standing cultural and ecological relationships with rivers.

a. Read a passage from a literary or environmental text—such as Edward Abbey's *The Monkey Wrench Gang* or Richard White's *The Organic Machine*—that deals with rivers and dams.
b. Analyze how the author frames the river: Is it a machine? A sacred place? A lost home?
c. Choose a hydropower project (e.g., Three Gorges, Glen Canyon, Columbia River) and research the cultural, historical, or spiritual significance of the river prior to damming.
d. Reflect on what was gained and lost—not just in material terms, but in narratives, memories, and identities tied to the river landscape.

References

Aiken, S.R. and Leigh, C.H. (2015). Dams and indigenous peoples in Malaysia: Development, displacement and resettlement. *Geografiska Annaler: Series B, Human Geography* 97 (1): 69–93.

Atkins, E. and Hope, J. (2021). Contemporary political ecologies of hydropower: insights from Bolivia and Brazil. *Journal of Political Ecology* 28 (1): 246–265.

Bin, D. (2021). Discussion on the development direction of hydropower in China. *Clean Energy* 5 (1): 10–18.

Cantor, A., Turley, B., and Maxfield, K. (2023). *Energy Storage and Environmental Justice: A Critical Examination of a Proposed Pumped Hydropower Facility in Goldendale*. Washington, DC: Antipode.

Diver, S., Oberholzer Dent, J.R., Sarna-Wojcicki, D. et al. (2024). Recasting Klamath Dam removal as eco-cultural revitalization and restorative justice through Karuk tribal leadership. *Water* 16 (16): 2295.

East, A.E. and Grant, G.E. (2023). A watershed moment for western US dams. *Water Resources Research* 59 (10): e2023WR035646.

Fearnside, P.M. (1988). China's three gorges Dam: "fatal" project or step toward modernization? *World Development* 16 (5): 615–630.

Fox, C.A., Reo, N.J., Fessell, B., and Dituri, F. (2022). Native American tribes and dam removal: restoring the Ottaway, Penobscot and Elwha Rivers. *Water Alternatives* 15 (1): 31–55.

Green, W.N. and Baird, I.G. (2016). Capitalizing on compensation: hydropower resettlement and the commodification and decommodification of nature–society relations in Southern Laos. *Annals of the American Association of Geographers* 106 (4): 853–873.

Hall, A. and Branford, S. (2012). Development, dams and Dilma: the saga of Belo Monte. *Critical Sociology* 38 (6): 851–862.

Harlan, T. (2023). Low-carbon frontier: renewable energy and the new resource boom in Western China. *The China Quarterly* 255: 591–610. https://doi.org/10.1017/S030574102200159X.

Hess, C.E.E. and Fenrich, E. (2017). Socio-environmental conflicts on hydropower: the São Luiz do Tapajós project in Brazil. *Environmental Science & Policy* 73: 20–28.

International Energy Agency (2024). Renewables 2023.https://www.iea.org/reports/renewables-2023 (accessed 11 November 2025).

Kelly, S., Heiret, Y.S., Guerra-Schleef, F. et al. (2024). Epistemic justice as energy justice: reflections from a transnational collaboration on hydropower and Indigenous rights. *Climate and Development* 16 (9): 784–797.

Lagendijk, V. (2016). Europe's Rhine power: connections, borders, and flows. *Water History* 8: 23–39.

Linton, J. and Budds, J. (2014). The hydrosocial cycle: defining and mobilizing a relational-dialectical approach to water. *Geoforum* 57: 170–180. https://doi.org/10.1016/j.geoforum.2013.10.008.

NASA (2007). *Rise of the Three Gorges Dam*. NASA.

Porter, T.M. (1996). *Trust in Numbers: The Pursuit of Objectivity in Science and Public Life*. Princeton University Press.

Ptak, T., Crootof, A., Harlan, T., and Kelly, S. (2022). Critically evaluating the purported global "boom" in small hydropower development through spatial and temporal analysis. *Renewable and Sustainable Energy Reviews* 163: 112490. https://doi.org/10.1016/j.rser.2022.112490.

Rampini, C. (2022). Climate and energy justice along the Brahmaputra river in Northeast India. *Environment and Planning E: Nature and Space* 5 (4): 1919–1946.

Reisner, M. (1993). *Cadillac Desert: The American West and Its Disappearing Water.* Penguin.

Roy, A. (2008). *The Shape of the Beast: Conversations with Arundhati Roy.* Penguin Books India.

Schramm, M.P., Bevelhimer, M.S., and DeRolph, C.R. (2016). A synthesis of environmental and recreational mitigation requirements at hydropower projects in the United States. *Environmental Science & Policy* 61: 87–96.

Schuyler, D. (2018). *Embattled River: The Hudson and Modern American Environmentalism.* Cornell University Press.

Souvannaseng, P. (2022). Liquidated: US/Japan-Chinese rivalry, financial crises, and explaining shifts in hydropower finance regimes in the Mekong. *Asian Perspective* 46 (1): 49–75.

Wittfogel, K.A. (1957). Oriental despotism: a comparative study of total power. New Haven, CT: Yale University Press.

Worster, D. (1992). *Rivers of Empire: Water, Aridity, and the Growth of the American West.* USA: Oxford University Press.

Further Reading

Roy, A. (2001). *Power Politics.* Boston, MA: South End Press.

Reisner, M. (1993). *Cadillac Desert: The American West and Its Disappearing Water.* Penguin.

White, R. (1996). *The Organic Machine: The Remaking of the Columbia River.* Macmillan.

Worster, D. (1992). *Rivers of Empire: Water, Aridity, and the Growth of the American West.* USA: Oxford University Press.

Chapter 7

Nuclear and Geothermal

Nuclear and geothermal electric power are the topics of Chapter 7. It may seem like a rather arbitrary grouping of energy resources, and maybe it is. However, both technologies rely on hot rocks to make electricity, and most of that heat comes from radioactive decay; geothermal from metals releasing heat inside the Earth and nuclear from fuel rods. They also have similar operating profiles as relatively "firm" resources and even utilize overlapping power-generating equipment, with steam turbines and condensers, and issues with hot wastewater depletion can sometimes make their operations sensitive to very warm summer temperatures. So, the grouping works to introduce key topics that the two sources of energy have in common. But there are important differences too. The chapter opens with a short history of nuclear power development and the associated yellow cake towns that developed in the United States and similar towns globally through the peak of the global nuclear weapons race. The next section describes the types of nuclear fuel processes, power plant types, and describes the different kinds of radiation and radiation dosage measurements used to measure the risk of exposure. Nuclear power is a low-carbon source of large quantities of energy in a handful of countries. In these countries, there is much concern that the older nuclear power plants across the country are increasingly expensive to operate. Geothermal power for electricity is much more regionally specific and limited, but it is common in many parts of the world. Both technologies use steam turbines, steam condensers, and cooling towers, and deliver power more or less consistently. It is also similar to nuclear power in minor, but important, ways such as producing large quantities of wastewater and using radiation as one of three heat sources on the Earth that allows this to be possible. Both technologies are capable of operating for long time periods.

Nuclear Power, Atoms for Energy

The origins of nuclear power date back to the discovery of radioactivity by Henri Becquerel in 1896. Nobel Prize-winning physicist Marie Curie was one of the first to study and develop applications for radioactive materials and is credited with developing ways to separate radium from other materials for research and application. Ernest Rutherford later won the Nobel Prize in Chemistry for developing the laws that govern radioactive decay of different elements, with the concept of the half-life, the time it takes for 50% of the atoms to split (Box 7.1). The first to demonstrate that a neutron could split an atom was Enrico Fermi in 1934, which kicked off a race to explore whether it was possible to split larger atoms and release large amounts of energy. After the discovery of fission in 1938, and amid concern that Germany might develop nuclear weapons,

Box 7.1 What Is the Energy in Radioactivity?

Several types of nuclear energy are released when the nucleus of an atom breaks apart or disintegrates—a process known as radioactive decay. This process emits nuclear radiation in various forms, including alpha particles, beta particles, gamma rays, and X-rays, each of which interacts differently with matter and has distinct implications for biological systems. Radiation is measured in multiple ways, depending on what aspect is being analyzed: (1) the rate of atomic disintegration (i.e., how many atoms decay per second), (2) the amount of energy released, or (3) the amount of energy absorbed by a material or living tissue. These differing measurement systems reflect the complex nature of radiation exposure. While low levels of background radiation are generally considered safe and occur naturally in our environment, high levels of exposure—especially over a short period—can be harmful or fatal, leading to cellular damage, cancer, or acute radiation sickness.

One of the challenges in understanding radiation is that different types of radiation have different penetrating abilities and biological effects. For instance, alpha particles can be stopped by a sheet of paper or skin, but are dangerous if inhaled or ingested. Beta particles penetrate further and can affect skin and internal organs.

Gamma rays and X-rays are highly penetrating and require dense shielding (like lead or concrete) to stop them. Moreover, not all radiation measurements are equal. Some focus purely on physical energy, while others incorporate a biological weighting factor to account for how damaging that radiation is to living tissues. For example, the Rad measures the absorbed energy, while the Sievert

(Sv) and Rem reflect biological impact, accounting for the type, target, and absorbed energy of the radiation.

Several different types of nuclear energy are released when a nucleus of an atom breaks apart or disintegrates. Nuclear radiation is measured based on rates of atomic disintegration, energy released, or energy received by an object. A small amount of low-level radiation exposure is probably alright, but high levels of radiation can have significant impacts on living organisms. Part of the confusing nature of radiation measurement is that different kinds of radiation—X-rays, gamma rays, and beta rays—have different implications for different materials. What follows are the different ways we approach radiation measurements. Note that some metrics measure the amount of radioactivity, energy, while other metrics represent a degree of energy absorbed or biological injury.

5 Bq—1 nucleus disintegration per second (s)
5 Ci—37,000,000,000 Bq (Beq)
5 Rad—radiation dosage of 1 joule (J) per kilogram (kg)
5 Rem—Roentgen equivalent mean dosage in Rads to cause some biological injury as one Rad of X-rays or gamma rays
5 Sievert (Sv) = 100 rem = 0.001 milli-Sievert (mSv)
5 X-rays and gamma rays: 1 rad = 1 rem = 10 mSv
5 Neutrons: 1 rad = 5–20 rem = 50–200 mSv
5 Alpha rays (He4): 1 rad = 20 rem = 200 mSv

There is little debate about the dangers of high-level radiation and health impacts. But there is less consensus and more questions about how low-level radiation impacts public health and workers (Perrow 2013; Boice et al. 2024).

the United States launched the Manhattan Project to make uranium into atomic weapons using laboratories and research teams across three primary sites Hanford, Washington, where plutonium would be made; Oak Ridge, Tennessee, where uranium fuel would be made; and Los Alamos, New Mexico, where the first bombs made of each fuel would be produced. All three places remain important to nuclear weapons and energy research, development, and demonstration. Scientists deployed what they knew about chain reactions in radioactivity decay to the development of weapons and eventually reactors for electricity and submarines.

The first nuclear-powered electricity generator was turned on in 1954. The Obninsk Nuclear Power Plant in the Soviet Union moved nuclear power from the military to civilian uses. It was a relatively small 5-MW power plant compared to today's reactors that can be measured in gigawatts. But it marked the introduction of a new workhorse resource for energy systems. Two years later, the first full-scale commercial nuclear power plant, Calder Hall, began delivering power to the United Kingdom's electric grid. The Shippingport Atomic Power Station opened in 1958 in Pennsylvania and became the first large-scale nuclear power plant in the world, as part of President Eisenhower's "Atoms for Peace" program. An experimental breeder reactor in Idaho was the first to produce electricity in 1951. For much of the period from 1970 to 2015, nuclear power produced about 20% of the United States' electricity demand.

Like many of the chaotic conceptions discussed in Chapter 2, nuclear power is many things at once. Nuclear power has been the topic of controversy, debate, misconstrued risk, failed design and engineering ideas, and a low-carbon energy source gaining favor again. Nuclear power is entangled with nuclear weapons, in history, through uranium and reactor fuel supply chains, and through concerns about weapons proliferation. Nuclear power has many dimensions that have different meanings to different epistemic communities, groups that have shared understandings and ways of describing the world. Nuclear power in some Indigenous communities in the western United States is associated with uranium mining. It is a decarbonization strategy for some energy modelers. With all these different ways of thinking about it, it is easy to see how nuclear power is a chaotic concept.

Nuclear power is one of the more hotly contested arenas in the decarbonization discourse. Once the superstar of the energy family in the 1950s and 1960s, the nuclear power industry in 2025 is saddled with many challenges in some parts of the world, but also successes that point to new opportunities. The COP28 summit in 2023 saw a proposal by 20 countries that seeks to triple the use of global nuclear power. Many older operating nuclear power plants struggle to compete with newer lower-priced sources. Newer projects face construction challenges, leading to significant cost overruns and delays in some parts of the world like the United States and Europe. Elsewhere, in places like China and India, projects seem to be built more efficiently, on time, and more or less at planned cost (Figure 7.1). With climate change motivating a quest for firm, low-carbon electricity resources, there is increased interest and even funding available to build new projects, including entrepreneurs and financiers pursuing next-generation nuclear facilities and technologies. Nuclear power plants simply produce so much low-carbon electricity from a relatively small facility that they may offer an opportunity to satisfy a large portion of future electricity demand. They have provided on the order of 20% of global power for many decades in some parts of the world, with 70% in France and up to 30% in Japan prior to the 2011 earthquake and tsunami (World Nuclear Association 2025).

Clean power is the term that encompasses a broad array of resources from renewables to nuclear and fossil resources with carbon capture. With climate change activism and advocacy now a global social movement, a significant portion of those engaged in policy and advocacy look favorably at clean energies like nuclear power because it produces a significant amount of electricity with very little life cycle greenhouse gases (GHG). There are multiple currents, but most share in common an

Figure 7.1 Rooppur nuclear power plant, Rooppur Pabna, Bangladesh. Source: Sanvi Ahmed Saim / Wikimedia Commons / CC BY-SA 4.0.

aspiration to rapidly move global energy sources away from fossil fuels. The global nuclear power production is at a 30-year low, with 440 nuclear reactors generating 9% of global electricity, but with 65 under construction that should bring that percentage back up in the relatively near future (World Nuclear Association 2025).

Nuclear energy's release of heat is a result of splitting uranium 235, one of several isotopes of that element found in rock at about 2–3%, depending on the origin of the ore. Most of the balance is nonfissile uranium, U-238. Each split uranium 235 atom releases heat, atomic fragments, and one or two neutrons that can cause other uranium 235 atoms to also split. A nuclear fuel is prepared to deliver a consistent rate of radioactive decay of uranium-235 when put in fuel rods in the core of a nuclear power plant. This heat is used to make steam, turn a turbine, and spin a generator. The reactor core is kept cool usually by circulating water in the tower. Passively cooled power plants, which are not yet commercially in operation, propose using gravity, density, or some other physical characteristic to ensure that the cores of reactors are kept cool without the need for an external energy source.

Most nuclear power plants are either boiling water reactors (BWRs) or pressured water reactors (PWRs). The major difference is in how the heat is transferred in the process of making steam. BWRs are designed to boil water in the reactor core and send that steam directly to a turbine. The technology was first researched by Argonne National Labs in the 1940s and eventually became the reactor sold widely throughout the world by General Electric Company. In 1957, the Vallecitos BWR built by General

Electric in Sunol, California, was the first in the world to deliver commercial electricity to the grid (Fennern 2018).

The Shippingport Nuclear Power Plant began operations shortly after in 1958 and became the widely adopted PWR. These power plants use a heat transfer fluid to produce the steam that spins the turbine, instead of directly "boiling the water" like a BWR. The technology was first developed for naval propulsion, but it became the most widely produced type, with designs led by Westinghouse (Cummins and Matzie 2018).

Opposition by environmental and antiwar organizations long linked nuclear power to nuclear winter, postnuclear war stories, and nuclear weapons accidents. The nuclear power plants at Seabrook in New England and Diablo Canyon in California (Box 7.2) were an early focus of protest for the antiwar movement that saw power and weapons as interconnected. The New England group called themselves the clamshell alliance, while the Diablo Canyon opponents called it the abalone alliance (Downey 1986; Wills 2012). In reality, the time of nuclear power's great slowdown was less to do with

Box 7.2 Diablo Canyon: Should California Keep Diablo Canyon Operating?

Diablo Canyon is a nuclear power plant located on the central coast of California that has been in operation since 1985. With a capacity of 2.24 GW, it remains the largest power plant in California nearly 40 years later. During the 2020 heatwaves, when California faced significant strain on its electrical grid—due to extreme temperatures and wildfires that compromised other parts of the energy system—Diablo Canyon played a critical role in maintaining grid stability for the state's nearly 40 million residents. However, the plant's future remains the subject of ongoing debate. Since 2016, Diablo Canyon has been slated for closure by 2025, largely due to its use of a once-through cooling system that draws in approximately 2.5 billion gallons of seawater per day, causing substantial marine life mortality through entrainment and impingement. The California State Water Resources Control Board, which oversees water quality and coastal resources, determined that this cooling method posed unacceptable ecological risks. The board's decision, challenged in court, was ultimately upheld after a lengthy legal battle.

Research led by Professor Pete Raimondi, a marine ecologist at UC Santa Cruz, has documented long-term ecological impacts from the plant's operations. Studies show shifts in marine species composition and warming of the local marine environment in Diablo Cove, attributable to thermal pollution from the plant's discharge (Steinbeck et al. 2007). These changes also have implications for "blue carbon" ecosystems—marine and coastal systems that sequester carbon and contribute to global carbon balance.

Despite the ecological concerns, growing energy demand and climate policy targets have reignited interest in Diablo Canyon as a source of firm, zero-carbon electricity. The plant's value was highlighted again in light of transmission bottlenecks between southern and northern California, a long-standing challenge for the state's grid. Although a third north–south high-voltage transmission line was completed in 2002, the prospect of a fourth is unlikely due to cost, permitting complexity, and land use constraints. Further complicating the debate is the Diablo Canyon Call Area, a site identified for potential offshore wind development. Developers have expressed interest in leveraging the plant's existing substation and coastal transmission infrastructure, adding another layer of strategic importance to the site. Public understanding of the plant's status is often shaped by a 2016 agreement involving PG&E, the Natural Resources Defense Council (NRDC), labor unions, and others, which proposed a phased shutdown of Diablo Canyon as part of a broader clean energy transition plan. While that agreement played a major role in setting the plant's original closure date, it is sometimes mistakenly viewed as the sole reason for the planned retirement. As of 2023, Diablo Canyon's operating life has been temporarily extended until at least 2030, supported in part by federal and state funding for grid reliability and climate resilience. This makes Diablo Canyon a compelling case study in the politics of infrastructure retirement, where valuable but aging capital assets may be devalued or revived amid shifting policy, economic, and environmental priorities (Nelson and Ramana 2023).

The broader lesson from Diablo Canyon is that infrastructure decisions are rarely simple. They are shaped by competing narratives, technical constraints, environmental impacts, policy trade-offs, and uncertainty about future conditions. In the case of nuclear power in California, the story is still being written—and like many in the energy transition, it remains multifaceted and deeply contested.

Three Mile Island and more to do with flattening electricity demand and cost overruns, which sidelined nuclear power for many decades in the United States and slowed its ascendance in developing countries. Despite years of stagnation, 19% of all electricity generated in the United States in 2018 was still from nuclear power, a percentage that continues to go down from a peak in the 1990s. Only two new nuclear plants in the United States were ordered after 1978, despite continued adoption globally. Most US nuclear power plants are beyond their expected lifetime, and a handful are up for retirement in the near future or have reported concern about economic viability. Concerns about losing these sources of electricity have prompted some energy policymakers to consider ways to make them more competitive or offer a bonus for low-carbon resources. If nuclear power plants are replaced with fossil fuel resources—coal and natural gas—the overall implications would raise GHG emissions. Some places like California are replacing nuclear power plants with renewable energy and energy storage. The question of nuclear power has no one-size-fits-all solution. Different countries are going to have to think through these questions in different ways and decide what path they want to take.

Chernobyl is the world's most widely known big nuclear power accident caused in part by design flaws and operator neglect. The Three Mile Island power plant accident occurred when an equipment failure resulted in the partial meltdown of a reactor. Fukushima Daiichi accident, the quake, and subsequent 14-m (45.9 ft) tsunami caused the main power to fail and the backup power system to be swept into the Pacific Ocean. There are numerous cases of accidents in addition to the accidents at Chernobyl, Fukushima Daichi, and Three Mile Island. Windscale was one of the first accidents in Europe (Garland and Wakeford 2007). Accidents such as the Church Rock disaster from a major uranium mining tailings pond breach that contaminated the Navajo Nation are illustrative of the far-reaching impacts from nuclear development that are far from power plants.

Nuclear power followed a steep learning curve with many criticality accidents in the early stages of development, including accidents that happened during fuel processing and also experimental and pilot reactor design. There were deadly lessons learned from the so-called "radium watch girls," of the Radium Watch Corporation, who were overexposed to the radioactive material while painting radium onto watches, were one of the first widespread exposures to workers in an occupational setting (Gunderman and Gonda 2015). There have also been long-standing cleanups and contamination incidents from yellow cake uranium mining towns to the places where fuels are processed. Hanford, Washington, in the United States is the largest cleanup site for radioactive and toxic waste—mostly produced from making plutonium for US nuclear weapons (Cram 2016). The lessons learned from the radium watch girls, and nuclear weapons testing from the Trinity test to Hiroshima and Nagasaki. These historic situations can provide lessons to improve the safe design and operation of nuclear power plants. Historically, it is uncommon to see the same exact = accident twice because the industry and facilities learn from other disasters and mistakes. The safety curve of complex machines can be steep.

Much of the conventional narrative about the history of nuclear power suggests that the expansion of the industry in the 1960s and 1970s was stymied primarily by public opposition and environmental activism. While it is true that antinuclear sentiment, fueled by concerns over radiation, safety, and environmental impacts, played a role in shaping public perception and political discourse, it is not entirely accurate to attribute the decline of nuclear power deployment solely to protest movements. Indeed, cost overruns, construction delays, and a flattening of electricity demand forecasts were arguably more significant factors in the slowdown of the nuclear industry in the United States. Many nuclear projects experienced massive budget increases and extended timelines, eroding investor confidence and public trust in the technology's feasibility. In addition, the energy crises of the 1970s and the advent of energy efficiency measures led to a revised outlook on future electricity demand, making large-scale nuclear investments less attractive. Globally, however, the trajectory of nuclear power has diverged. The nuclear industry in the United States and much of

Europe has remained relatively stagnant or even declined in some areas, due in part to regulatory complexity and market competitiveness from renewables. By contrast, countries like China, Russia, India, and the United Arab Emirates are actively expanding their nuclear capacity and collectively constructing dozens of new nuclear power plants. In these contexts, nuclear energy is framed as a critical component of energy security, decarbonization, and technological modernization.

A question on the minds of many energy thinkers today is whether nuclear power is reliable or needed to ensure a reliable grid. Nuclear power plants have a high-capacity factor, meaning they run a very large portion of the year. Some term this "clean firm power," noting that it is not subject to the same intermittencies with renewables. At the same time, global warming is causing some nuclear power plants to go offline when source water temperatures are too high to provide the power plant with adequate cooling. In 2019, a July heatwave shut down 10% France's installed nuclear capacity due to river temperatures running too warm. In other parts of the world, coastal power plants face sea-level rise, and debris carried by floodwaters has at times clogged cooling systems as well. Nuclear power plants built and designed in the 1950s and 1960s are sometimes located in places before we considered potential impacts from climate change, posing an adaptation–mitigation dilemma (Kopytko and Perkins 2011).

The focus on deploying low-carbon energy resources led to a brief nuclear renaissance by the time the United Nations' Intergovernmental Panel on Climate Change (IPCC) was awarded the Nobel Prize in 2007. But in some places the enthusiasm was short-lived. In March 2011, the global nuclear renaissance discussion paused as the Fukushima Daiichi Nuclear Power Plant in Japan underwent a meltdown from the damages caused by a tsunami that occurred after a 9.0 magnitude earthquake. The disaster sparked international debates on the usages of nuclear power generation and the place that it holds. Countries such as China and the United Kingdom remained steadfast in commitment to increase nuclear power, while others like Germany and Japan took the opposite approach and began reexamining their national policies on nuclear energy (Brumfiel and Cyranoski 2011); Germany's policy was ratcheted up to phase out nuclear much earlier than anticipated, leading to criticism because they left old coal-powered power plants open, some burning the highly carbon-intensive lignite coal. The Fukushima accident had similar consequences to accidents at Three Mile Island and Chernobyl, negatively changing the public opinion regarding nuclear energy. It did so globally, and locally, as many nuclear power plants in Japan faced social opposition during efforts to restart them (Visschers and Siegrist 2013). Although, notably, Fukushima did not slow power plant approvals in the United States or China and India.

Uranium Supply Chains

Uranium makes up only two parts per million of the Earth's crust. A single pound of uranium contains as much energy as three million pounds of coal. Uranium is much more energy-dense than other energy sources. Uranium mining is an important

Box 7.3 Uranium Supply Chains

The typical nuclear power plant generates on average about 20 metric tons of high-level radiation waste annually (Kessides 2010). The nuclear industry produces 2,300 metric tons each year and has created around 70,500 metric tons since its inception (Davis 2012). The United States has no central repository for nuclear waste, though there have been plans to build one in the Yucca Mountains near the atomic test site an hour west of Las Vegas. The United States has invested billions into the project to date, and there are calls to revive the project. California law requires a national nuclear repository to be opened before allowing any new nuclear power plants in the state. In 2024, several Silicon Valley technology companies were proposing to keep operating and revive shuttered nuclear power plants to meet new energy demands for artificial intelligence.

Source: Adapted from KQED (2025).

supply chain impact consideration for nuclear power, often ignored in climate action conversations today (Box 7.3). In the United States, environmental justice impacts have occurred on Native American lands, including the Church Rock disaster on Navajo lands in 1979, where a tailing pond dam collapsed and caused the largest unintentional release of radioactive material in US history (Tsosie 2015). What makes uranium mining for nuclear power complicated is that it shares the same supply chain as nuclear weapons. Even if there were no more nuclear power plants, there would still be uranium mined in some places because of the dependency on the metal for nuclear warheads. Maybe that will change someday, but that day seems off.

Mining impacts remain a key area of environmental damage in most assessments of the life cycle impacts of nuclear power. The world's leading producers of uranium ore are Canada, Kazakhstan, Namibia, Australia, and Uzbekistan, which collectively mine 85% of global supplies. In the United States, uranium is mined mainly across the western states in sandstone deposits along with coal, such as Wyoming and the Four Corner regions. The United States was the first to develop extensive uranium mining, with many so-called yellowcake towns scattered across the American West around the time of the Manhattan Project to develop nuclear weapons. Uranium oxide (UO_2) fuel pellets are assembled into rods commonly made of a zirconium alloy. The pellets heat the rods, which are the source of heat for the steam turbine (Figure 7.2). Some nuclear fuel rods incorporate reused uranium and plutonium in a mixed-oxide fuel (MOX). This fuel can be made from weapons-grade plutonium for nuclear disarmament. Fuel rods in a nuclear power plant need to be replaced about every other year on average. The waste fuel rods are transported for fuel recovery or stored onsite surrounded by water, which shields radiation. The concern with some nuclear accidents has been with the loss of this water during an accident.

Figure 7.2 Nuclear fuel pellets. Source: Nuclear Regulatory Commission / Flickr.

Uranium is a mined ore that is subject to many of the problems faced by mining communities and workers. Tailing piles can contain low-level radioactive waste or acids that leach into the groundwater. Some frontiers for global uranium supplies include Namibia, currently the fourth largest uranium producer in the world. Research in political ecology describes how communities facing the prospects of a global uranium rush are shaped by local resistance or accommodation to mining efforts (Conde and Kallis 2012). The authors of that research remind us that uranium mining is the forgotten source of nuclear power, not nuclear power plants.

In the United States, uranium is subject to political debates about whether it should qualify as a critical mineral, a designation that allows mining companies to bypass very basic environmental best practices and community engagement that is crucial to ensuring that mining companies have the social license to operate. The Energy Act of 2022 does not allow uranium to be listed as critical because it is an energy fuel.

Making Nuclear Power Safer, Cheaper, and Waste-Free

Nuclear power plants can provide important jobs, community revenues, and financial support to local communities and help boost the local economy. A full appreciation and assessment of the financial and community costs associated with decommissioning a nuclear power plant depends on many factors. What does decommissioning actually look like from the perspective of jobs, community revenues, or justice? Many nuclear power plant jobs can be highly skilled, and they may be organized labor in some parts of the world. But these jobs also create other economic activities to employ people across various activities the support the plant and community.

Cost is perhaps the most compelling answer to the common question: *If nuclear power is so great, why isn't there more of it?* The high expense of nuclear energy—particularly during the construction phase—has been a major barrier to its widespread deployment. A significant portion of nuclear project costs stems from complex engineering requirements and lengthy construction timelines, which are often extended by design changes and the rigorous regulatory reviews those changes trigger. These reviews, while essential for ensuring safety, are sometimes conducted with an overabundance of caution, leading to further delays and cost overruns. In addition to construction, there are other long-term financial obligations that must be considered in any discussion of nuclear energy development. These include the substantial costs associated with decommissioning aging plants and the safe storage or disposal of nuclear waste, which can span decades and require sustained public and private investment.

Most of the recent global growth in nuclear power is occurring in China, which has aggressively expanded its nuclear fleet as part of its energy security and decarbonization strategies. In contrast, nuclear power plant construction in the United States largely occurred between the 1960s and the early 1990s, after which new plant orders essentially stopped. Approximately 95% of the nuclear reactors operating in the United States in 2025 were built during that earlier period, meaning that even the youngest plants were already over 30 years old. While the slowdown in US nuclear construction is often attributed to public opposition—particularly stemming from the environmental and anti-war movements—this explanation overemphasizes their role and simplifies a far more complex situation. It is true that high-profile nuclear accidents significantly influenced public perception and policy debates. These include (1) Windscale (the United Kingdom, 1957)—a reactor fire that released radioactive contamination; (2) Kyshtym (the USSR, 1957)—a major accident at a nuclear weapons material reprocessing facility near Chelyabinsk, ranked among the worst nuclear disasters; (3) Saint-Laurent (France, 1969)—a partial core meltdown in a graphite gas reactor; and (4) Three Mile Island (the United States, 1979)—a partial meltdown in Pennsylvania that became the most serious accident in US commercial nuclear history, and Chernobyl (USSR, 1986)—a catastrophic explosion and meltdown with global repercussions. These incidents, particularly Three Mile Island and Chernobyl, undoubtedly played an important role in shaping public distrust and in amplifying calls for stronger regulation.

However, the primary reasons for the slowdown in US nuclear development were economic and institutional. Key factors included flat or declining electricity demand in the 1980s and 1990s, which undercut the need for new baseload capacity, cost overruns, and lengthy construction delays on existing nuclear projects. This made utilities and regulators wary of future investments, regulatory backlash that shielded ratepayers from bearing the brunt of rising project costs, reducing financial incentives for utilities, low fossil fuel prices, particularly cheap oil and natural gas, which made alternative energy sources more economically attractive.

In sum, the slowdown of nuclear power development in the United States is a multifaceted story. It reflects not only public concerns over safety and environmental impacts but also deeper economic, regulatory, and market dynamics that reshaped utility commitments to nuclear energy.

There are some ideas promoted in the second generation of reactor technologies and fissile fuels. Small-modular reactors are proposed as a solution to the challenges of cost overruns. The basic idea is to make nuclear reactors manufactured products where cost reductions are possible from economies of scale. Some proponents of nuclear power propose thorium reactors in place of uranium- and plutonium-based ones, as thorium is more abundant in Earth's crust than uranium, does not produce the same amounts of high-level waste, and poses lower weapons proliferation risks. However, to produce electricity from thorium, it is first converted to protactinium by bombarding it with neutrons; this protactinium decays to form uranium 233. Uranium 233 is not found in appreciable quantities in Earth's crust and was discovered during the Manhattan project; scientists say it has more attributes that make it even more suitable than uranium 235 for making weapons. Currently, India operates the only thorium-based nuclear power plant in the world, the 30 kW Kalpakkam Mini Reactor.

The holy grail of nuclear power is sustaining a fusion reaction, similar to the process used to make energy in the stars and our sun. The basic idea of fusion research is to produce energy greater than the amount needed to create it. Most fusion research has focused on deuterium–tritium fuel. The National Ignition Facility in Lawrence Livermore National Laboratory in Livermore, California, is one of the leading places where burning plasma has been achieved (Zylstra et al. 2022). The site is funded by the US Department of Energy and uses lasers to compress isotopes of hydrogen.

Geothermal Hot Rocks for Electricity and Heat

Geothermal energy is sometimes referred to as a form of indirect nuclear power—and in a sense, that's true. Geothermal power harnesses heat stored beneath the Earth's surface, primarily by tapping into underground reservoirs of hot water or steam that have been naturally heated by contact with hot rocks deep within the Earth's crust. The sources of Earth's internal heat are generally considered to be threefold: residual heat from planetary formation, including gravitational compression, frictional heating from the movement of tectonic plates and the viscous flow of the mantle, radioactive decay of isotopes such as uranium-238, thorium-232, and potassium-40 in the Earth's mantle and crust. Of these, radioactive decay is thought to account for a significant portion—estimates suggest between 45% and 90% of the heat in the Earth's interior comes from this source, depending on geological models.

The Earth is estimated to lose heat at a rate of approximately 40–50 TW globally, which translates to an average geothermal heat flux of about 0.08–0.1 W/m^2 at the surface—though this can be significantly higher in tectonically active regions such as Iceland, the Philippines, or the western United States.

In addition to being mostly generated in Earth's center, geothermal energy is accessible everywhere on Earth if it is accessible and not too deep. The sources of this heat include friction from fluids moving in Earth's core, radiative decay of elements in the mantle, and pressure from the planet's weight. Geothermal hot spots that make it to Earth's surface include hot springs and the famous "Old Faithful" geyser at Yellowstone National Park. Many places along the Ring of Fire in the Pacific Ocean use geothermal resources.

The first uses of geothermal power for heat date back to antiquity. Electric power generation from geothermal did not arrive until the twentieth century. There are three geothermal power plant technologies used to convert steam coming from underground into electricity—dry steam, flash steam, and binary cycle. One of the most efficient types of geothermal technology used today is the dry steam geothermal power plant. Dry steam plants apply steam directly to a turbine, which drives a generator that generates electricity. The type of geothermal power plant is dictated by the nature of the resource. The Geysers complex—comprising around 22 geothermal power plants located in the Mayacamas Mountains of Northern California—is the world's largest single source of geothermal electricity, with an installed capacity of over 700 MW; though, because of plant retirements and geothermal reservoir depletion, it is about half of the peak it once was.

Geothermal power plants share supply chain components with several other energy technologies that rely on steam turbines for electricity generation, including concentrated solar power, biomass, and nuclear energy. These systems all require turbines, condensers, heat exchangers, and generators, forming a common industrial backbone. In addition, geothermal projects also draw heavily from the oil and gas industry, particularly in the drilling, casing, and well-completion technologies. The exploration and development of geothermal reservoirs rely on high-temperature drilling rigs, mud circulation systems, and subsurface imaging techniques, many of which were pioneered or adapted from hydrocarbon extraction. This overlap in infrastructure and expertise means that geothermal energy sits at a unique intersection between renewable and extractive industries—offering opportunities for cross-sector innovation, but also facing similar challenges such as well failure risks, high capital costs, and resource uncertainty at depth.

There are several different designs for geothermal power plants depending on the thermal resource they are working with. Flash steam plants are the most common type of geothermal power generation plant in operation today. Like dry steam plants, they use a fluid that is under ground at temperatures greater than 360 °F and pump it under high pressure into a tank. The tank at the surface is held at a much lower pressure, causing some of the liquid to vaporize at a rapid rate. From it, the vapor turns the turbine connected to an electricity generator to generate electricity for usage. Binary-cycle geothermal power generation plants are different from dry steam and flash steam systems technology because the fluid from the geothermal reservoir never comes in contact with the turbine or generator. Binary-cycle power plants use a closed-loop system. Like the two other power plants, the binary-cycle power plants virtually emit nothing to the atmosphere. It is predicted that a significant proportion of geothermal electricity in the future could come from binary-cycle plants because most of the geothermal energy is below 300 °F, a resource level that favors binary cycle.

Geothermal energy offers valuable opportunities for direct heating applications, particularly by utilizing low-grade geothermal resources—reservoirs with moderate to low temperatures (typically below 150 °C or 300 °F). While these lower-temperature sources may not be suitable for electricity generation, they are highly efficient for heating purposes and can be harnessed for space heating, agricultural uses, and industrial processes.

One of the most promising applications is in district heating systems, which deliver heat to multiple buildings or even entire neighborhoods through a centralized network of insulated pipes. These systems can capture and distribute waste heat from geothermal plants or use low-enthalpy geothermal sources directly. Because geothermal heat is both consistent and renewable, it provides a stable and low-emissions alternative to heating systems based on fossil fuels. Several countries have successfully adopted this model. In Iceland, for example, over 90% of homes are heated by geothermal energy, primarily through district heating systems. Similar systems are in place in parts of France (e.g., the Paris Basin), Germany, China, and in a few locations in the western United States, including Boise, Idaho, which has operated a geothermal district heating system since the 1980s. As cities look for ways to decarbonize heating—which can account for a significant portion of urban energy use—geothermal district heating offers a compelling solution, especially in areas with accessible subsurface heat and supportive policies. Additionally, integrating geothermal heat with building retrofits, heat pumps, or solar thermal technologies can further improve efficiency and sustainability.

The potential of global geothermal energy growth is significant. Sites along the Pacific "ring of fire" or other places near active volcanic areas offer significant potential for electricity generation. Geothermal energy provides more than 15% of electricity in several countries, including Iceland, Kenya, El Salvador, Costa Rica, and the Philippines (Box 7.5). Iceland famously has a geothermal facility at the Blue Lagoon is also a popular swimming attraction for tourists. Baths are located directly in the shadows of the power generation facility. One major use of geothermal power in Iceland is the production of aluminum. While geothermal energy is widely accessible if you go deep enough, the economics of geothermal energy are far more complicated. Geothermal reservoirs can be of different quality that can pose site- and rock-specific challenges or be limited by water or other uses. One innovation in this space is called enhanced geothermal systems (EGS). With this technology, the heat reservoir rock is fractured and cracked to maximize the surface area of hot rock as water is passed over it (Figure 7.3). The development of EGS technology could make a much larger resource available, though this EGS process may be linked to increased potential for earthquakes and may be too expensive. This is one reason EGH technologies borrow innovations from natural gas industries like slant drilling and fracturing rock layers could bring enhanced geothermal to be a widespread clean firm energy resource across a wider area (Aljubran and Horne 2025).

Like most things that require taking material or fluid from Earth's crust, geothermal power plants must manage sulfur emissions. When a volcanic eruption in Hawaii

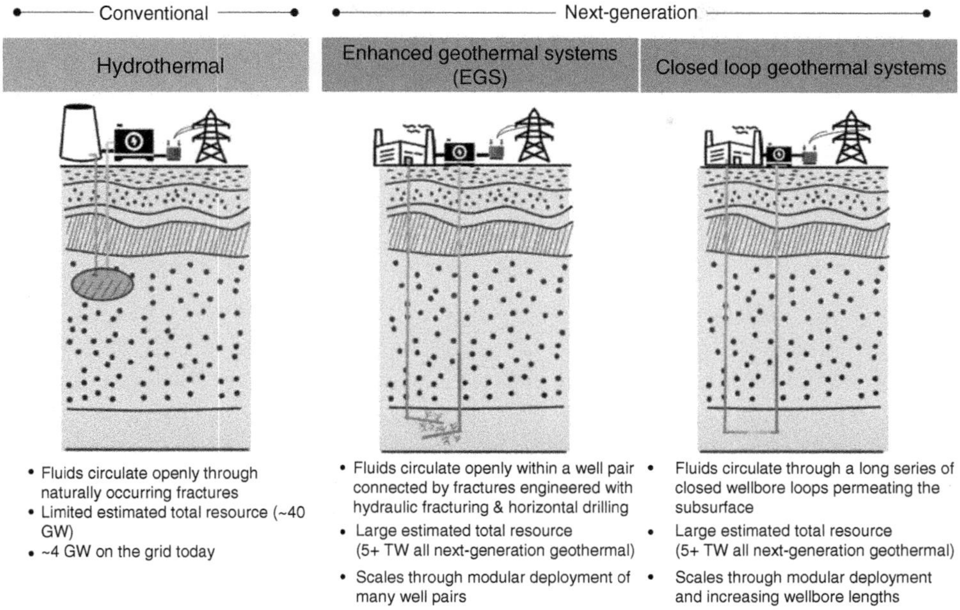

Figure 7.3 Conventional and next-generation geothermal power plant designs. Source: United States Department of Energy, 2024. https://commons.wikimedia.org/wiki/Category:Geothermal_power_plants#/media/File:Geothermal_technology_overview.png / Public Domain/last September 18, 2025.

led to lava encroachment at a geothermal facility on the Big Island in 2019, there was concern about deadly hydrogen sulfide emissions as the fire encroached. Where geothermal brines are rich in metals, some propose extracting them, notably lithium, nickel, rare earth elements, and uranium (Box 7.4).

Low-grade geothermal heat just a few feet below ground can also be used more directly to heat buildings and homes. Ten feet down below the Earth, the ground temperature is consistently between 10 and 16 °C. This resource can be tapped to provide cooling in the summer and heating in the winter for homes and large buildings. Geothermal heat pumps work by burying a loop of pipes called heat exchangers just below the ground surface. Filled with water or antifreeze solutions, pumps will circulate the heated fluid from inside to outside and the other way around, saving money on air-conditioning during winter and summer. Geothermal heat pumps will be discussed in a later chapter on the built environment. This use of geothermal energy takes low-grade heat from the earth to balance either the warm or cool temperatures in a house. They can also be used in greenhouses for agricultural operations.

Trace elements, especially heavy metals, are often found in the wastewater discharge from geothermal power plants. This might be construed as an opportunity as some have proposed extracting important elements like lithium and rare earths from geothermal wastes. Cooling water and subsidence issues can also be challenges for these projects in certain locations. Other sites may find it challenging because of the presence of rare ecosystems or endangered species (Box 7.6 and Figure 7.4).

Box 7.4 Extraction of Elements from Geothermal Brines and Nuclear Waste

In some parts of the world, geothermal power plants extract not only heat but also valuable minerals and heavy metals from subsurface brines. These geothermal brines—especially those associated with high-temperature, high-salinity geothermal reservoirs—can contain strategically important elements for the energy transition, including lithium, manganese, cobalt, rare earth elements, and even uranium. Among these, lithium has garnered particular attention due to its essential role in lithium-ion batteries used in electric vehicles, grid-scale energy storage, and consumer electronics. Extracting lithium and other metals from geothermal brines relies on the development of direct lithium extraction (DLE) technologies, which aim to separate lithium from complex mineral mixtures without the need for large evaporation ponds. These technologies often utilize membranes, sorbents, ion exchange systems, or advanced filtration techniques to isolate lithium and other salts efficiently.

A major site of interest is the Imperial Valley of California, near the Salton Sea, where geothermal operations already provide significant renewable electricity. Several companies and research initiatives are now working to co-develop DLE technologies alongside geothermal power plants, aiming to create integrated systems that produce both clean energy and critical minerals. If proven economically and environmentally viable, these projects could position the region as a domestic source of lithium, helping to reduce reliance on international supply chains dominated by countries like China, Australia, and Chile. The co-production of minerals and energy from geothermal brines represents a promising frontier in resource extraction—one that could help address both clean energy goals and the supply security of critical materials (U.S. Department of Energy 2021). However, challenges remain in scaling up extraction technologies, managing brine reinjection, and addressing potential impacts on groundwater and ecosystems.

Source: Adapted from Britton et al. (2024).

Box 7.5 Geothermal Power in Kenya

There are a few countries in the world where geothermal contributes meaningful amounts of renewable energy to the grid; Iceland, the Philippines, and Kenya. Kenya is an eastern African country with low rates of electricity access; around 25% have access to electricity. The country has been developing a power grid as well and uses a substantial amount of geothermal power for electricity.

The history of geothermal development in Kenya starts with the Great African Rift system of tectonic activities that gives rise to Mt. Kilimanjaro and causes volcanoes in the eastern part of the continent. The first geothermal power plant for electricity was explored in the 1950s when geothermal fields were drilled at Olkaria, Kenya, and power plants were built in the 1970s (Mwangi 2010). The opportunities for low-pollution electricity attracted the interest of the World Bank, who soon after invested in numerous geothermal facilities in the region. Some have come to question the sustainability of some of Kenya's geothermal sites. Some call geothermal over-extraction "heat mining" to denote how heat is literally removed from the heat basin. Geothermal power is also used in agriculture, where it is used to heat greenhouses in some of the regions with geothermal resources.

Box 7.6 Geothermal Versus the Endangered Toad

In Nevada's Dixie Valley, a unique environmental and energy development conflict has emerged around a tiny amphibian: the Dixie Valley toad (*Anaxyrus williamsi*). This species, discovered and formally described only in 2017, is found nowhere else on Earth—its entire habitat is confined to a single spring-fed wetland complex in the Dixie Valley region, an ecologically isolated area in the Great Basin. In the same year, Ormat Technologies, a leading developer of geothermal power plants, proposed a project to develop a geothermal energy facility in the valley with a planned capacity of up to 120 MW. While geothermal energy is a key part of Nevada's clean energy strategy and a critical technology for the global energy transition, the proposed project raised serious ecological concerns due to its proximity to the toad's sole habitat. In response, the U.S. Fish and Wildlife Service (USFWS) listed the Dixie Valley toad as endangered under the Endangered Species Act in 2022, granting the species legal protection and prompting a temporary halt to the geothermal project. The listing was based on concerns that the geothermal development could alter groundwater dynamics—potentially drying or degrading the delicate spring and wetland system the toad relies on. Alongside endangered status, the USFWS also designated areas of critical habitat, ensuring regulatory oversight of land use and development in the toad's range.

This case illustrates the emerging tensions between biodiversity conservation and the push for renewable energy infrastructure, particularly in ecologically sensitive areas. It underscores how the energy transition, while necessary for mitigating climate change, can still create local environmental trade-offs—especially in regions where rare species or cultural resources overlap with high-value energy resources. Beyond its ecological dimension, the Dixie Valley geothermal project also raised concerns among Indigenous communities and conservationists, as it could affect geophysical features such as hot springs, which hold spiritual and cultural significance. Geothermal development can sometimes reduce surface spring flows or lower reservoir temperatures, leading to long-term degradation of both natural ecosystems and sacred landscapes.

In cases like this, decision-makers must grapple with difficult questions: Are there alternative siting options that could avoid harm to critical habitat? Can adaptive management or technological solutions reduce environmental risks? How do we balance urgent climate goals with long-term ecological stewardship? The Dixie Valley toad is a symbol of the complex and layered challenges in building a clean energy future that also respects biodiversity, cultural values, and local voices.

Figure 7.4 The Dixie Valley Toad's only habitat was proposed for a geothermal power plant, but the company proposed its project elsewhere. Source: USFWS Pacific Southwest Region / Flickr.

Highlights/Takeaways

- Nuclear power plants provide tremendous quantities of power from relatively small power plants, but face economic challenges as well as public policy questions around what to do with aging or old nuclear power plants.
- Nuclear power has critical challenges in uranium supply chains, some of which reproduce environmental inequality.
- Costs of nuclear and geothermal both face obstacles owing to the unique resource or reactors designs.
- The use of geothermal resources can provide firm, renewable energy as well as other resources like critical metals. But it will likely face challenges where there are unique environments or groundwater issues.

Questions for Review

1. What should countries with aging nuclear power fleets facing costly repairs do with aging nuclear power plants in the United States? In Germany? Japan? France? These closures could lead to increased electricity GHG emissions, particularly from natural gas. But are there opportunity costs? If new nuclear power plants are built, and uranium supply chains continue to develop, what can be done to protect those who live alongside the commodity chains? Future power plants will need to be smaller and more flexible to operate.
2. Geothermal and nuclear resources are important to some electricity grids because they offer power resources that are available throughout the day. A power plant that can operate at high percentages for long periods is said to have a high-capacity factor.
3. Many energy technologies and resources require thinking about trade-offs when it comes to development. What are some key trade-offs with nuclear or geothermal technology? How would you navigate these trade-offs if a proposed project pitted conservation versus energy development?

Exercise 7.1

Geothermal energy harnesses heat from inside the Earth, while nuclear power relies on geologic materials like uranium and thorium.

a. What are the two primary sources of Earth's internal heat?
b. Create a poster or infographic that illustrates the geological processes involved in the production of both geothermal and nuclear energy.

Your design should address the following layers:

1. The Resource: Where does the energy originate geologically, and how is it accessed?
2. Transformation: What processes are used to convert it into electricity? What other natural resources (e.g., water, land) are involved, and how are they affected?
3. Societal Impact: What are the political and economic changes that accompany the development of these energy resources? Who benefits, who bears the costs, and how does the surrounding environment or community change?

Use diagrams, brief text, and color to clearly communicate the science and broader implications.

Exercise 7.2

Research a nuclear power plant that has recently closed, is scheduled for closure, or has been proposed for shutdown. Try to find one in your region or country, but any plant globally is acceptable.

a. What were the main concerns or reasons behind the proposed or actual closure (e.g., economics, safety, public opinion, environmental regulations)?
b. What was the community's historical relationship with the plant (jobs, tax revenue, environmental concerns)?
c. Who were the key stakeholders or decision-makers, and how did they weigh the trade-offs?
d. In your opinion, was the decision to shut it down justified? Explain your reasoning using evidence from your research.

Exercise 7.3

Projects like uranium mines and geothermal plants can provide critical energy resources, but they may also cause irreversible harm to ecosystems or communities—especially when they impact rare species, culturally significant lands, or sacred Indigenous sites.

a. Research an example where an energy development project (either nuclear or geothermal) conflicted with a sensitive ecological or cultural resource.
b. How could the project have been designed differently to reduce or mitigate harm?
c. What role did the community or affected group play in the decision-making process?
d. Reflect: Should communities have the right to veto or block a project if it threatens their way of life or environment? Why or why not?

References

Aljubran, M.J. and Horne, R.N. (2025). Techno-economics of geothermal power in the contiguous United States under baseload and flexible operations. *Renewable and Sustainable Energy Reviews* 211: 115322. https://doi.org/10.1016/j.rser.2024.115322.

Boice, J.D., Cohen, S.S., Mumma, M.T. et al. (2024). Mortality from leukemia, cancer and heart disease among U.S. nuclear power plant workers, 1957–2011. In: *The*

Million Person Study of Low-Dose Radiation Health Effects, (eds. Boice, J.D. et al.), 141–62. CRC Press.

Britton, A., Olmedo, L., Torres, C.A., and Blair, J.J.A. (2024). Hydrosocial imaginaries of green extractivism: water-energy transitions and geothermal lithium development at the salton sea in imperial valley, California. *The Extractive Industries and Society* 20: 101567. https://doi.org/10.1016/j.exis.2024.101567.

Brumfiel, G. and Cyranoski, D. (2011). Fukushima deep in hot water: rising levels of radioactive liquid hamper clean-up effort. *Nature* 474: 135–137.

Conde, M. and Kallis, G. (2012). The global uranium rush and its Africa frontier. Effects, reactions and social movements in Namibia. *Global Environmental Change* 22 (3): 596–610.

Cram, S. (2016). Wild and scenic wasteland: conservation politics in the nuclear wilderness. *Environmental Humanities* 7 (1): 89–105.

Cummins, W.E. and Matzie, R. (2018). Design evolution of PWRs: shipping port to generation III+. *Progress in Nuclear Energy* 102: 9–37.

Davis, L.W. (2012). Prospects for Nuclear Power. *Journal of Economic Perspectives* 26 (1): 49–66.

Downey, G.L. (1986). Ideology and the clamshell identity: organizational dilemmas in the anti-nuclear power movement. *Social Problems* 33 (5): 357–373.

Fennern, L. (2018). Design evolution of BWRs: Dresden to generation III+. *Progress in Nuclear Energy* 102: 38–57.

Garland, J.A. and Wakeford, R. (2007). Atmospheric emissions from the windscale accident of October 1957. *Atmospheric Environment* 41 (18): 3904–3920.

Gunderman, R.B. and Gonda, A.S. (2015). Radium girls. *Radiology* 274 (2): 314–318.

Kessides, I.N. (2010). Nuclear power: Understanding the economic risks and uncertainties. *Energy Policy* 38 (8): 3849–3864.

Kopytko, N. and Perkins, J. (2011). Climate change, nuclear power, and the adaptation–mitigation dilemma. *Energy Policy* 39 (1): 318–333.

KQED (2025). Will California revive nuclear power to feed AI energy demands?," February 2, 2025. https://www.kqed.org/news/12024817/ai-is-bringing-nuclear-power-back-from-the-dead-maybe-even-in-california (accessed 12 November 2025).

Mwangi, M.N. (2010). Environmental and socio-economic issues of geothermal development in Kenya. *GRC Bulletin* 4: 24–35.

Nelson, S. and Ramana, M.V. (2023). Managing decline: devaluation and just transition at diablo canyon nuclear power plant. *Environment and Planning A: Economy and Space* 55 (8): 1951–1969.

Perrow, C. (2013). Nuclear denial: from Hiroshima to Fukushima. *Bulletin of the Atomic Scientists* 69 (5): 56–67.

Steinbeck, J.R., Hedgepeth, J., Raimondi, P. (2007). Assessing power plant cooling water intake system entrainment impacts. San Luis Obispo, CA. Http://Www.Energy.ca.Gov/2007publications/CEC-700-2007-010/CEC-700-2007-010.PDF (accessed 12 November 2025).

Tsosie, R. (2015). Indigenous peoples and the ethics of remediation: redressing the legacy of radioactive contamination for native peoples and native lands. *Santa Clara Journal of International Law* 13: 203.

U.S. Department of Energy (2021). Can geothermal energy solve the lithium shortfall?https://www.energy.gov/eere/geothermal/articles/can-geothermal-energy-solve-lithium-shortfall#:~:text=Hot%20salty%20water%2C%20or%20geothermal,from%20thousands%20of%20feet%20underground (accessed 12 November 2025).

Visschers, V.H.M. and Siegrist, M. (2013). How a nuclear power plant accident influences acceptance of nuclear power: results of a longitudinal study before and after the Fukushima disaster. *Risk Analysis: An International Journal* 33 (2): 333–347.

Wills, J. (2012). *Conservation Fallout: Nuclear Protest at Diablo Canyon.* University of Nevada Press.

World Nuclear Association (2025). Information library. https://world-nuclear.org/ (accessed 12 November 2025).

Zylstra, A.B., Hurricane, O.A., Callahan, D.A. et al. (2022). Burning plasma achieved in inertial fusion. *Nature* 601 (7894): 542–548.

Further Reading

Bersano, A. and Segantin, S. (2024). History of nuclear power plants development. In: *Nuclear Power Reactor Designs* (ed. J. Wang, S. Talabi, and S.B.Y. Leon), 3–40. Academic Press.

Bragg-Sitton, S.M. (2024). Achieving a net-zero future: the role of nuclear energy (No. INL/MIS-24-76549-Rev000). IdahoNational Laboratory (INL), Idaho Falls, ID (United States). MIT: The Future of Nuclear Power.

Huenges, E. and Ledru, P. (ed.) (2011). *Geothermal Energy Systems: Exploration, Development, and Utilization.* John Wiley & Sons.

Rhodes, R. (2018). Why nuclear power must be part of the energy solution. *Yale Environment* 360: 19.

Chapter 8

Solar, Wind, Waves, and Tides

Earth's energy flows originate primarily from the sun, which drives a vast range of atmospheric, oceanic, and biological processes. Solar radiation fuels fundamental weather systems, including wind patterns, ocean currents, and the water cycle, all of which can be harnessed as sources of renewable energy. A smaller but still important contribution comes from the gravitational interaction between the Earth and the moon, which generates the tidal forces that move vast volumes of water around the planet on predictable cycles. This chapter explores several of these repeating natural energy flows—including solar, wind, wave, and tidal energy—and examines their potential to be captured, converted, and deployed for human use. These flows, though intermittent in nature, can be transformed into thermal, kinetic, or electrical energy through various technologies designed to work with Earth's dynamic systems. While often grouped under the umbrella of "green energy," these renewable energy sources are not entirely impact-free. Their development depends on materials, infrastructure, and land use decisions that have ecological, social, and political consequences. Understanding the material implications of harvesting these flows—such as the metals and minerals required, land use trade-offs, or localized environmental impacts—is essential to building a truly sustainable energy future. Nonetheless, when compared to fossil fuels, these renewable flows offer the promise of cleaner air, lower greenhouse gas (GHG) emissions, and more resilient energy systems. Harnessed responsibly, they represent not just a technological solution, but an opportunity to reimagine our relationship with the planet's natural systems—working with Earth's rhythms rather than extracting from them.

Harnessing Solar Energy Flows

Solar energy has become one of the most iconic technologies of the low-carbon energy transition. It produces fewer emissions than most other energy sources and offers a unique compatibility with residential energy demand, as well as the potential for integration into the built environment and infrastructure. If deployed in ways that displace dirtier energy sources, solar power could help address many of the environmental harms associated with conventional energy production—such as air and water pollution and the generation of hazardous waste.

The primary technology for electricity generation from the sun is photovoltaics (PVs) (Box 8.1), though concentrated solar power (CSP) technologies also contribute to clean electricity production. While these technologies have minimal emissions during operation, most of their environmental impact occurs upstream—in the supply chain, including materials extraction, manufacturing, and disposal. This creates significant challenges for a just transition (Mulvaney 2019; Fthenakis et al. 2009).

A just transition to solar power would emphasize: (1) Improved energy access to enhance livelihoods and protect vulnerable populations; (2) Displacement of fossil fuel pollution, which disproportionately affects marginalized communities; and (3) The prevention of resource exploitation, ensuring that resource-dependent communities are not stripped of local benefits while supplying energy services to distant users.

Box 8.1 The Photovoltaic Effect

The first stage in the PV commodity chain is mining. One of the most critical materials is high-purity quartz, which is used to produce silicon metal—the foundation for semiconductor-grade silicon used in both electronics and photovoltaic technologies.

While countries such as Brazil, China, Madagascar, South Africa, Canada, the United States, and Venezuela possess large reserves of industrial quartz, much of this material is not pure enough for the stringent requirements of glass and silicon manufacturing. Only a small fraction qualifies as high-purity quartz, which must contain extremely low levels of impurities such as iron, aluminum, and titanium. Most high-purity quartz is extracted through open-pit mining, which allows direct access to quartz veins with minimal contamination. The Spruce Pine Mining District in North Carolina (United States) is currently the world's leading source of high-purity quartz, supplying much of the global demand for semiconductor and photovoltaic applications.

In addition to silicon, photovoltaic technologies rely on a variety of ferrous and non-ferrous metals, precious and rare metals, and specialty minerals. The specific materials required vary depending on the semiconductor technology used. For example, crystalline silicon PV requires silicon, silver (for conductive paste), and aluminum (for frames and contacts). Thin-film technologies, such as CdTe (cadmium telluride) or CIGS (copper indium gallium selenide), use critical and often less abundant materials like cadmium, indium, gallium, and tellurium. The extraction and processing of these raw materials represent a resource- and energy-intensive phase of the PV life cycle and have significant implications for both environmental sustainability and supply chain ethics.

- Crystalline silicon photovoltaic, which comprise ~95% of the market share, and are traced to extractive industries from high-purity quartz, glass-quality quartz, copper, metallurgical silicon, silver, aluminum, cassiterite (tin), and industrial diamonds.
- Cadmium telluride (CdTe) thin-film PVs require glass-quality quartz, tellurium, cadmium, molybdenum, or tin.
- Copper indium gallium selenide (CIGS) thin films, which require copper, indium, gallium, selenium, tin, molybdenum, and cadmium or zinc.
- Perovskites are emerging solar cell technologies based on different compounds of tin or lead. Early applications of perovskites include stacking atop crystalline silicon solar cells.
- Heterojunction crystalline silicon solar cells could move from using a layer of amorphous silicon to thin films of perovskites.
- Heterojunction gallium arsenide cells remain limited in application due to expense and are primarily satellite or telecommunications equipment, but require gallium, arsenic, germanium, phosphorus.
- Dye-sensitized solar cells are typically made of plastics but require indium tin oxide, iodine, and the very uncommon metal ruthenium.
- Plastic solar cells are made from organic compounds but use indium tin oxide as the transparent conductor.

Yet in public discourse, detailed evaluations of solar energy often stop at basic environmental metrics such as energy payback time (EPBT) or greenhouse gas (GHG) emissions. Less attention is given to social, occupational, or community-level impacts along the solar value chain.

There are significant potential benefits for communities living near aging or combustion peaker power plants—typically powered by oil or natural gas—if these facilities are replaced by solar-generated electricity. Such a transition could reduce NO_x emissions, particulate matter, and other pollutants that disproportionately affect fenceline communities. However, the occupational health and safety challenges in solar manufacturing facilities, often located far from installation sites, remain an underexamined issue (Figure 8.1).

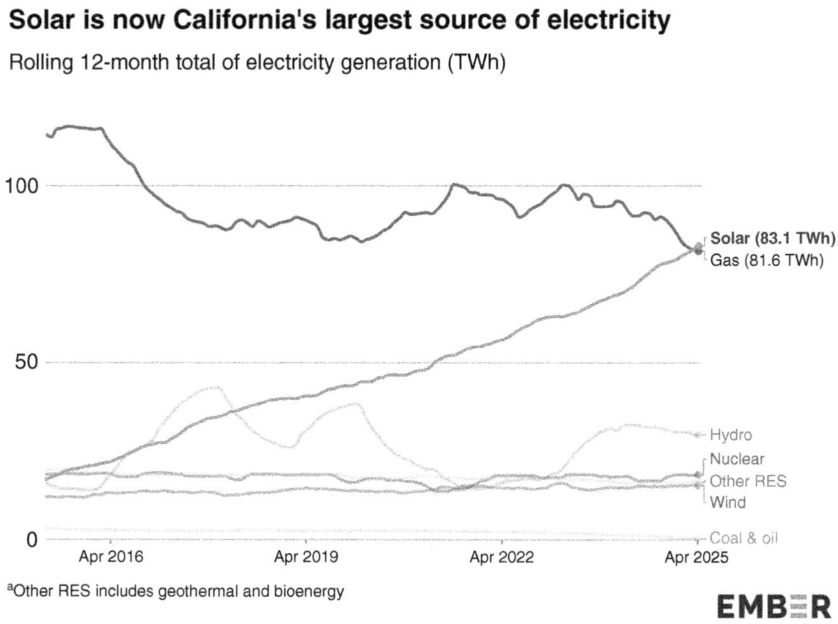

Figure 8.1 Solar is the leading source of electricity in California. Source: PV magazine/Ember/Solar becomes top source of electricity in California/https://pv-magazine-usa.com/2025/07/09/solar-becomes-top-source-of-electricity-in-california//CC BY 4.0/Last accessed on September 18, 2025.

Although life cycle assessments (LCAs) generally confirm that PVs have far lower GHG emissions per unit of energy than fossil fuels, they also reveal environmental and social impacts that are unevenly distributed. A critical exploration of solar materials, supply chains, and disposal practices is essential to identify pathways that address environmental injustice in climate mitigation strategies. As with other energy sectors, environmental inequality in solar energy manifests in numerous forms—across geography, class, labor, and ecosystems.

The scaling up of PV technologies, particularly crystalline silicon and thin-film PV, is likely to introduce new environmental and social challenges across global supply chains, from material extraction to end-of-life waste (Figure 8.2). The sourcing of critical minerals and metals such as copper, tellurium, indium, silver, and rare earth elements will become increasingly important to examine. Copper, in particular, and its related metals like tellurium and cobalt, will be embedded in nearly all aspects of the electrified economy, raising urgent questions about justice in copper mining, especially in regions where Indigenous lands or biodiverse ecosystems are affected (Lee et al. 2020). For crystalline silicon PVs, environmental impacts include quartz mining for silicon feedstock, waste generation during polysilicon refining, the use of lead-based solders in panel assembly, and large quantities of chlorine- and fluorine-based chemicals used in wafer and cell production. Addressing these impacts through

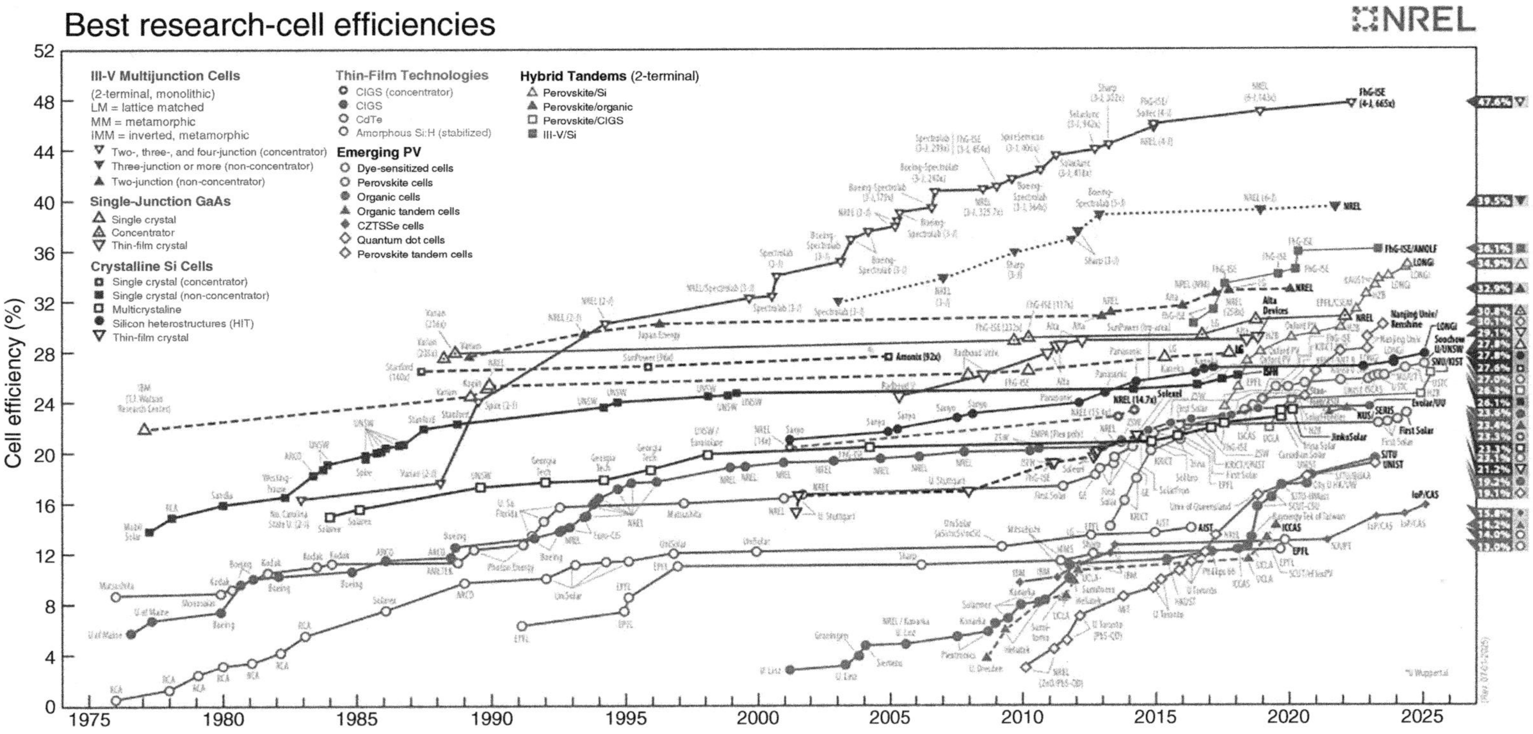

Figure 8.2 Best efficiencies for photovoltaics by technology. Source: National Renewable Energy Laboratory/Best Research-Cell Efficiencies/https://commons.wikimedia.org/wiki/File:Best_Research-Cell_Efficiencies.svg/last accessed on September 18, 2025.

better materials management, regulatory frameworks, and corporate accountability is essential if solar energy is to fulfill its potential not only as a clean energy solution, but also as a driver of a more equitable and sustainable energy system.

Tracing impacts from securing supplies of metals can be extremely complicated. Some metals are tertiary and secondary ore products, and others already blend in recovered and recycled metals. These complications make it difficult to generalize about mining in practice. Quartz for solar quality glass—with low iron and aluminum content—is another, less pure, high-purity quartz. Terawatt-scale PVs production will require "an unprecedented expansion" in glass manufacturing, necessitating the construction of thousands of glass factories (Burrows and Fthenakis 2015). Glass requires numerous inputs in addition to quartz, including soda ash, sodium sulfate, dolomite, limestone, and coal or charcoal for coke.

For the workhorse of the industry in 2020, crystalline silicon PVs, the initial refining turns quartz into metallurgical-grade silicon, a substance used to harden steel and other metals like silicon. That process occurs in large furnaces, which often use coal or charcoal and energy to coke the material to silicon. Both coal and charcoal production are the source of carbon for "coking" silica to be a silicon metal. The next step—turning metallurgical-grade silicon into polycrystalline silicon, a form of pure silicon—creates the very toxic compound silicon tetrachloride. Making polysilicon involves combining hydrochloric acid with metallurgical-grade silicon to turn it into trichlorosilanes. The trichlorosilanes then react with added hydrogen, producing polysilicon along with liquid silicon tetrachloride—3 or 4 tons of silicon tetrachloride for every ton of polysilicon. Most manufacturers recycle this waste to make more polysilicon. Recapturing silicon from silicon tetrachloride requires less energy than obtaining it from raw silica, so recycling this waste product can save manufacturers money. The refining process for polysilicon is the most energy stage of production. A much smaller PVs industry was able to obtain pure silicon from chipmakers, which routinely rejected the fraction of their wafers that did not meet the computer industry's standards. But the boom in PVs demanded more than semiconductor-industry leftovers, and new polysilicon manufacturers built out new refineries in China.

Photovoltaic cell manufacturers purify chunks of polysilicon to form bricklike ingots and slice the ingots into wafers. Then they intentionally introduce impurities into the silicon wafers, creating the essential architecture that produces the photovoltaic effect. At this stage, manufacturers rely on hydrofluoric acid to clean the wafers, remove damage from sawing, and texture the surface to better collect light. Hydrofluoric acid works great for all these things, but when it touches human skin, this highly corrosive liquid can destroy tissue and decalcify bones. So handling hydrofluoric acid requires extreme care, and it must be disposed of properly. But accidents do happen and are more likely in places that have limited experience in manufacturing semiconductors or that have lax environmental regulations. In August 2011, a factory in China's Zhejiang province owned by Jinko Solar Holding Co., one of the largest photovoltaic companies in the world, spilt hydrofluoric acid into the nearby Mujiaqiao River, killing hundreds of fish. And farmers working adjacent lands, who used the contaminated water to clean their animals, accidentally killed dozens of pigs (Mulvaney 2019).

Thin-film PVs typically rely on one of several toxic heavy metals and metal compounds, most notably cadmium-based semiconductors (Fthenakis et al. 2009). Where heavy metal exposures are most likely, there are a number of safeguards that need to be taken into account to ensure that these clean tech and green technologies improve environmental quality overall. Thin films, despite requiring toxic metals and metal compounds, also require fewer energy and material inputs, so there could be important tradeoffs with energy and toxics that require careful comparisons or simply better overall manufacturing standards.

The waste industries are the final stage in the solar commodity chain. The number and complex actors involved in management and governance, and the "situated intersectionalities between actors shape both the implementation and experience" of waste (Cornea et al. 2017). Waste occurs throughout every stage of the production process, but for renewables, it often implies the disposal of the device itself. End-of-life electronic and electrical equipment waste (e-waste) is one of the fastest-growing global waste streams. In 2019, 53.6 metric tons of e-waste were disposed of (Forti et al. 2020); these volumes are only expected to grow. Because PVs can have 30-year lifetimes, it is expected to be another decade before it becomes a major waste stream, reaching a cumulative 80 million metric tons by 2050 (International Renewable Energy Association 2016). Photovoltaic waste would comprise about 10% of the 2050 overall e-waste flows (Heath et al. 2020).

The heavy cost of extractive industries makes waste reduction, recovery, and recycling critical to global equality and sustainability. Recovered materials include glass cullet, silicon wafers and granulates, silver, indium, tin, molybdenum, zinc, copper, aluminum, steel, tellurium, cadmium, selenium, gallium, ruthenium, and plastic components. The current focus remains on the easiest material targets—aluminum, glass, silver, silicon, and copper—but specific materials recovered and recovery rates vary by disposition method and specific recovery facilities. Photovoltaics can be mostly glass by weight (>80% or more). Broken glass is called cullet, and in theory, the material can be recycled indefinitely, with reasonable losses. Recycling glass into high-value feedstocks for glass faces several challenges, but can lower the melt-point for float glass, avoiding combustion emissions. Some end-of-life modules go to landfills, others to hazardous waste disposal sites, and others to materials recovery, and even within that category, not all smelters recover the same materials. The most energy-intensive process of making a photovoltaic module is purifying silicon from quartz to silicon metal to polysilicon, so recovering silicon can have environmental benefits (Heath et al. 2020). Establishing take-back and collection systems can facilitate secondary markets for reused photovoltaic modules.

Utility-Scale Solar Energy

What are the implications of solar for where they are installed? These can vary significantly in impacts to communities and ecosystems depending on the land use change or if the project is integrated into the built environment. Solar power may lead to cleaner air for fenceline communities through peaker power plant displacement (Krieger et al. 2016) or could have impacts on biodiversity (van de Ven et al. 2021). Further down the road, communities' exposures to refinery pollution may be attributed to vehicle electrification.

Led by foreign direct investment in agricultural land, from multinational foodstuffs to biofuels and now to wind and solar, the transfer of land to energy production has exacerbated food insecurity in developing countries (Baka 2011). Land use change from energy sprawl will continue to be a major global challenge affecting global ecologies and communities (McDonald et al. 2009). Land in the American West, and throughout the world, will come under increasing land pressures, increasingly becoming a source of land use conflict (Mulvaney 2017). There are already reports of marginalized communities and land users being forced to leave land or significantly alter access to resources and land. In Morocco, there are tribal communities forced to leave lands to make solar farms safe investment vehicles, which corroborates the nation-state's vision of growth and developer influence over state power (Rignall 2016). Even in India, where projects are discursively framed around benefits to gender, there are examples emerging of this claim being exploited and further undermining women's access to better livelihoods (Stock and Birkenholtz 2019; Stock 2020). Land use conflicts are often long-standing ones about land ownership, rights, and land titling, which can lead to violence or forced resettlements (Bebbington and Bury 2013). Other users of land, including communal users—marginalized or indigenous communities—should consider in natural resource law and policy how these spaces are used as cultural resources for traditional knowledge or self-determination to avoid land use conflicts. Impacts to indigenous cultural resources as damaged include artifacts, geoglyphs, intaglios, and the unearthing of ancient cremation sites.

Solar farm is a colloquial term used to describe a utility-scale solar power plant that harnesses energy from the sun. These are also commonly referred to as solar parks or large-scale solar installations. There are two main types of utility-scale solar technologies used to generate electricity. PV power plants convert sunlight directly into electricity using semiconductor materials that absorb photons and release electrons. CSP uses mirrors or lenses to concentrate sunlight to produce heat, which is then used to generate steam that drives a steam turbine. Solar farms can take various forms: fields of photovoltaic modules (solar panels), heliostats—flat mirrors that track the sun and reflect sunlight onto a central receiver, or parabolic troughs—curved mirrors that focus sunlight onto a pipe containing a heat transfer fluid. These systems are widely viewed as key strategies for decarbonizing electricity generation due to their potential to produce large amounts of clean, renewable energy at scale.

Utility-scale solar farms are typically connected to the transmission grid, allowing them to supply electricity to entire regions or multiple cities. This distinguishes them from distributed solar systems, such as rooftop or small-scale installations, which are usually connected to the local distribution grid and serve individual homes or buildings. Though utility-scale and distributed solar may use similar photovoltaic technologies, only the large, centralized installations are referred to as "solar farms." These systems can vary significantly in size. Some occupy tens of acres and generate a few megawatts (MW) of electricity; others span thousands of acres and produce hundreds of MW. The land-to-power output ratio varies by technology and geography, but a typical estimate is around 7–10 acres per MW of capacity, depending on panel efficiency and spacing requirements.

The environmental benefits of solar farms come from the displacement of dirtier energy on the grid. Some studies suggest that replacing coal, natural gas, and petroleum-based fuels with solar electricity could reduce air pollution by 90% (Fthenakis et al. 2008). The climate benefits of solar electricity can be even greater with PVs emitting twenty times fewer GHG emissions per unit of energy than fossil-fueled electricity. Solar farms also offer economic benefits from jobs and tax revenues, such as sales tax, property tax, and employment tax, as well as associated economic activity, especially during power plant construction.

Because of the diffuse nature of solar insolation and terrestrial sites of construction of solar farms, they can have considerable impacts on landscapes when placed in areas of ecological, agronomic, or cultural resource significance. In the western United States, solar farms have been controversial because of impacts on public lands and land use change in protected areas (Hernandez et al. 2015). These socio-ecological conflicts sometimes pit groups working toward climate change mitigation and renewable energy versus groups advocating the protection of wildlife habitat and biodiversity conservation (Mulvaney 2017). Solar farms can have direct impacts on landscapes and habitats when projects require displacing wildlife. With very large-scale projects, solar farms can also significantly fragment habitat by impeding the movement of species like bighorn sheep or elk across the landscape. There are some data and observations that suggest that there may be a lake effect or polarized light mechanism that attracts aquatic birds to solar farms in desert regions (Kosciuch et al. 2020).

Solar farms can undermine community vulnerability where marginalized people are displaced or lose access to land where they make their livelihoods. Research from India suggests that solar farm siting has led to the displacement of farmers or landless peoples (Stock and Birkenholtz 2020). Elsewhere in India, questions about procedural justice issues have been raised, where it appears locals have had little input on the impacts from developing what at the time was India's largest solar park (Yenneti and Day 2015). Similar patterns are emerging elsewhere, for example on the Yucatan Peninsula in Mexico, where the social benefits of solar farms are coming into question in a similar context of dispossession (Reyes Maturano 2021). Procedural injustices often occur when there is a lack of public participation or consent in the development of solar farms. The process of development for many infrastructures lacks quality public participation and input in part because of its "decide-announce-defend" nature. Companies decide where a project will be located, they announce it, and then defend it during what is typically the participation period. Improvements in public consultation, participation, and most importantly consent and even ownership are known to best mediate these conflicts, although some are more intractable than others.

Not all solar farms result in negative land use impacts. In fact, many are developed on disturbed or underutilized lands, such as brownfields, industrial sites, salt-contaminated cropland, abandoned mines, and retired landfills. These locations offer significantly less ecologically damaging alternatives to developing on high-quality habitat or open space. In some cases, solar farms are even being designed to support ecological restoration, including efforts to enhance pollination, biodiversity, and ecosystem services by providing habitat and forage for native species.

In the future, solar farms may increasingly be integrated with agricultural production systems to create synergistic land use models. Known as agrivoltaics, this approach combines photovoltaic arrays with farming activities. For instance, solar panels can be spaced and arranged to allow rows of crops to grow underneath or between them. Some projects are co-sited with apiaries to support honeybee populations or elevated to allow tractors and harvesting equipment to operate beneath the panels. Another emerging model is "ranch-voltaics," which integrates solar infrastructure with grazing lands, allowing livestock to roam under and around solar panels. In addition, floating solar farms—sometimes called "floatovoltaics"—place solar panels on bodies of water such as reservoirs, aquaculture ponds, or wastewater treatment basins, helping to reduce land competition while sometimes improving water quality or reducing evaporation.

As the global energy transition accelerates, the number of solar farms is expected to increase substantially. However, without thoughtful planning and inclusive engagement, these projects may run into land use conflicts, particularly given the diffuse and land-intensive nature of solar energy. Solar farms that are co-developed with community input and minimize ecological harm are far more likely to avoid controversy and build long-term local support.

By contrast, solar development in sensitive habitats can reduce ecosystem services, such as pollination, erosion control, and carbon sequestration, provided by native flora and fauna. As a result, the location, design, and integration of solar infrastructure must be approached with careful consideration of ecological and social landscapes, ensuring that clean energy does not come at the expense of biodiversity or community well-being (Grodsky and Hernandez 2020). In some parts of the world, solar power has resulted in the loss of land access for marginalized groups (Stock and Birkenholtz 2020).

Scaling up production for photovoltaic technologies—both thin films and crystalline silicon—will entail new environmental and social impacts. The quest for just transitions to solar power does not involve reinventing the wheel. There are already frameworks and concepts to explore the impacts of production systems that could be drawn from to elaborate on the entanglements and socio-ecologies enrolled on this project of decarbonization along the energy continuum (Hernández 2015). Here, no particular patterns of labor exploitation take hold along the extraction-to-waste chain. Instead, patterns are shaped by the social lives of materials as they move through space, where they are endowed with meaning, but also reveal socio-ecological connections that may contradict the dominant claims (Appadurai 1988).

Just transitions might also look for opportunities to repair and restore communities historically marginalized by economies. One example would be using community-owned solar revenues as reparations to begin to restore communities from past harms and injustices (Luke and Heynen 2020). Other creative and innovative ideas will be needed, and specifically, there will be a need to interrogate assumptions about "least-cost" and "cheaper" solar. Global capital tends to take the easiest routes to cheapness, and that can readily reinforce existing inequalities across the solar power commodity chain. "Land-sparing" opportunities may also help facilitate just outcomes by improving the quality of land use change or making infrastructure more sustainable (Hoffacker et al. 2017). The move toward just solar power will require attention to these dynamics and an emphasis on differentiating good solar power from better solar power.

Distributed Energy Resources (DERs) Powered by Solar

Distributed energy resources (DERs) are decentralized energy technologies and devices located close to the point of consumption. These include residential rooftop solar panels, battery energy storage systems, demand-responsive appliances, smart inverters, electric vehicle (EV) chargers, microgrids, and virtual power plants (VPPs). Collectively, DERs offer significant benefits to both electricity customers and grid operators. These include avoiding costly grid infrastructure expansion, improving reliability and resilience, reducing transmission losses, and enhancing energy independence at the household or community level.

However, the deployment of DERs faces numerous obstacles, particularly related to interconnection delays, permitting, and inspection requirements. These processes are often slow and inconsistent, as they vary widely by local jurisdiction, utility practices, and building or electrical codes (O'Shaughnessy et al. 2022). These hurdles can delay or discourage adoption, especially for households lacking time, resources, or technical assistance.

One of DERs' key advantages is their compatibility with existing human-made structures—such as rooftops, parking lots, and public infrastructure—which allows them to generate clean energy without requiring new land development. This can result in co-benefits beyond electricity generation, such as mitigating the urban heat island effect. For example, installing solar canopies over parking lots can lower surface temperatures and provide shaded areas for cars and people. In France, a recent policy mandates that all parking lots with 80 or more spaces be covered by solar canopies, illustrating how land-efficient DER deployment can be scaled through regulatory innovation. Despite their promise, DERs are unevenly deployed due to structural and socio-technical barriers. In many areas, aging grid infrastructure lacks sufficient circuit hosting capacity to support new DER interconnections. In California, these constraints disproportionately affect Black-identifying households and disadvantaged communities, contributing to inequities in DER access and benefits (Brockway et al. 2021).

To realize the full potential of DERs, several policy reforms and technical enablers are needed: Equitable net-metering compensation to ensure fair value for exported DER power, targeted investments and incentives to support DER adoption in low-income and environmental justice (EJ) communities, and regulatory reform to ease the deployment of community microgrids and non-utility power sharing.

A key example is California Public Utilities Code Section 218, commonly referred to as the "over-the-fence rule." This regulation prohibits the sharing of electricity across property lines by non-utilities, even for neighborhood-based microgrids. As a result, several promising microgrid projects have been delayed, downsized, or canceled altogether (Sustainable Systems Research Foundation 2021). Reforming this provision could open the door to utility cooperatives, community-owned microgrids, and other local energy solutions that challenge traditional monopoly utility models. DERs also compete with utility-scale energy projects, especially when aggregated and deployed as VPPs—networks of distributed solar panels, batteries, and flexible loads that can operate like centralized power plants. Under Federal Energy Regulatory Commission (FERC) Order 2222, these aggregated DERs are permitted to participate in wholesale electricity markets, a transformative step toward decentralized grid management. However, state and utility implementation of FERC 2222 has been slow and uneven, with some jurisdictions yet to begin rulemaking processes. Unlocking the potential of DERs will be crucial to enabling a range of decarbonization technologies, including EV charging, home solar and battery systems, and heat pump adoption. These technologies not only help reduce emissions but also support grid flexibility and resilience, especially in the face of climate-related disruptions.

Solar metrics

One important concept in energy studies and life cycle analysis is EPBT. This concept is grounded in the idea of embodied energy—the total amount of energy required to extract raw materials, manufacture, transport, and install a product or piece of equipment. In the context of renewable energy technologies, EPBT helps us understand how long it takes for a system to generate the same amount of energy that was invested in its production. Put simply: if we know how much energy it takes to make a product, we can ask how long it takes that product to "pay back" that energy through its operation. This duration is referred to as the EPBT.

In the scientific and engineering literature, EPBT for solar PV systems typically ranges from less than one year to around four years, depending on the technology and installation location. For example, First Solar, a leading manufacturer of thin-film PV modules, claims that its modules can achieve EPBTs as short as six months under optimal conditions. Geography plays a key role in EPBT: sunnier locations generate more electricity, which shortens the payback time, while cloudier or higher-latitude regions may experience longer paybacks. Solar panels manufactured in China had 30% longer EPBTs than those made in Europe (Liang and You 2023). Additionally, the carbon footprint of Chinese-made panels was estimated to be twice as high, due largely to the carbon intensity of China's coal-heavy electricity mix used in manufacturing processes.

This leads to considering an extension of the energy payback concept—carbon payback time. Instead of focusing on energy, this concept considers the GHG emissions embedded in a product's life cycle and asks: how long does it take for a renewable energy system to "offset" the GHG emissions produced during its manufacture? However, carbon payback time is more location-dependent and context-sensitive than EPBT. For instance, installing a PV system in a region where electricity is already produced with low emissions (e.g., hydropower or nuclear) will displace fewer emissions, resulting in a longer carbon payback time. Conversely, installing PV in a coal- or gas-heavy grid will displace more emissions, leading to a faster carbon payback. Both energy and carbon payback times are valuable tools for understanding the life cycle efficiency and environmental impact of renewable energy systems. They also underscore the importance of considering regional contexts, supply chain emissions, and grid carbon intensity when evaluating clean energy technologies and their potential to mitigate climate change.

Wind Power

Wind power is primarily driven by the sun's uneven heating of the Earth's surface, which creates temperature and pressure differences in the atmosphere (Box 8.2). The Earth's rotation and the Coriolis effect also contribute to wind patterns, though to a lesser extent. Wind energy is initially harvested as mechanical energy

Box 8.2 Estimating Wind Turbine Power and Betz' Limit

Power = $0.5' \rho' C'$ v3′ *A* Sweep

*v*3: meters per second—doubling of the box
Wind speed results in an eightfold increase in power.
ρ: 1.2 kg/m^3 (kilograms per cubic meter) high-density air results in more power (altitude and temperature).
*A*s: m^2 (square meters) area swept by the blade; a slight increase in blade length increases the area greatly.
C: efficiency coefficient, no units; different types of wind turbines have different maximum theoretical efficiencies (Betz limit $\approx$ 0.593 or 16/27 ths) but usually between 0.4 and 0.5.

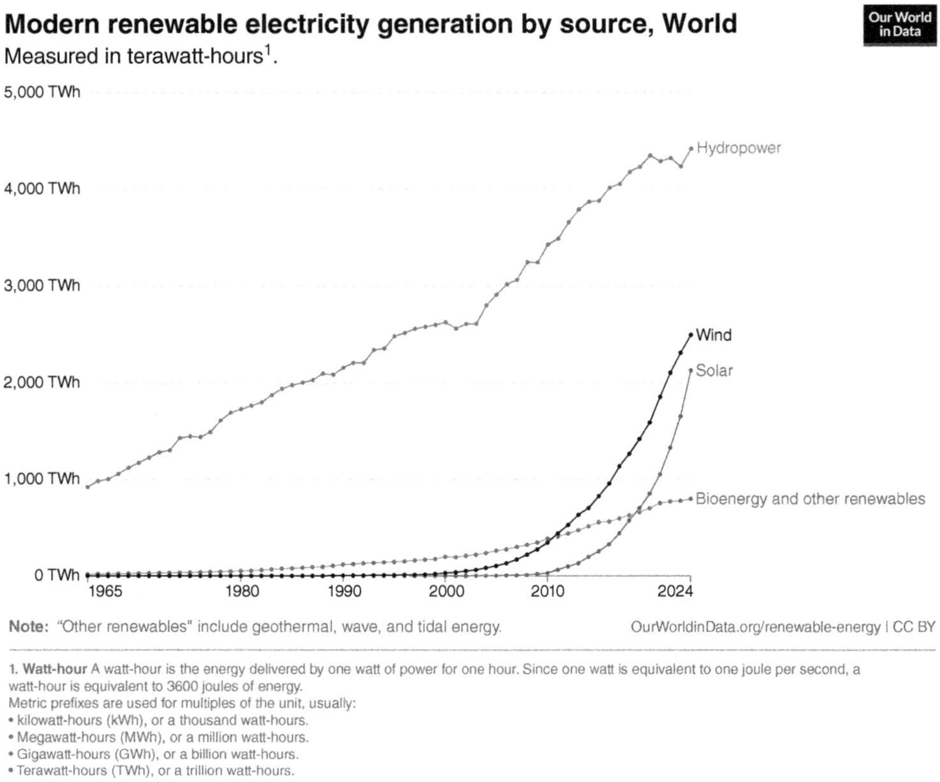

Figure 8.3 Wind and solar are the world's fastest growing renewable energy supplies. Source: Modern renewable electricity generation by source, World/Our World in Data/https://ourworldindata.org/grapher/modern-renewable-prod/last accessed on September 18, 2025.

(e.g., via rotating blades), which is then converted to electricity using a generator. Historically, humans have harnessed wind for navigation (sailing ships), milling grain, and pumping water. Wind was among the last "prime movers" before the rise of the fossil fuel era. The use of wind for electricity generation dates back to the late nineteenth century, when Charles F. Brush constructed a 60-ft windmill in Cleveland, Ohio, in 1888, which powered a 12-kW dynamo—one of the first documented uses of wind to generate electricity.

In recent decades, wind energy has experienced dramatic growth, emerging as a cornerstone of the global renewable energy transition (Figure 8.3). Cumulative global wind power capacity surpassed 1,000 gigawatts (GW) in 2023, according to the Global Wind Energy Council (GWEC). Much of the growth in global renewables over the past two decades has come from utility-scale wind farms, as turbine technology has improved in size, efficiency, and cost-effectiveness.

According to the U.S. Department of Energy, the average turbine capacity has increased significantly—from around 1.5 MW in 2010 to 2.5–3.0 MW in 2020, with some of the largest onshore turbines now exceeding 5 MW and offshore turbines reaching nearly 15 MW.

Figure 8.4 Offshore windfarm. Source: Ralph Greig / Wikimedia Commons / CC BY-SA 2.0

China leads globally with around 35% of cumulative installed wind capacity, followed by the United States (17%), Germany (10%), India (6%), Spain (4%), and other countries such as the United Kingdom, France, Brazil, Denmark, and Canada, each contributing smaller but significant shares (Global Wind Energy Council 2023). In the United States, Texas is the top wind energy producer, generating over 22% of its electricity from wind as of 2022. The wind-rich corridor stretching from Texas and New Mexico to Montana and the Great Lakes region is sometimes dubbed the "Saudi Arabia of Wind."

Wind power is most abundant in the mid-latitudes, particularly in regions dominated by prevailing westerlies in the Northern and Southern Hemispheres. However, one major obstacle to expanding wind power is the distance between high-quality wind resources and major population or demand centers, requiring investment in transmission infrastructure.

To address wind's intermittency, many wind farms are now being paired with battery storage systems, allowing wind energy to become more dispatchable and provide grid flexibility.

Offshore wind offers several advantages over onshore wind. Offshore locations tend to have stronger, more consistent winds and fewer obstructions, leading to higher efficiency and up to 50% more power output for the same turbine compared to land-based systems (Figure 8.4). Offshore wind has been operating in Northern Europe since the 1980s, with the first U.S. offshore wind farm opening in 2016 off Block Island, Rhode Island. Social science research from the Block Island project has highlighted the importance of community trust, transparent engagement, and early involvement in the siting process to ensure successful deployment and public acceptance.

Offshore wind could play a crucial role in decarbonizing coastal regions where land availability is constrained and grid congestion is common. However, developing offshore wind requires substantial investment in ports, staging areas, subsea cables, and grid interconnection infrastructure. These activities may disturb benthic (seafloor) ecosystems, present navigation hazards, and raise public safety and environmental concerns.

For regions with deep coastal waters, where fixed-bottom turbines are impractical, floating wind turbines are emerging as a promising solution. The three main floating turbine platforms under consideration are: Spar-buoy: Anchored by heavy ballast below the water surface, offering stability in deep water and minimizing impact from surface waves; semi-submersible: Buoyancy-driven, with a wide base that allows stability in moderate depths. These platforms can be assembled at port and towed to sea, reducing reliance on expensive marine vessels; Tension-leg platform (TLP): Uses taut mooring lines for vertical stability. Although more complex to install, TLPs offer superior stability and are suited for deeper waters.

Wind farms can affect avian and bat populations, particularly through collisions with turbine blades. The Altamont Pass Wind Resource Area in California has long been cited as an early case of high bird mortality, especially among golden eagles. The area lies between the San Francisco Bay estuary and the Pacific Flyway, placing turbines in a major avian migration corridor. Many analysts attribute the high mortality to poor siting and older turbine designs. In recent years, the project has been repowered: hundreds of older, small turbines were replaced with fewer, more efficient, and better-sited turbines. This case underscores the critical role of ecological monitoring, adaptive siting, and data-informed decision-making in reducing wildlife impacts.

Wind power—onshore and offshore—is a central pillar of the global energy transition. Technological innovation, paired with careful siting, policy support, and environmental safeguards, will be essential for expanding wind's role in a low-carbon, sustainable energy future. As turbine capacities increase and floating technologies evolve, wind energy will become more versatile—but its deployment must be socially equitable, ecologically responsible, and strategically planned to avoid replicating the environmental harms of past energy systems.

Wave and Tidal Power

Wave energy is generated by the movement of the ocean's surface, influenced by various forces including wind, storms, gravity, and surface tension. One of the key drivers of wave formation is the fetch of a storm—the distance over which wind blows across the ocean surface. Even regions far from storm centers can experience large waves as swell energy travels thousands of miles, particularly across the Pacific Ocean. Wave energy can be harnessed using technologies that extract energy from surface motion, pressure fluctuations below the surface, or entire wave forms. Some technologies remain stationary and let waves pass over them, while others move in response to wave motion.

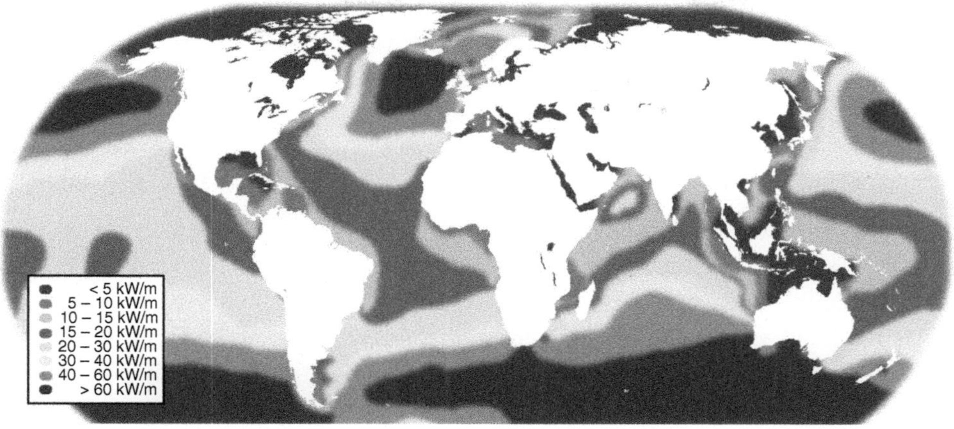

Figure 8.5 The global distribution of wave energy showing it tends to be more powerful at lower and high latitudes.

Wave energy potential is greatest in the mid-latitudes (Figure 8.5), where trade winds and local winds combine, making areas like the North Sea, the northern Pacific (California to Alaska), New Zealand, South Africa, and southern South America prime locations for wave energy development.

Wave energy devices can be installed in various locations and are often classified by distance from shore: Shoreline: Close to land, easier to maintain, connected to grid infrastructure; nearshore: In shallower waters, some energy is lost due to seabed friction; and offshore: In deeper waters, offering higher energy yield but with greater construction and maintenance challenges. Three common converter types include Point Absorbers: Buoy-like devices that float and move vertically with wave motion. These are non-directional and often deployed in nearshore environments (Rusu and Onea 2017). Attenuators: Long floating devices aligned with the wave direction, generating energy as they flex along their length. Notable examples include the Salter Duck, Bristol Cylinder, Pelamis Anaconda, and Clamshell designs. Terminators or overtopping devices, which are positioned onshore or nearshore, these devices act like breakwaters, funneling wave water into a reservoir. The stored water flows through low-head turbines, similar to small hydroelectric plants.

Wave energy is still in its early stages of commercialization. Costs are high due to the technical complexity, harsh ocean conditions, and lack of mature supply chains. Bringing projects forward will require: Government subsidies and tax credits, research and development funding, private-public partnerships, and regulatory support and permitting frameworks. One major cost driver is the need for underwater transmission cables, which are expensive and complex to maintain. To reduce these costs, some propose co-siting wave energy systems with offshore wind farms or floating solar ("floatovoltaic") arrays. These hybrid systems can share infrastructure, reduce transmission costs, and improve output reliability by combining complementary energy sources (Rusu and Onea 2017) (Box 8.3).

Box 8.3 Ocean Energy

Ocean energy—encompassing technologies like tidal, wave, and ocean thermal energy conversion (OTEC)—is an emerging but promising frontier in the renewable energy landscape. According to research led by Pacific Northwest National Laboratory (PNNL), ocean energy holds the potential to provide clean, resilient power to coastal, island, and remote communities, where energy costs are often high and supply is vulnerable to disruptions. These marine energy systems complement more established renewables like solar and wind by offering more consistent and potentially base-load energy generation, especially in regions with strong tidal or wave activity. PNNL's work emphasizes not only the technical advancement of marine energy converters—such as tidal turbines, point absorbers, and oscillating water columns—but also the need for robust environmental monitoring tools to ensure that these systems do not adversely affect marine ecosystems. Researchers are developing technologies to track and mitigate impacts on marine wildlife, helping accelerate permitting and responsible deployment. In addition to electricity, ocean energy can support the "blue economy" by powering offshore aquaculture, desalination, and ocean observation systems. Though still in early stages, with challenges around cost, scale, and grid integration, PNNL and other institutions argue that ocean energy can play a meaningful role in decarbonizing coastal infrastructure, diversifying renewable portfolios, and creating synergies with maritime industries. As research continues and technologies mature, ocean energy may become a valuable asset in achieving climate and energy goals, particularly in coastal regions with strong marine resources.

Wave energy projects must be carefully designed to minimize impacts such as disruption to marine life and habitats, interference with navigation and shipping routes, conflicts with commercial and recreational fishing, and visual impacts and coastal view degradation. Marine sanctuaries and protected areas are generally unsuitable for wave energy development due to strict conservation regulations. Marine spatial planning, incorporating the expertise of ecologists, oceanographers, social scientists, and local stakeholders, will be essential for responsible deployment.

Improved wave forecasting is crucial to integrate wave power into energy systems. Seasonal variation is significant—winter typically provides stronger wave conditions than summer.

There are two main forecasting approaches, using either physics-based models (e.g., spectral wave models that balance wave energy equations): Better for long-term forecasts or time series models (e.g., neural networks correlate historical variables using deep learning): More accurate for short-term forecasting (Reikard et al. 2015). Wind and wave conditions are closely related, and integrated forecasting systems will be key to optimizing hybrid energy platforms.

Tidal energy, as the name suggests, harnesses the ebb and flow of ocean tides, primarily driven by the gravitational pull of the moon, with a smaller contribution from the sun. Tidal energy is predictable and can be harvested in two main ways. Tidal stream generators are turbines placed in natural tidal flows, such as estuaries or narrow channels, where water moves in a dominant direction. Tidal barrages are structures similar to dams that impound water during high tide and release it during low tide, generating power as the water flows through turbines. The Severn Estuary in the United Kingdom is a historic and often-cited site for proposed barrage development. The power output increases with the square of the tidal range, meaning greater tidal differences produce exponentially more energy than simply increasing basin size. While still in early stages, wave and tidal energy represent promising frontiers for clean, predictable, and renewable electricity generation. However, they face considerable technological, economic, and ecological challenges. Continued research, public investment, stakeholder collaboration, and policy innovation will be essential to realizing their full potential.

Takeaways/Highlights

- Most renewable energy flows originate from the sun. The sun drives the majority of Earth's renewable energy flows—including solar, wind, wave, and indirectly tidal (which is primarily lunar)—by creating recurring thermal and kinetic patterns that life and human infrastructure can harness.
- Solar energy is foundational to the low-carbon energy transition. Solar PVs produce significantly fewer emissions than fossil fuels and offer compatibility with both utility-scale installations and DERs. However, challenges remain across the supply chain, including mining impacts, energy-intensive refining processes, and end-of-life waste management.
- Wind power is a maturing and scalable solution. Wind energy has seen massive global expansion, especially in China, the United States, and Europe. Offshore wind, in particular, offers higher efficiency and capacity due to stronger and steadier winds, though it comes with increased infrastructure costs and environmental planning requirements.
- Energy and carbon payback time matter. A critical metric in renewable energy systems is the EPBT—the time it takes for a technology to generate the energy used to make it. Similarly, carbon payback time estimates how long it takes to offset embedded GHG emissions. Both metrics highlight the importance of installation location and the carbon intensity of the grid being displaced.
- Just transitions require more than clean energy. A truly just transition to solar power would include improving energy access for marginalized communities, minimizing occupational risks along the supply chain, and ensuring community input, ownership, and benefit—especially where Indigenous, rural, or displaced populations are impacted.

Questions for Review

1. What is the PVs effect? How does it work? Explain the photovoltaic effect to someone who might not be familiar. How does the energy get collected and move through the system.
2. Why is there land use conflict over solar and wind projects? List all that you can think of that may be sources of contestation or social resistance.
3. What are the prospects of getting energy from the ocean? What are the opportunities and key challenges?

Exercise 8.1 Developing Solar Energy Farms Responsibly

In this assignment, you will explore and summarize the controversies over siting solar power plants. Your essays should be 500 words, and you will also collect key attributes about the power plant.

Part 1: Reflect on your observations of the controversies over siting large-scale utility solar power plants. Begin your essay by describing why large-scale projects have been pursued, and describe some of the controversies these projects have engendered opposition.

Part 2: You will each be assigned a solar power plant. Provide a brief background and key information about the project. Who is building the project? Who is leading the review? How big is it? How much land? Whose land? You will review the environmental impact reports (skim, as these are long), agency websites, and news stories. Find the public comments for the project and describe the key issues that are raised. Are there any solutions to mitigating the primary concerns? What is the current status of the project, and what are the future prospects? Use maps in the environmental impact statement to identify where the project is located and launch the Desert Renewable Energy Conservation Plan databasin site analyst. http://drecp.consbio.webfactional.com/survey Click on the site or nearby, what is the general landscape context? Which species are on site, and what kinds of impacts could be expected? This map can help you locate your project: http://www.energy.ca.gov/maps/renewable/renewable_development_drecp.pdf.

Part 3: Explore the websites of the Solar Programmatic Environmental Impact Statement (http://solareis.anl.gov) and/or Desert Renewable Energy Conservation Plan (http://www.drecp.org). What are the goals of each program, and how can these planning processes help identify appropriate sites for solar energy deployment? Write about how you might apply key development principles to utility scale solar project planning efforts. What is important and why? Describe what sustainability means, and evaluate how these projects meet (or do not meet) your own interpretations of and expectations for sustainability.

Format of assignment

Introduction—one paragraph, short description of the general issues of siting solar

Key information—include the following in a table or list

- *Location, developer*
- *Output (MW)*
- *Size (acres)*
- *Water use (if available)*

Short utility scale solar project description—one paragraph

Species impacted and major environmental concerns—one or two paragraphs

Environmental justice or cultural resource considerations—one paragraph

How does the project you investigate reflect principles of sustainable or ecosystem-oriented project design? One paragraph

Conclude with a paragraph with your impressions of the project, what do you think? How might you have done things differently if you were to propose this project again? What alternative sites would you consider?

Solar Power Plant Projects

Use the internet or try this National Environmental Policy Act repository to find a utility-scale solar project. https://eplanning.blm.gov/eplanning-ui/home.

Exercise 8.2 Life Cycle Assessment of Photovoltaic Technologies

In this assignment, work with a life cycle inventory for a photovoltaic technology. Each team will be assigned a different photovoltaic product made in different parts of the world.

Photovoltaic modules manufacturers

A. Mono-crystalline silicon modules made in Xinjiang, China
B. Mono-crystalline silicon modules in Washington, United States
C. Mono-crystalline silicon modules in Iceland
D. CdTe thin-film modules made in Malaysia (CdTe)

1. Review the Excel sheet and put your team's names on the copy your team will submit. Skim over tables 1 and 2 to understand the various components used in all four technologies and the EPA Tool for the Reduction and Assessment of Chemicals and Other Environmental Impacts (TRACI) indicators for each of the inventory inputs.

2. Look over your assigned production system. These are tables 3A, 3B, 3C, and 3D. Notice that most of the inventory and emissions factors have been completed in Table 3. The emissions for each material are obtained by multiplying the "kg of substance/m^2" by each impact category "kg pollutant/kg substance." The result is kg of pollutant per square meter. We will focus on GHG emissions, which in TRACI is called Global Warming Air.
3. Complete this missing information in the cells highlighted in yellow. This information is in the first two tables highlighted in orange. Once finished, try to complete the rest of the PV modules.
4. Review your team's PV module results on the "graphs" worksheet.
5. Discuss any environmental hotspots that are driving the outcome. What can be addressed to make these PVs with less energy or GHG emissions?
6. Prepare for final submission.

Exercise 8.3 Your Region's Renewable Electricity Fleet

Goal: Build an Excel calculator that estimates the resources for electric power generation in 2050. Write a short essay describing the Excel table and focus on one key consideration for this future energy system.

Deliverables: 500–1,000-word essay plus Google Sheet. Upload the essay to canvas and add a link to the Google Sheet in the comment section of the submission.

Podcasts: https://xenetwork.org/ets/episodes/episode-161-expanding-transmission/

https://xenetwork.org/ets/episodes/episode-46-100percent-renewables/
https://www.volts.wtf/p/transmission-fortnight-burying-power?s=r
https://xenetwork.org/ets/episodes/episode-146-why-local-solar-costs-less/

PART ONE

Two energy system integration researchers (Professors Mark Delucchi & Mark Jacobson) have argued that global energy needs could be met with wind, water, and solar (WWS) power. For this assignment, you will read the California version of their analysis and critically assess their argument. This paper and the prior summary version for the United States have been posted on the canvas site in *Module 7* as Jacobson et al. (2014).

Write a 500–1,000 word summary of the central ideas presented in their analysis and answer the questions below (1-in margins, single-spaced, 12-point

Arial or Times font). References do not count toward the page count, but *use references.* Your essay should have three parts.

First, describe the reasons for a WWS approach and its physical constraints. Summarize the key technologies proposed and how they fit into California's future. What are the key geographies? What are the key challenges or constraints? Is the WWS argument convincing, or are there assumptions that require further investigation? Missing technologies that are key?

Next, focus on some of the integration issues, as well as the policy and economic issues. What is the upshot of their argument? What are some of the key justifications for their conclusions? If the barriers they suggest do in fact exist, how might they be overcome?

Use the Jacobson et al. (2014) paper to begin to identify key concepts that might be important to your particular project(s). Somewhere in your write-up, you should include some key takeaway points that should inform your individual or group projects.

How has this analysis been used in other research? Search for any articles that thoroughly engage with their analysis. For example, see "No quick switch to low-carbon energy" by Kramer and Haigh (2009) in *Nature* (https://www-nature-com.libaccess.sjlibrary.org/articles/462568a). Did these authors agree with Delucchi and Jacobson? Disagree? What did they contest or agree with? How about the podcast "Is 100% renewables realistic?" What are the major shortcomings of the analyses?

PART TWO

Open a Google Sheet and make a table like table 2 Jacobson et al. (2015). Populate with the information found in the columns under (1) energy technology (2) rated power, (3) percent of 2050 power demand met by technology, and (4) technical potential nameplate capacity.

Once you've copied these columns from the Jacobson table, add a column for capacity factor and estimate energy use. *Reminder: capacity factor is the percent of time the power plant is running. . .* Find some data on capacity factors here: https://www.eia.gov/electricity/monthly/epm_table_grapher.php?t=epmt_6_07_b

To estimate energy use multiply the nameplate capacity of the electricity fleet by the capacity factor times the number of hours in a year (8,760 hours).

$$\text{Energy} = \text{Power} \times \text{Time} = \text{Power} \times 8760\,\text{hours} \times \text{capacity factor}$$

Next, think about how one might improve it to capture more of the sustainability benefits or tradeoffs we have. List new themes/calculations you might add, like land use, material requirements.

$$\text{Land} - \text{space} = \text{power}\left(\text{Watts}\right) / \text{power density}\left(\text{W}/\text{m}^2\right)$$

Power density, for example can estimate land use. Use the table below to estimate the size of the energy footprint for each, using the table from van Zalk and Behrens (2018). The median power density (W/m^2) is in parentheses on the left of the chart. To get from terawatts to watts, you need to multiply by 1,000,000,000,000/one trillion/1×10^{12}. Then divide by the power density.

Add two columns with water withdrawals and water consumption. This paper has good water use coefficients for water withdrawals and consumption. Take the medians and multiply by the energy generation.

References

Appadurai, A. (1988). *The Social Life of Things: Commodities in Cultural Perspective.* Cambridge University Press.

Baka, J. (2011). Biofuels and wasteland grabbing: how India's biofuel policy is facilitating land grabs in Tamil Nadu, India. *International Conference on Global Land Grabbing, Institute of Development Studies.* University of Sussex, Brighton, United Kingdom, 6–8 April 2011.

Bebbington, A. and Bury, J. (2013). *Subterranean Struggles: New Dynamics of Mining, Oil, and Gas in Latin America.* University of Texas Press.

Brockway, P.E., Sorrell, S., Semieniuk, G. et al. (2021). Energy efficiency and economy-wide rebound effects: A review of the evidence and its implications. *Renewable and Sustainable Energy Reviews* 141: 110781.

Burrows, K. and Fthenakis, V. (2015). Glass needs for a growing photovoltaics industry. *Solar Energy Materials and Solar Cells* 132: 455–459.

Cornea, N., Véron, R., and Zimmer, A. (2017). Clean city politics: an urban political ecology of solid waste in West Bengal, India. *Environment and Planning A* 49 (4): 728–744.

Forti, V., Balde, C.P., Kuehr, R., and Bel, G. (2020). *The Global E-Waste Monitor 2020: Quantities, Flows and the Circular Economy Potential.* United Nations University (UNU)/United Nations Institute for Training and Research.

Fthenakis, V.M., Kim, H.C., and Alsema, E. (2008). Emissions from photovoltaic life cycles. *Environmental Science & Technology* 42 (6): 2168–2174.

Fthenakis, V., Wang, W., and Kim, H.C. (2009). Life cycle inventory analysis of the production of metals used in photovoltaics. *Renewable and Sustainable Energy Reviews* 13 (3): 493–517.

Global Wind Energy Council. (2023). *Global Wind Report.* www.gwec.net

Grodsky, S.M. and Hernandez, R.R. (2020). Reduced ecosystem services of desert plants from ground-mounted solar energy development. *Nature Sustainability* 3 (12): 1036–1043.

Heath, G.A., Silverman, T.J., Kempe, M. et al. (2020). Research and development priorities for silicon photovoltaic module recycling to support a circular economy. *Nature Energy* 5 (7): 502–510.

Hernández, D. (2015). Sacrifice along the energy continuum: a Call for energy justice. *Environmental Justice* 8 (4): 151–156.

Hernandez, R.R., Hoffacker, M.K., Murphy-Mariscal, M.L. et al. (2015). Solar energy development impacts on land cover change and protected areas. *Proceedings of the National Academy of Sciences of the United States of America* 112 (44): 13579–13584.

Hoffacker, M.K., Allen, M.F., and Hernandez, R.R. (2017). Land-sparing opportunities for solar energy development in agricultural landscapes: a case study of the Great Central Valley, CA, United States. *Environmental Science & Technology* 51 (24): 14472–14482.

International Rewnewable Energy Association (2016). *End-of-Life Management: Solar Photovoltaic Panels*. International Renewable Energy Agency, IEA-PVPS.

Jacobson, M.Z., Delucchi, M.A., Ingraffea, A.R. et al. (2014). A roadmap for repowering California for all purposes with wind, water, and sunlight. *Energy* 73: 875–889.

Kosciuch, K., Riser-Espinoza, D., Gerringer, M., and Erickson, W. (2020). A summary of bird mortality at photovoltaic utility scale solar facilities in the Southwestern US. *PLoS One* 15 (4): e0232034.

Kramer, G.J. and Haigh, M. (2009). No quick switch to low-carbon energy. *Nature* 462: 568–569.

Krieger, E.M., Casey, J.A., and Shonkoff, S.B. (2016). A framework for siting and dispatch of emerging energy resources to realize environmental and health benefits: case study on peaker power plant displacement. *Energy Policy* 96: 302–313.

Lee, J., Bazilian, M., Sovacool, B., and Greene, S. (2020). Responsible or reckless? A critical review of the environmental and climate assessments of mineral supply chains. *Environmental Research Letters* 15 (10): 103009.

Liang, H. and You, F. (2023). Reshoring silicon photovoltaics manufacturing contributes to decarbonization and climate change mitigation. *Nature Communications* 14 (1): 1274.

Luke, N. and Heynen, N. (2020). Community solar as energy reparations: abolishing petro-racial capitalism in New Orleans. *American Quarterly* 72 (3): 603–625.

McDonald, R.I., Fargione, J., Kiesecker, J. et al. (2009). Energy sprawl or energy efficiency: climate policy impacts on natural habitat for the United States of America. *PLoS One* 4 (8): e6802.

Mulvaney, D. (2017). Identifying the roots of Green Civil War over utility-scale solar energy projects on public lands across the American Southwest. *Journal of Land Use Science* 12 (6): 493–515.

Mulvaney, D. (2019). *Solar Power: Innovation, Sustainability, and Environmental Justice*. University of California Press.

O'Shaughnessy, E., Dong, S., Cook, J.J. et al. (2022). Effects of local permitting and interconnection requirements on solar PV installation durations. *Energy Policy* 161: 112734.

Reikard, G., Robertson, B., Buckham, B. et al. (2015). Simulating and forecasting ocean wave energy in western Canada. *Ocean Engineering* 103: 223–236.

Reyes Maturano, I. (2021). Social dispossession, the real 'benefit' of green projects in Yucatan. *Development* 65 (1): 63–70.

Rignall, K.E. (2016). Solar power, state power, and the politics of energy transition in pre-Saharan Morocco. *Environment and Planning A* 48 (3): 540–557.

Rusu, E. and Onea, F. (2017). Joint evaluation of the wave and offshore wind energy resources in the developing countries. *Energies* 10 (11): 1866.

Stock, R. (2020). *Deus ex mitigata*: denaturalizing the discursive power of Solar India. *Environment and Planning E: Nature and Space* 4 (2): 354–382. https://doi.org/10.1177/2514848620908166.

Stock, R. and Birkenholtz, T. (2020). Photons vs. firewood: Female (dis) empowerment by solar power in India. *Gender, Place & Culture* 27 (11): 1628–1651.

Stock, R. and Birkenholtz, T. (2021). The sun and the scythe: energy dispossessions and the agrarian question of labor in solar parks. *The Journal of Peasant Studies* 48 (5): 1–24.

Sustainable Systems Research Foundation (2021). *A Political Analysis of Microgrids & California's Over-the-Fence Rule.* https://sustainablesystemsfoundation.org/wp-content/uploads/2021/04/Political-Analysis-of-Microgrids-and-Over-the-Fence.formatted-converted-5.pdf

Watts, M. (2001). Petro-violence: community, extraction, and political ecology of a mythic commodity. In: *Violent Environments* (eds. N.L. Peluso and M. Watts), 189–212. Ithaca, NY: Cornell University Press.

Yenneti, K. and Day, R. (2015). Procedural (in) justice in the implementation of solar energy: the case of Charanaka solar park, Gujarat, India. *Energy Policy* 86: 664–673.

van de Ven, D.-J., Capellan-Perez, I., Arto, I. et al. (2021). The potential land requirements and related land use change emissions of solar energy. *Scientific Reports* 11: 2907.

van Zalk, J. and Behrens, P. (2018). The spatial extent of renewable and non-renewable power generation: A review and meta-analysis of power densities and their application in the US. *Energy Policy* 123: 83–91.

Further Reading

Ayaburi, J., Bazilian, M., Kincer, J., and Moss, T. (2020). Measuring "reasonably reliable" access to electricity services. *The Electricity Journal* 33 (7): 106828.

Bedi, H.P. (2021). Solar power for some? Energy transition injustices in Kerala, India. *Environment and Planning E: Nature and Space* 5 (3): 1146–1163. https://doi.org/10.1177/25148486211046963.

Cha, J.M. (2020). A just transition for whom? Politics, contestation, and social identity in the disruption of coal in the Powder River Basin. *Energy Research & Social Science* 69: 101657.

Cohen, J.J., Reichl, J., and Schmidthaler, M. (2014). Re-focussing research efforts on the public acceptance of energy infrastructure: a critical review. *Energy* 76: 4–9.

Devitt, D.A., Young, M.H., and Pierre, J.P. (2020). Assessing the potential for greater solar development in West Texas, USA. *Energy Strategy Reviews* 29: 100490.

https://digitalscholarship.unlv.edu/cgi/viewcontent.cgi?article=1818&context=sls_fac_articles.

Dunlap, A. (2018). Counterinsurgency for wind energy: the Bíi Hioxo wind park in Juchitán, Mexico. *The Journal of Peasant Studies* 45 (3): 630–652. https://doi.org/10.1080/03066150.2016.1259221.

Haggerty, J.H., Haggerty, M.N., Roemer, K., and Rose, J. (2018). Planning for the local impacts of coal facility closure: emerging strategies in the US West. *Resources Policy* 57: 69–80.

Hamidi, M. (2016). Indonesian female factory workers: the gendered Migration policy in Malaysia. *People: International Journal of Social Sciences* 2 (1): 643–664.

Jenkins, K., McCauley, D., Heffron, R. et al. (2016). Energy justice: a conceptual review. *Energy Research & Social Science* 11: 174–182.

Martinez, N. (2020). Resisting renewables: the energy epistemics of social opposition in Mexico. *Energy Research & Social Science* 70: 101632.

Mayer, A., Hazboun, S.O., and Howe, P.D. (2021). For the love of sun and wind? Proximity to renewable energy facilities and support for renewable power across time and space in the United States. *Energy Research and Social Science* 73: 101910.

Moore-O'Leary, K.A., Hernandez, R.R., Johnston, D.S. et al. (2017). Sustainability of utility-scale solar energy–critical ecological concepts. *Frontiers in Ecology and the Environment* 15 (7): 385–394.

Morello-Frosch, R., Pastor, M., and Sadd, J. (2001). Environmental justice and Southern California's "riskscape" the distribution of air toxics exposures and health risks among diverse communities. *Urban Affairs Review* 36 (4): 551–578.

Newell, P. and Mulvaney, D. (2013). The political economy of the 'just transition'. *The Geographical Journal* 179 (2): 132–140.

Ngai, P. (2016). Incomplete subjects: circular migration and the life and death struggles of the migrant workers in China. In: *Impact of Circular Migration on Human, Political and Civil Rights* (ed. C. Sole, S. Parella, T.S. Marti, and S. Nita(Schroder)), 175–194. Springer.

Sareen, S. and Kale, S.S. (2018). Solar 'power': socio-political dynamics of infrastructural development in two Western Indian states. *Energy Research & Social Science* 41: 270–278.

Sovacool, B.K., Ali, S.H., Bazilian, M. et al. (2020). Sustainable minerals and metals for a low-carbon future. *Science* 367 (6473): 30–33.

Stevis, D. and Felli, R. (2020). Planetary just transition? How inclusive and how just? *Earth System Governance* 6.

Sward, J.A., Nilson, R.S., Katkar, V.V. et al. (2021). Integrating social considerations in multicriteria decision analysis for utility-scale solar photovoltaic siting. *Applied Energy* 288: 116543.

Yenneti, K., Day, R., and Golubchikov, O. (2016). Spatial justice and the land politics of renewables: dispossessing vulnerable communities through solar energy megaprojects. *Geoforum* 76: 90–99. https://www.sciencedirect.com/science/article/abs/pii/S0016718515303249.

Chapter 9

Metals and Minerals

The energy transition will be highly metal- and mineral-intensive. While including oil and coal in the total mass of extracted materials may show a net reduction in overall mining activity, the demand for specific metals—such as copper, aluminum, lithium, nickel, and cobalt—is expected to increase substantially. As gaseous and liquid fuels currently transported by pipeline, rail, ship, or truck are replaced by electrified systems, new landscapes and sites of mineral production will emerge to support this transformation. Some analysts project that meeting the infrastructure needs of a 1.5 °C-compatible future will require a five- to sixfold increase in the production of key metals and minerals. This will involve not only expanded mining and brine extraction but also improved recycling systems to recover critical materials for low-carbon technologies. Understanding the material foundations of the energy transition is essential, as different socio-technical pathways enroll distinct materials, ecosystems, labor regimes, and geopolitical dynamics into the production of our energy systems. The choices made about which technologies to pursue—and where and how their inputs are sourced—will have significant implications for environmental justice, biodiversity, and local livelihoods. This chapter highlights the importance of interdisciplinary approaches to navigating the tradeoffs of the transition. It challenges the notion that mining expansion is simply a necessary cost of decarbonization and argues that each mining proposal must be contextually assessed, with attention to local environmental, social, and governance conditions. By anticipating the material demands of future energy systems, land use planners, resource managers, and policymakers can make more informed decisions about where and how to develop while avoiding the repetition of extractive patterns that have historically harmed people and ecosystems.

Mineral and Metal Intensive Energy Transition

Extractive industries remain the backbone of the material economy, supplying the natural resources that underpin industrial production, infrastructure, and now, the renewable energy transition. Unlike fossil fuel-based systems—where energy is extracted and combusted—renewable energy is harvested directly from flows such as sunlight, wind, and waves. However, the major environmental and resource impacts of renewable systems lie not in their operation, but in the production, siting, and disposal of the devices and infrastructures that make renewable energy possible. This shift in impact location underscores the importance of comprehensively analyzing the supply chains and extractive industries linked to renewable energy (Box 9.1). Technologies like solar photovoltaics, wind turbines, batteries, and electric vehicles (EVs) require large quantities of minerals and metals, including quartz, bauxite (for aluminum), cassiterite (for tin), copper, lithium, and rare earth elements such as neodymium, dysprosium, and praseodymium. As demand for these materials rises,

Box 9.1 Critical Metals and Minerals: Foundations of Energy Transitions

Critical metals and minerals form the material foundation of the global energy transition, as they are essential to the manufacture of technologies that will harness, generate, store, and transmit clean and renewable energy (Bazilian 2018). These materials are embedded in everything from solar panels and wind turbines to EVs, battery storage systems, and power transmission infrastructure. In particular, batteries are central to decarbonization strategies, as they enable the storage of intermittent energy from solar and wind, support grid flexibility, and power the electrification of transportation. A wide range of metals and minerals are needed to produce modern batteries, particularly lithium-ion batteries, which dominate today's EV and grid storage markets. Key materials include lithium, cobalt, nickel, manganese, graphite, aluminum, copper, zinc, iron, vanadium, lead, and chromium. Each of these plays a different functional role: lithium is used in cathodes; cobalt and nickel improve energy density and cycle life; graphite is the most common anode material; and copper is critical for current collectors and conductivity. The high concentration of value and performance in a small volume of these materials has led to what some call the "metal intensity" of clean energy systems.

This demand raises sustainability concerns, including geopolitical supply risks, environmental degradation, labor conditions in mining regions, and price volatility. In response, research laboratories and battery developers are investing in the exploration of next-generation battery chemistries that could reduce reliance on scarce or expensive materials. Some promising alternatives include sodium-ion batteries (which use abundant sodium instead of lithium), zinc-air batteries, and solid-state batteries, which may one day be constructed with non-metallic or polymer-based components, such as glass or organic plastics. These emerging technologies aim to deliver safe, high-performance energy storage using widely available, low-cost, and environmentally benign materials. While many of these technologies are still in the experimental or pilot phases, they represent an important frontier for reducing the environmental and social footprint of battery production, ensuring that energy transitions are not only low-carbon, but also more equitable and sustainable in terms of material inputs.

new landscapes are being brought into production, transforming them from rural, remote, or even wild ecosystems into industrial extraction zones (Sovacool et al. 2020).

In some commodity systems, increased demand means enrolling new territories, communities, and ecosystems into the global extractive economy. Meeting the material requirements of the energy transition may benefit from innovations in metallurgy, chemical processing, and recycling, as well as more stringent environmental and labor standards across the mining and refining sectors. To illustrate the material intensity of renewable infrastructure, consider that building a 100 MW wind farm requires an estimated: 30,000 tons of iron ore, 50,000 tons of concrete, and 900 tons of non-recyclable plastics (primarily for turbine blades)—all of which are mined, transported, and manufactured using fossil fuels. Virtually all projections of a clean energy transition point toward increased material throughput over time, particularly for the production of batteries, photovoltaics, wind turbines, EVs, and transmission lines.

These trends suggest a boom in extractive activity and an expansion of mining into new geographies and sensitive ecosystems. For context, EVs contain over six times the amount of metal and mineral content as internal combustion engine (ICE) vehicles (Box 9.2). While it's true that ICE vehicles burn large volumes of petroleum over their lifetimes, and that petroleum itself is a mined resource, the shift from combustion fuels to electrified systems does not eliminate material intensity—it reshapes it. Comparisons between ICE and EV impacts can obscure the fact that new mining introduces new environmental and social risks that cannot simply be offset by reductions in fossil fuel use. Rather than focusing narrowly on material substitution, the more meaningful questions revolve around how to pursue best practices in mining,

Box 9.2 Battery Chemistries, Metals, and Manufacturers

A metric ton (MT) of lithium can supply enough material to produce batteries for approximately 50 EVs, depending on the specific battery chemistry, size, and vehicle range. Lithium-ion batteries have become a cornerstone of both EV development and stationary energy storage, making lithium one of the most strategically important minerals in the clean energy transition. At the most basic level, lithium-ion batteries consist of three key components: a cathode, anode, and electrolyte. Each of these components depends on specific metal combinations, often governed by intellectual property protections, patents, or licensing agreements. Different battery chemistries offer different performance characteristics—such as energy density, cycle life, charging speed, and thermal stability—and often involve trade-offs. For instance, a battery with higher energy density may also carry a greater risk of thermal runaway or fire, while others prioritize safety and longevity over capacity. The table below outlines some of the most prominent lithium-ion battery chemistries, the leading manufacturers behind them, and the types of vehicles or applications in which they are commonly used (Table 9.1). Because this is a rapidly evolving industry, ongoing innovation is likely to produce new chemistries, shifting manufacturer alliances, and emerging technologies with novel performance capabilities.

Table 9.1 Battery manufacturers and the battery chemistries they use, and the main or exclusive buyers of their batteries.

Battery OEM	Battery chemistry	Auto manufacturer/application
BYD	LFP	Ford, BYD
Contemporary Amperex Technology Co., Limited (CATL)	LFP, NMC, sodium ion	Tesla, Toyota, Peugeot, Hyundai, Honda, BMW, VW, Volvo
LG energy solution	NMC, LFP	General Motors, Daimler, Tesla
Panasonic	NCA, NMC, LFP	Tesla, others
Samsung	LMO/NMC	Ford

processing, and recycling; how to minimize harms to communities and ecosystems; and how to ensure the energy transition is not built on extractive injustice. The transition must be planned with careful attention to governance, equity, and environmental limits, recognizing that the sustainability of renewable energy depends as much on its materials as on its emissions profile.

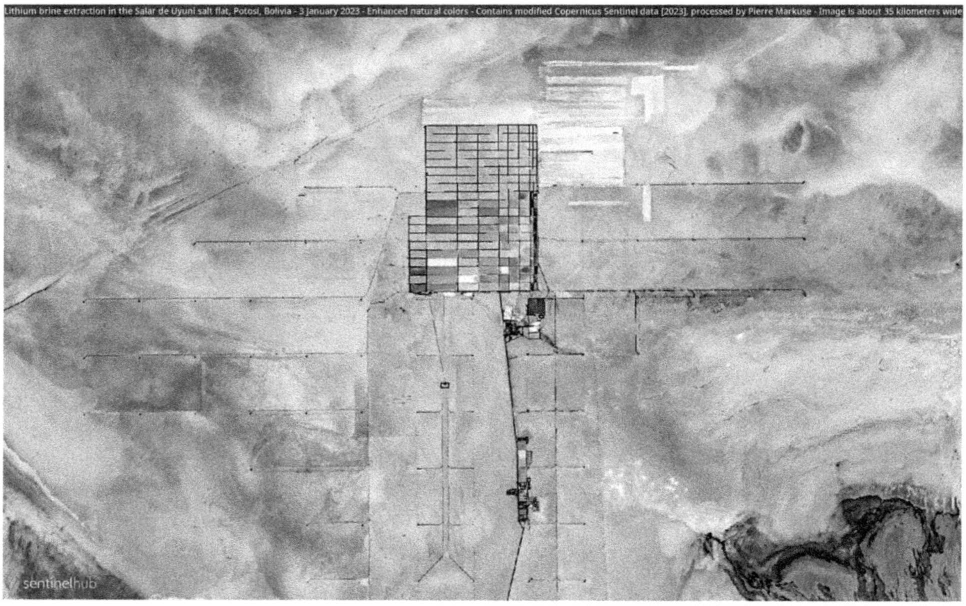

Figure 9.1 Lithium extraction from brines is one source of the key battery ingredient. This method involves drilling wells and using evaporation ponds to process the lithium salts. Source: Pierre Markuse / Flickr.

The future role of lithium-ion batteries in energy systems will depend heavily on cost trends in energy storage. In 2010, the average price of lithium-ion batteries was over \$1,100 per kilowatt-hour (kWh). By 2019, prices had fallen by roughly 87%, reaching an average of \$156/kWh, due to advancements in manufacturing, economies of scale, and increased demand from EVs and renewable energy integration. Battery prices have since surpassed the long-anticipated milestone of dropping prices below \$100/kWh, with some sources reporting average global pack prices around \$139/kWh, and lithium iron phosphate (LFP) batteries reaching as low as \$100–120/kWh, particularly in China. However, the trajectory toward cheaper storage is no longer as predictable. Supply chain disruptions, inflation, and increased shipping and raw material costs—especially for lithium, cobalt, and nickel—have introduced new price volatility (Figure 9.1). In 2022, for example battery prices temporarily rose for the first time in over a decade, driven largely by soaring material costs. But that was only a temporary blip on the otherwise steady fall in prices.

This raises important questions for analysts and planners: Will battery prices continue to decline with every doubling of production, as predicted by learning curves and economies of scale? Or will resource constraints, geopolitical tensions, and environmental costs lead to plateaus or even reversals in cost reductions? These uncertainties are central to energy modeling and long-term policy planning, where assumptions about future battery costs significantly influence projections for EV adoption, grid storage capacity, and renewable energy feasibility. It's also important to note that not all battery price estimates are equivalent. Some data reflect the cost of battery cells only, while others include the entire battery pack, which adds costs for management systems, cooling, enclosures, and more. When comparing battery prices across sources, it's essential to understand whether the figures refer to cells or complete systems to ensure valid comparisons.

Cobalt is a bluish-gray metal valued for its unique electrical and thermal properties, making it an important component in high-performance alloys, magnets, and especially rechargeable batteries. Approximately 50% of global cobalt demand was used in the production of battery cathodes, primarily for lithium-ion batteries used in EVs, portable electronics, and energy storage systems (Cobalt Institute 2000). Around 70–75% of the world's cobalt supply is mined in the Democratic Republic of the Congo (DRC), which has long dominated global production. Of that, an estimated 15–30% is produced through artisanal and small-scale mining (ASM)—a labor-intensive process often carried out under hazardous conditions, with limited oversight or worker protections. The mechanization of cobalt extraction in the DRC typically involves the arrival of large-scale mining companies, often operating through public-private partnerships with the state. While these operations may bring infrastructure and investment, they also tend to result in displacement, environmental degradation, and increased tensions with local communities (Radley and Geenen 2021). These dynamics highlight the complex relationship between resource extraction, state governance, and community well-being, especially in regions where economic dependency on mining coexists with deep structural inequality.

Cobalt mining has become emblematic of the complex relationship between natural resource extraction and human rights abuses, particularly in the DRC. The DRC, a former Belgian colony, remains deeply entrenched in cycles of poverty and exploitation, much of it tied to the global demand for critical minerals. The country lies along Africa's Copper Belt, a geologically rich zone recognized by the mining industry for its vast reserves of copper and cobalt. In many mining operations, cobalt is not the primary target but rather a by-product or tertiary product of copper mining.

The southeastern city of Lubumbashi, near the border with Zambia, is the center of the DRC's mining industry. While the DRC produces around 3% of the world's copper, it supplies over 50% of the global cobalt demand—a metal essential for lithium-ion batteries used in EVs, grid storage, and consumer electronics, all key components of the global energy transition. Two of the largest cobalt producers operating in the country are Glencore and China Molybdenum Company. Approximately 15–20% of the cobalt mined in the DRC comes from ASM, which is often informal, unregulated, and conducted under harsh and unsafe conditions. However, ASM also provides critical income for impoverished and marginalized communities. Completely eliminating ASM from the supply chain could result in greater industry consolidation, benefiting large firms while excluding small-scale miners who rely on this work for survival. Some analysts and practitioners advocate for the formalization and regulation of artisanal mining as a pathway toward a more just and equitable transition (Byemba 2020).

Human rights concerns in the DRC's mining sector are significant. Reports suggest that up to 40,000 children may be working in mining operations, often in hazardous conditions without adequate protection or legal oversight. The mining boom has also raised concerns about tax avoidance and corruption. Although some governments have explored raising taxes or tariffs on mineral exports to increase domestic revenue, weak governance and loopholes in accounting practices frequently allow multinational mining companies to avoid paying their fair share (Radley 2024).

In response to these ethical concerns, some battery manufacturers have pledged to stop sourcing cobalt from the DRC, while others are shifting to cobalt-free battery chemistries. By 2022, an estimated 10% of lithium-ion batteries sold globally contained no nickel, manganese, or cobalt, reflecting the rapid rise of LFP batteries. While, these chemistries are less energy-dense than nickel manganese cobalt (NMC) batteries, ongoing technological improvements have brought their performance closer to expectations for many use cases.

Cobalt, however, is not the only metal with social and environmental concerns. Copper, bauxite (for aluminum), iron ore, and manganese also raise serious issues at the intersection of mining, human rights, and environmental health. For example, manganese mining in Burkina Faso, Ghana, Mozambique, and Tanzania has been linked to public health problems, though it is often difficult to disentangle these effects from broader underdevelopment and inadequate healthcare systems (Farnham et al. 2022). Bauxite, used in aluminum for lightweight EV parts and battery casings, will also become increasingly important to the energy transition.

Moreover, battery supply chains involve a complex network of materials, including steel, graphite, plastics, and numerous other industrial inputs. Smelters and refineries—often located far from extraction sites—are where traceability becomes most opaque, making it difficult to verify ethical sourcing and environmental compliance. As the demand for critical minerals intensifies, ensuring transparency, accountability, and justice across the entire supply chain will be essential to building a truly sustainable and equitable clean energy future (Table 9.2).

Mining is central to the energy transition, but it involves not only the extraction of valuable materials—it also produces significant volumes of waste. Alongside the desired metals, mining generates overburden, tailings, and low-grade ores, all of which require careful management. For example, copper ore typically contains only

Table 9.2 Example of a materials list for a nickel-manganese-cobalt lithium-ion battery.

Graphite	52 kg
Aluminum	35 kg
Nickel	29 kg
Copper	20 kg
Steel	20 kg
Iron and carbon	18 kg
Nickel or other	2 kg
Manganese	10 kg
Cobalt	8 kg
Lithium	6 kg
Iron	5 kg

about 1% copper, meaning that 99% of the extracted material becomes waste or is processed for secondary and tertiary products. Similarly, graphite ore has a yield of around 8%, with the rest ending up as tailings or overburden (USGS 2022).

As the world shifts from molecules (fossil fuels) to electrons for transportation and other energy services, demand for metal-intensive technologies is soaring. Among these, copper is especially critical. It is used extensively in EVs, power transmission, solar panels, and wind turbines. Copper demand is expected to rise by two to three times under decarbonization scenarios, with 3–5 tons required per megawatt of renewable energy capacity (Valenta et al. 2019). A Rio Tinto executive once called copper "the oxygen of decarbonization." Copper demand, even for combustion engines, is on the rise with increased electronics and electrical equipment. However, it takes two and a half times more copper per EV than combustion engine vehicles and electric wires and cables are also copper intensive.

Meeting the material demands of electrification is not straightforward. Although rising copper prices in recent years have incentivized investment, developing new "greenfield" mines typically takes 10–15 years. Many of the world's top-producing copper mines, such as Escondida in Chile and Morenci in Arizona, have been operating for over a century. Chile remains the largest global producer, supplying about 25% of the world's copper. The Morenci mine, the largest in the United States, has a long history of labor disputes, often related to racial inequality and wage conflicts (Kluger 2016; Watt 1956). In addition to long development times, ore grades are declining. For instance, typical nickel ore contains only about 1.5% nickel, and copper ore averages just 0.7% (Henckens and Worrell 2020), meaning more material must be processed for the same yield—raising energy use, emissions, and waste.

Aluminum is another critical material for the energy transition, used in lightweight vehicles, battery casings, and solar panel frames due to its lightness, corrosion resistance, conductivity, and malleability. Aluminum is primarily derived from bauxite (Al_2O_3), a group of aluminum hydroxide minerals. Production of 1 ton of aluminum requires the amount of energy consumed annually by a typical US household. Aluminum is widely used because it is more light-weight and flexible than harder and heavier metals. One important case study of bauxite mining and alumina production is Jamaica (Young 1965). Only 15% of mined bauxite becomes aluminum after refining, but 40% of bauxite can be processed into finished aluminum products, making it a relatively high-yield operation compared to other metals (Government of Canada 2023). One key advantage is that the aluminum industry is already electrified, making it well-positioned for low-carbon production, particularly in regions with access to renewable energy. Moreover, aluminum is highly recyclable, and recycling aluminum saves up to 95% of the energy required to produce primary aluminum.

Tin, though required in smaller quantities, is a crucial component of solder, which is used in electronics, energy devices, and solar power systems. Most tin is derived from cassiterite, which is classified as a conflict mineral in many regions due to its role in funding armed groups and contributing to human rights abuses. The ethical sourcing of tin and other minor but essential metals remains a major concern for clean energy supply chains. As global demand for critical minerals accelerates, balancing the imperatives of resource extraction, environmental protection, and social equity will be essential to building a truly just and resilient clean energy future.

Mining and Extractives Industries

Primary demand spans all major categories, including ferrous metals such as iron, manganese, nickel, chromium, cobalt, and molybdenum; non-ferrous metals aluminum, lead, zinc, copper, and tin, precious and rare metals and minerals; and energy and specialty minerals. Most metals can be obtained from primary, secondary, or tertiary mining or metallurgy from bauxite, copper, tin, zinc, and lead. Harnessing the opportunity to tap into product waste streams—turning the linear natural resource basis of the economy into a circular one—may be critical to the planet's health. It could also have national security implications, as recycling can reduce metal imports. Countries like the United States have long stockpiled minerals and metals on the grounds of economic or national security. Of the 35 critical minerals listed by the USGS, 14 were 100% dependent on imports for supply, and 14 more were greater than 50%.

The global commodity chain approach can help us trace back up the supply chain to understand the industry geographies and socio-ecological networks that bind together modern global production networks. Geographers describe mining and natural resource sectors as existing mostly on the periphery of the global economy, where labor is cheap and manufacturing is a dominant workforce activity. Add to this very developed mining sectors in some countries like Canada, Australia, and the United States, where there are more industrialized and mechanized mining activities. The 1980s saw a particularly rapid expansion of mining into developing countries and countries that lacked institutions for environmental governance, leading to many damaging projects with particular impacts on indigenous communities (Ballard and Banks 2003).

Other materials include aggregate and sand for construction, which dominate the list of mining activities. Mining locations are remote; they are often in lands that have conservation value because of the lack of prior development or suburban impacts. In the United States, a large portion of mining activities are near Native American reservations and communities. Communities living near mining activities have more tailings with more water pollution, more dust more exposure. They may be exposed to more flooding risk or seismic hazards because of the mining activities. Often these are lands that are dependent upon people for their livelihoods or recreation.

Rare earth elements are a class of metals that have also been in the discourse around critical metals (Box 9.3). In late 2021, China announced it was merging three separate suppliers of rare earth elements into one megafirm. The company would control 62% of China's supply of rare earths. This will allow for greater purchasing power and influence over the global sector. China controls on the order of 60–70% of rare earth

Box 9.3 Rare Earth Elements

A just transition to wind power raises similar questions about supply chains with product systems reaching back to mining, extractives, and plastics manufacturing. One key area of focus for wind turbine supply chains is rare earth elements—17 metals usually found together at the bottom of the periodic table—three of which are used in wind turbine motors. Rare earth elements are a group of 17 lanthanide elements: lanthanum, cerium, praseodymium, neodymium, promethium, samarium, europium, gadolinium, terbium, dysprosium, holmium, erbium, thulium, ytterbium, lutetium, scandium, and yttrium. They are critical metals for defense, renewable energy, and electric engines. The extractive activities that produce rare earth elements have been linked to water contamination and worker health and safety hazards ranging from California to Malaysia to northern China, where today much of the production takes place. Extracting and processing rare earth oxides requires solvents and produces large volumes of waste and mining tailings that also require careful handling, management, and waste stewardship. Often these sites of extraction far upstream are furthest from view, and there is limited understanding of their emissions, waste, or community or labor impacts. These supply chains have also been at the center of geopolitical contests for nations seeking to secure these critical elements. These issues extend beyond rare earths, as China controls 80% of the world's refining and mining of clean energy raw materials.

mining and 90% of rare earth processing industries (Figure 9.2). China peaked its market share in 2014 at 86%; with newer operations coming online in Malaysia and elsewhere, that number has fallen to the mid-60% (Figure 9.3).

We might think of the system, instead of just being relational chains of economic activity, to also emphasize social-ecological relationships that are altered or shifted by these dynamics. The emphasis on non-economic actors seems like it is the most important contribution, as well as a very detailed understanding of the missing links in battery supply chains

The manufacturing of rare earth components for wind turbines and key associated equipment has shifted over time from its modern origins in the United States, Denmark, Germany, and Spain. Today, by far the leading manufacturing nation is China. Around the same time period that saw this transition in manufacturing, the supply chain evolved as well, with much of the supply chain also moving to China. In the mid-twentieth century, nearly all global rare earth elements were produced at a mine at Mountain Pass, California, with the manufacturing of color televisions driving most of the demand. By 2010, 97% of rare earth element production occurred in China, giving that nation a virtual monopoly on a material that many see as a critical input not only for wind turbines, but also for EVs and technologies for national defense and military applications.

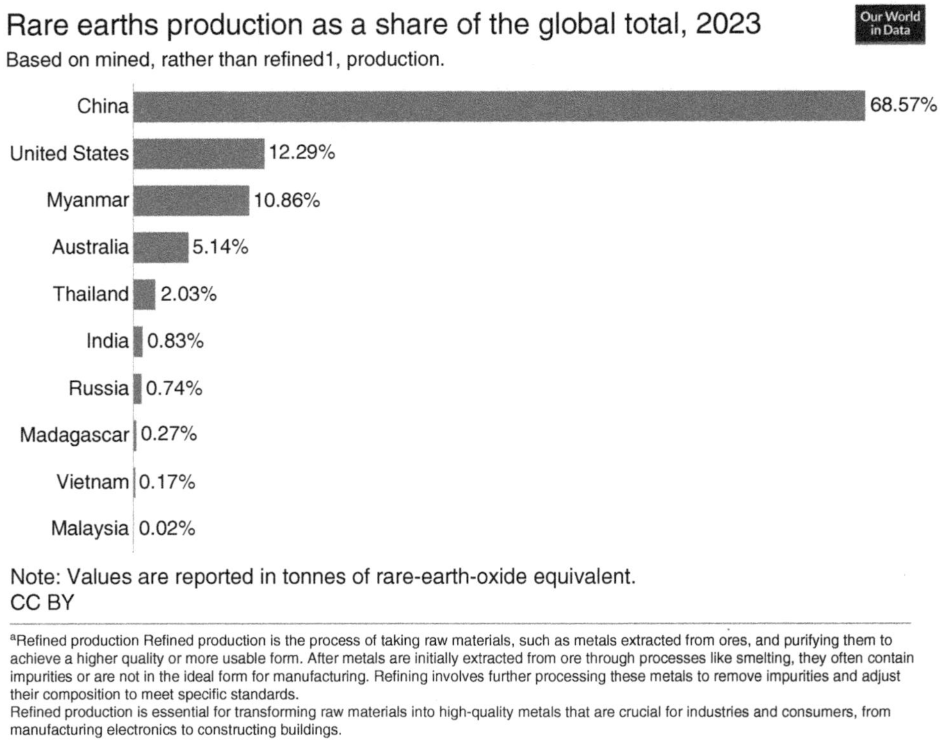

Figure 9.2 Critical minerals are often extracted or processed in geographically limited places. Rare Earth elements used in permanent magnets for wind turbines and electric motors are concentrated in a few places. Source: Rare earths production as a share of the global total, 2023/Our World in Data/CC BY 4.0.

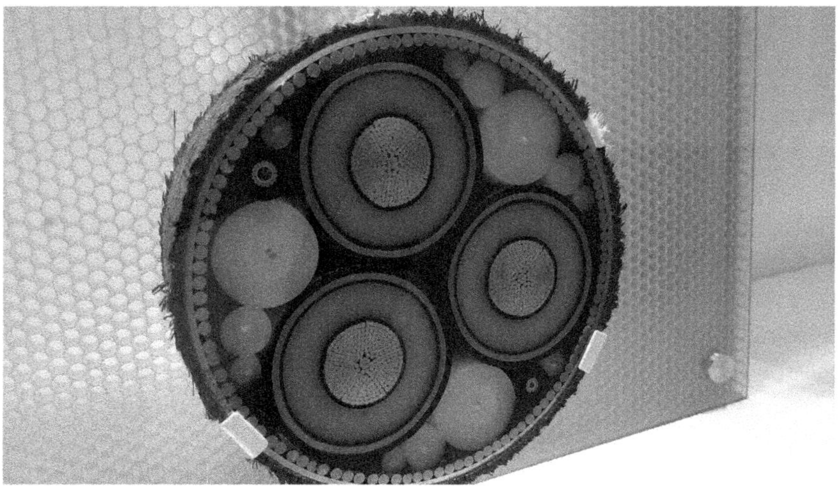

Figure 9.3 A cross-section of an undersea transmission power line for an offshore wind farm. Eduardo Sanchez / wikimediacommons / Public domain.

Given the special electromagnetic properties of these elements, these supply chains have also become embroiled in international disputes (Klinger 2018). Routine shipments from China to Japan and the United States were blocked when the government of China began to restrict exports of rare earth elements (and also tungsten and molybdenum). Whether the conflict produced a real or perceived scarcity, the issue escalated to a World Trade Organization dispute settlement tribunal ruling against China's export restrictions, and they were dropped in 2015.

Some efforts are made to recycle these metals or find alternative sources (such as from coal waste) that would eliminate the need for these metals, which exhibit important electrical and magnetic properties. One area that is receiving interest is deep-sea mining, with some suggesting that interest in rare-earth element extraction may become a motivator of conflict and disputes in the South China Sea. Extracting rare earth elements from coal and coal ash waste might be an opportunity for retraining and reinvestment in regions facing losses in employment and tax revenue because of the transition away from coal. The U.S. Department of Energy, among others, are looking to invest in infrastructure to recover rare earth elements from coal ash waste. Using the remnant waste products of the fossil fuel era as materials to make new clean tech seems like an important prospect for the circular economy and just transitions (Bagdonas et al. 2022).

Who bears the burden of mining has a complicated answer, as some of these activities are done under poor conditions with little more than survival as the reward, while other parts of the sector are heavily mechanized, bring communities economic activity, and sometimes even have jobs backed by organized labor and collective bargaining. Resisting mines in the developed world can lead to a race to the bottom, where mining companies move to or further develop places with little, weak, or no regulation or oversight of environmental or social impacts.

Lead times for mines can be decades in the making, as in some places the process of obtaining permits can take years of procedural action and data collection.

One key challenge in civilization's quest for metals and minerals is the declining ore of natural resources. Quality ores have diminished over the past century; for example copper ores are down in the twentieth century. Better mining and processing have made up for some of the inefficiencies associated with production from lower-quality ores.

The world produced 88 thousand MTs of lithium. To meet the projected demand increase from EVs, world production will need to reach 1.5 million MT by 2025. There are 200 mega-factories for battery production scheduled to be operating by 2030. Many countries are adding requirements that many of the battery's components be sourced domestically. Lithium extraction is done either by hard rock surface mining or through brine extraction pumped from below and evaporated in open-air ponds (Table 9.3). Lithium carbonate is one common form of the metal taken from brines in aquifers that have high levels of the dissolved compound (Figure 9.1). The brines are next evaporated in a series of ponds until they reach appropriate concentrations. At this stage, the other dissolved metals and impurities are removed.

Table 9.3 Lithium resources for feedstocks to make lithium-ion battery cathodes.

Resource	Chemical route	Extraction method
Lithium brines	Lithium carbonate	Brine extraction
Spodumene	Lithium hydroxide	Hard rock mining
Direct lithium extraction	Lithium hydroxide/carbonate	Geothermal brines

Lithium hydroxide is the secondary product of brine extraction, but where direct lithium extraction (DLA) is used, it could also lead to the production of lithium carbonate. The extraction of lithium water, which accumulates in salts in arid areas, can lead to groundwater stress or sinking groundwater tables if done unsustainably. It can take over a half-million gallons of water to produce a ton of lithium (Blair et al. 2024a). It is expected that hard rock mining sources of spodumene will continue to grow market share as lithium-ion batteries are set to power vehicles and the electricity grid, among other electronics.

The flow rates of extraction and processing can be a limiting factor of production. The evaporative methods that rely on sunlight for processing a limited by evaporative processes, which are augmented by the additives that make the colorful ponds seen in South American lithium industries. A major producer of lithium on the Atacama Salt Flat, Albemarle, was fined for over-pumping groundwater by the government of Chile under their groundwater monitoring, which might be essential for any water basin where water or brine extraction occurs (Blair et al. 2024b). The so-called lithium triangle and western Nevada have similar landscape characteristics and are arid regions susceptible to groundwater depletion.

The global production network for lithium from the case of Bolivia represents an interesting case study because many of its commercial prospects have not been developed, as they have in neighboring countries in the region (Bos and Forget 2021). Bolivia has the world's largest lithium reserves, but Chile and Argentina produce far more lithium, with Bolivia mostly producing pilot-scale quantities. The country's nationally owned mineral resources have led many investors to eschew deals with the country to develop these resources. Bolivia, for its part, has developed a state-owned lithium extraction company. Like many countries with resources do when they want to take greater value from their resources, Bolivia has a national strategy to also take a greater role in the refining and chemical processing stages of lithium production. The development outcomes of the extraction of any metals and minerals should keep in mind the importance of development outcomes, not merely the foreign direct investment or revenue flows. These not only result in people sharing more in the benefits associated with these projects, but also can extend social license to develop in other situations as other communities take note.

LFP batteries hold promise for avoiding cobalt, manganese, and nickel in EVs and stationary storage batteries. But phosphorus, too is a material in high demand and that can be subject to price volatility and supply chain disruption. Only five countries produce 85% of global phosphorus, raising concerns about global phosphorus demand for lithium-iron-phosphate batteries (Spears et al. 2022).

Other battery technologies may avoid lithium altogether. Sodium flow batteries have been used in Japan since the 1990s, but sodium-ion batteries may be suitable for many higher energy density applications. Aluminum ion batteries, manganese-based (Li et al. 2021), zinc-air batteries, graphene ion batteries, and others on the horizon could obviate much of the need for lithium. Each is another battery storage technology that is getting attention from investors and policy-makers, and one major lithium-ion battery manufacturer began to produce sodium-ion batteries for one of its popular EVs in 2025. These batteries may also be less fire-prone.

Metals and minerals are not the only materials with questions about sourcing. One question unresolved is where specialized plastics and new fiber composites will come from. The solution here may require the development of new or substitute materials, labor, and land for renewable energy. For renewables like biofuels, the production of the fuel itself can be the largest environmental impact, especially where it requires large amounts of land or causes water quality impacts (Figure 9.4).

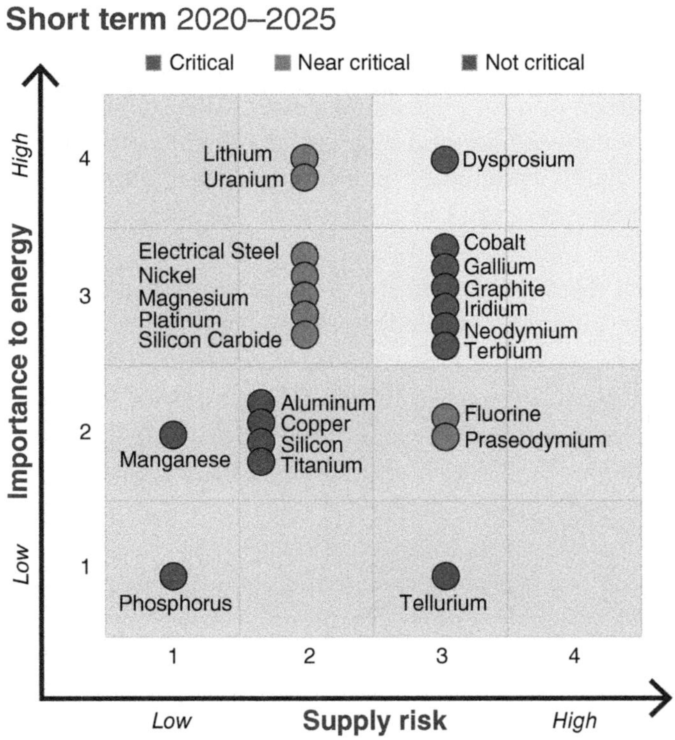

Figure 9.4 U.S. Department of Energy "criticality" map that tries to classify major metals and minerals for energy transition based on their supply risk and importance to energy industries. Source: U.S. Department of Energy/https://www.energy.gov/cmm/what-are-critical-materials-and-critical-minerals/Public Domain/last accessed on September 18, 2025.

The global transformation of the world by mining and mineral extraction cannot be understated in some parts of the world. The global terraforming of coal and copper mines left behind pits that look like stadiums and amphitheaters cut in the hillside. Mountaintop removal (MTR) for coal is a practice that is becoming less common because of cost and regulatory interventions to protect riparian areas and forests. But still transformed over 500 mountaintops of the Appalachia range.

These changes brought by mining caused physical but also social change. Geographers have documented dispossession in mining communities around the world, including on the Bolivian Altiplano (Perreault 2013). Mining companies can fracture communities by varying some in long-standing disputes over land titling, or worse, outright take the lands away from communities. Developing mines proceeds best when mining companies can develop the social license to operate, and this can mean meeting early and often with key stakeholders and modifying to accommodate local concerns. Community benefit agreements often also show a mining development's commitment to ensuring some of the revenues stay local (Blair et al. 2025).

Iron will also be a critically important metal to energy transitions as well. Iron was a forest product when it relied on charcoal for smelting. Iron ore production today is mostly for steel and is produced across the world, with leading producers being Brazil, China, Malaysia, and Australia. Similar to other mining activities, major impacts from developing and mining iron ore are from land use change, tailings, and spoils, which can contaminate water supplies and riparian ecosystems with trace heavy metals (Zabowski et al. 2001). Pollution from these activities contributes to increased ecological risks and increased risk of cancer to people living in nearby communities (Diami et al. 2016).

Nickel is a metal that is important to several chemistries of lithium-ion batteries that has also made international headlines (Box 9.4). Nickel concentrates are also important for several types of lithium-ion batteries, as they strengthen the energy a battery cathode is able to store (see Chapter 2, how batteries work). One strategy to avoid cobalt has been to increase the amount of nickel compared to manganese and cobalt in NMC batteries. Nickel-rich cathodes have been accompanied by increasing energy densities.

Competing industrial uses of nickel compounds include use in steel production, smelting and refining metals for alloys, and electroplating materials. The world's largest nickel producer is Indonesia. Nickel mining via conventional practices is both energy-intensive and generates tailings and wastewater. In 2022, nickel prices rose after Russia's invasion of Ukraine, as sanctions were expected to affect global nickel supply chains.

Gold is required for some of the inputs used to make inverters. Gold mining is one of the most damaging and least efficient in terms of metal to tailings of all types of mining. Artisanal gold mining in Nigeria caused a major public health crisis when many miners were exposed to high levels of lead (Bartrem et al. 2022). There are many new gold mines opening up across Nevada and California.

Box 9.4 The London Metals Exchange and the Commodities Traders

Commodities traders play an outsized yet hidden role in our world. The London Metals Exchange is a futures trading exchange for commodities like steel, nickel, oil, and nearly all the most important extractive materials. Futures markets for metals and grains have existed for a hundred years. These derivatives are basically short bets that the price of a commodity will rise or fall. They serve as a sort of insurance for farmers, as futures markets allow them to lock in the price far ahead of time, ensuring they do not lose much. Periodically the metals exchanges have historic moments due to global crises or something to do with supply chains. There was a tin crisis in 1985 after an international tin council that had stabilized prices for decades suddenly became insolvent. The Russian invasion of Ukraine in 2022 caused a nickel crisis, leading the London Metals Exchange to close trading and even cancel a great number of contracts. This is shaping up to be an important moment in the direction of the materiality of lithium-ion batteries, as many car and battery makers are announcing shifts away from nickel and cobalt batteries, such as Tesla and Volkswagen.

Land

Developing metals and minerals often means terraforming land. The United Nations estimates that 40% of the global terrain has been devalued, mainly by agriculture, but also mining, urbanization, and suburban sprawl. They say that if nothing changes, an area the size of South America will be damaged by 2050. A shift away from subterranean fossil fuels to renewables collected from devices and infrastructures on the Earth's surface will come with tradeoffs, and these drive demand for new natural resources and lands rich in renewable energy resources (Sovacool et al. 2020) (Figure 9.5).

One important challenge that comes with the development of new mining activities has to do with the ownership and titling of land. Often, historic processes of dispossession of colonization can shape land ownership and occupancy, and mining companies may find themselves in a situation where the lands are contested (Box 9.5). Even in the developed world, we see these contestations. Mining companies have been applying for placer claims for lithium deposits numbering in the tens of thousands. Two prominent fights over Red Mountain or Thacker Pass on the Nevada and Oregon border, and another dispute over a buckwheat plant that only occurs on a few hundred acres at the Rhyolite Ridge, are leading examples of these controversies manifesting in the industrialized world. These impacts are despite claims that onshoring supply chains to the global north would minimize the impacts of extractive industries by shifting them toward sites with purportedly better governance (Riofrancos 2023).

Mining is one of the most environmentally and socially disruptive activities undertaken by human civilization, causing significant damage to ecosystems and

Figure 9.5 The largest open pit mine is the world is Mina de Chuquicamata, in Calama, Chile. Cancillería del Ecuador / Flickr.

Box 9.5 Mining and Land Dispossession

Dispossession refers to the process of removing individuals or communities from their land, resources, or rights, often through legal, economic, or coercive means (Perreault 2013). This concept is central to understanding the colonial histories and ongoing injustices faced by Indigenous peoples around the world. During the era of European conquest and colonization, Indigenous communities across the Americas, Africa, Asia, and Oceania were systematically dispossessed of their ancestral territories—lands that were often integral to their cultural, spiritual, and economic life. The imposition of European legal systems and concepts of private property often disregarded Indigenous land tenure systems, treating unceded or communally held lands as "empty" or "unused." This legacy continues today, particularly in the context of mining and natural resource extraction. Many modern land conflicts cannot be fully understood without acknowledging the historical and ongoing contestations over land ownership, access, and sovereignty (Willow 2016). Dispossession is not only a historical phenomenon but also an active process that unfolds through the granting of concessions, the enforcement of state mineral laws, and the undervaluing or erasure of Indigenous and customary land rights.

In one notorious case in Ecuador, land titles were stripped from Indigenous communities in the Amazon without their knowledge or consent to make way for resource development. In other cases, even when surface land rights are recognized, conferring subsurface mineral rights to companies or the state can trigger displacement and dispossession. This is particularly problematic in countries where mineral rights are legally separated from surface land rights, allowing extraction to proceed without meaningful consent from the people who live on or use the land.

These patterns of dispossession often result in cultural disruption, environmental degradation, and loss of livelihood, and can generate long-term social conflict. They raise critical questions about who controls land and resources, who benefits from their development, and how historical injustices are addressed—or perpetuated—in the era of green energy and global decarbonization. Addressing these issues requires not only legal reform, but also a commitment to Free, Prior, and Informed Consent (FPIC), restorative justice, and the recognition of Indigenous land governance systems.

Source: Adapted from Shade (2015).

posing serious hazards to human health and safety. From a global perspective, the distribution of mining's benefits and burdens is highly uneven, reflecting broader patterns of unequal development and resource dependency. Both industrialized and rural countries engage in mining, but the intensity of environmental degradation, labor conditions, and regulatory enforcement varies widely by region. Some countries have stricter environmental and labor protections, leading to fewer emissions and safer working conditions. Others, often with weaker governance or less regulatory oversight, face more severe pollution, workplace risks, and community displacement.

Interestingly, when mining executives are asked where they prefer to operate, many cite locations such as Nevada and Utah in the United States. These regions are known for their stable regulatory environments, predictable permitting processes, and infrastructure access, making them attractive despite their relatively higher labor and environmental standards. This reveals a paradox: companies may prefer to operate in places where the rule of law is strong and risks of social and political conflict are lower—even if that means higher costs—highlighting how regulatory clarity and stability can be as valuable as resource abundance.

Metal and mineral commodity chains

To say that metal and mineral commodity chains can be hard to untangle is an understatement. There is increased interest in understanding and improving the socio-ecological impacts of global production systems through our understanding of linkages in global commodity chains. The commodity chain approach is used by geographers, anthropologists, and sociologists to identify what shapes world-economic spatial inequalities. Mapping how firms are connected through space can reveal patterns influenced by the labor process, industry organizations, science, regulation, political power, and culture.

The commodity chain approach has much in common with what is often described as a value chain. The more positively framed term in concept suggests firms construct value over space through sourcing and contracting arrangements. This approach aims to understand how these activities are organized and governed, where value is added or extracted across the supply chain as materials move from raw materials to finished products. While several research communities use the commodity chain concept, they have in common an aspect of unveiling that helps reveal the empirical realities of making things (Box 9.6). Production systems that produce solar power will not be immune to the dynamics of labor and land exploitation. Ensuring that technological solutions to environmental problems do not have unintended outcomes will help society enact a just transition to solar power.

Box 9.6 Metal and Mineral Supply Chains, Monitoring and Assurance Programs

One of the key strategies for improving environmental and social outcomes in the mining sector has been the development of certification schemes and voluntary guidance frameworks. These initiatives aim to encourage best practices, promote transparency, and minimize harm in the sourcing of critical minerals used in energy technologies. A prominent example is the Organization for Economic Cooperation and Development's Due Diligence Guidance for Responsible Supply Chains of Minerals from Conflict-Affected and High-Risk Areas, which provides guidance to companies—including miners, smelters, refiners, and downstream purchasers—on how to identify and mitigate human rights abuses and environmental risks in their supply chains. A leading initiative is the Responsible Minerals Initiative (RMI), a multi-stakeholder effort involving companies across the electronics, automotive, and clean energy sectors. RMI promotes supply chain accountability by auditing smelters and refiners and encouraging companies to source only from certified conflict-free operations. Its broader mission includes improving mining safety, working conditions, and community public health, while ensuring that the economic benefits of mining can still support local development. Major automotive manufacturers have increasingly acknowledged their responsibility in this space. For example, one auto company emphasized its commitment to "improving artisanal cobalt mining conditions as well as the living conditions of people in surrounding communities" (Responsible Minerals Initiative, 2020). This reflects growing recognition that companies must move beyond passive sourcing to take an active role in shaping mining practices—particularly in regions like the DRC, where a significant portion of cobalt is produced through ASM under difficult conditions. While voluntary initiatives are not a substitute for binding regulations, they are an important component of a more just and sustainable transition to clean energy. However, their effectiveness varies, and critics note that without independent oversight, stakeholder inclusion, and clear enforcement mechanisms, these frameworks can fall short of delivering on their promises. For certification efforts to have a meaningful impact, they must be inclusive of affected communities, grounded in transparent data, and integrated with broader efforts to reform global mineral governance.

There is much to be said about the many unique and rare metals that are used to make some technologies that will be core to the energy transitions. But the material that seems at first pass most critical to the future of energy looks to be copper. The growth of copper production is poised to grow in the coming years. Vanadium is also set to grow, especially if emerging vanadium flow batteries come online. This material is currently sourced from coal, oil, and gas.

Recycling and Mining Waste for Materials

Economies currently extract most natural resources in a take-make-waste economy. Many sustainability thinkers have been saying for years that in the future landfills might be seen as future mines to recover long-lost materials. While that playing out is unlikely, there are many potential sources of metals available in waste flows that might be harvested to augment supplies overall or displace resources. Developing a value chain for various critical metals can also help buffer supplies that might be vulnerable to disruption. Analysts at the energy consulting company McKinsey see some key materials for use in the valuable energy transition industry could exceed $400 billion, and the value of battery recycling alone is poised to be over $39 billion per year by 2030 (McKinsey & Company and Global Battery Alliance 2023).

Batteries age based on the up and down cycling of the charge, which over time accumulates impurities that reduce the storage and transport of electric charge. In theory, large portions of the materials in batteries could be recovered through various recycling and recovery means. Studies on the recycling of lithium-ion batteries suggest they could be a significant source of the metal and mineral supply once end-of-life batteries go into retirement. The National Renewable Energy Lab estimated that 60% of the future 2040 cobalt supply could be met by recovered end-of-life materials (and this was before there were signals that cobalt-based batteries could decline in market share).

New laws and regulations are being developed to ensure batteries are recycled, including a requirement that manufacturers put funding into a mandatory recycling target for raw materials. The European Union's battery law, passed in 2022, sets recycling targets for cobalt and nickel of 90% by 2027 and 95% by 2031, and lithium at 50% by 2027 and 80% by 2031. These recycling rules favor clean electricity use in battery production. Any manufacturer making batteries must report the life cycle CO_2 footprint from mining to recycling or disposal. The standard sets a threshold to meet. Mandatory due diligence rules mean that companies, be they car original equipment manufacturers or battery makers, are responsible for conducting due diligence along the supply chains of lithium, nickel, cobalt, and graphite. These rules will apply to all battery categories from 2025 after a 24-month implementation period. Companies will have to pay attention to a series of non-exhaustive environmental and social risks, including water, biodiversity, workers' rights, child labor, etc., as well as community rights (including Indigenous peoples). Companies will have to set up grievance mechanisms as part of their due diligence policies. However, the obligations are not extended to manufacturers placing batteries on the market. The European Commission will prepare guidelines for companies and set up a management scheme. Will these standards reshape the industry to the highest sustainability achievements? Or will they just reshuffle the deck and continue business as usual.

Asteroid and Deep Sea Mining

Deep-sea mining for nickel nodules is seen as another frontier for critical metals development, but comes into direct conflict with ocean conservation goals and fishing or aquaculture interests. Prospects for deep sea nickel and cobalt mining off Southeast Asia have received a great deal of attention despite some buyers like Volkswagen saying they were committed to keeping metals from deep sea mining out of their supply chains. These are new areas where resources will be opened up, meaning there is a need for a new section on deep-sea mining, asteroid mining, and governance.

Electrification and energy transition will bring great attention to efforts to mine the metals and minerals needed to modernize and clean up the energy system. At the same time, mining companies are never sure to miss opportunities like these and arguably could use pretense of energy transition and electrification to mine everything they can sell. As much as a discourse of tradeoffs in energy transition, we need a vision for alternatives that helps bring attention to alternative arrangements or unexpected outcomes.

New technologies, devices, infrastructures, and supply chains will produce the new energy economy. What do these new systems of production look like? Where do the resources come from, and what does that mean for communities? What labor is used to make them, and where is the land used to deploy these devices and infrastructures? Just transitions research is already showing important themes to explore for emerging renewable energy inequalities in, for example the supply chains of metals that will be required for decarbonization (Sovacool et al. 2020). A new movement for looking to repair things and design products for the environment will play an important role too.

Highlights/Takeaways

- The energy transition will be metal- and mineral-intensive because of the types of materials needed for renewable and clean energy generation and use.
- Developing new mines will be new sites of land use conflict, and new mines look for these new geologies in places with the right infrastructures and political economies for development.
- Community benefits, careful planning, better value of waste tailings, and efficiency in ore recovery will all be critical to ensuring these developments do not reproduce inequality.

Questions for Review

1. Metals and minerals will be in great demand in decarbonization and energy development scenarios. Some metals will need to be extracted at rates many times over current consumption patterns.
2. Choose a particular metal that will be critical to energy transition and explain the impacts of extracting those metals. Locate where it is made in place.
3. How do supply chain dynamics and geopolitical control over mineral refining shape global access to critical materials for the energy transition? Give examples from one or more countries.
4. What are the trade-offs between eliminating artisanal mining for safety and formalizing it to improve livelihoods and ensure community benefit? Use cobalt or another relevant example.
5. Discuss the role of certification schemes (e.g., OECD Due Diligence, RMI) in addressing environmental and human rights risks in mining. What are their limitations?
6. Explain how battery chemistries influence demand for certain metals. Compare at least two chemistries and their implications for sustainability and sourcing.
7. Why is it important to consider both the energy intensity and the social impacts of metal extraction when evaluating the sustainability of clean energy technologies?
8. What are the major forms of mining waste, and how do low ore grades affect environmental and energy footprints? How might this influence planning for future mines?

Exercise 9.1 Ethical Sourcing of Battery Chemistries

Imagine you are part of a battery manufacturing team. Consider the different roles of a battery chemist or designer and a procurement specialist responsible for sourcing raw materials. Choose one battery chemistry (e.g., NMC, LFP, or solid-state). Identify the key metals and materials that constitute this battery.

Where are these materials typically sourced from? What are the major producing countries or regions? What are the ethical or environmental challenges associated with sourcing these materials (e.g., human rights issues, pollution, artisanal mining)? As someone in charge of this supply chain, how would you proceed to address or mitigate these challenges? Consider approaches like certification schemes, recycling, or material substitution.

Exercise 9.2 How Much Power Can be Added Based on the Production of Photovoltaic Metals?

Using available global production of gallium, estimate how much electricity could be generated annually from one year's worth of gallium-based photovoltaic semiconductors.

Use the following assumptions:

- Power conversion efficiency: 25%
- Solar irradiance (AM 1.5): 1,000 W/m^2
- Annual global gallium production: 220 MTs
- Minimum material intensity: 0.00005 MTs/m^2
- Capacity factor: 0.30

Questions:

- How many square meters of PV panels could be made from 220 MTs of gallium?
- What is the installed capacity (MW) of those panels?
- How much electricity would be generated in a year (in TWh)?

Exercise 9.3 Estimate the Overburden of Mining

An average EV contains approximately 23 kg of copper. An ore grade of 0.6% means 1,000 kg of rock yields 6 kg of copper → ~167 kg rock per 1 kg copper. Multiply accordingly to find the total overburden for 10 million EVs. Assume a copper ore grade of 0.6%.

Estimate how many kilograms of rock (non-copper material) must be processed to obtain 1 kg of copper. Then calculate the overburden required to supply copper for 10 million EVs (assume 10 million were made).

References

Bagdonas, D., Enriquez, A., Coddington, K. et al. (2022). Rare earth element resource evaluation of coal byproducts: a case study from the Powder River Basin, Wyoming. *Renewable and Sustainable Energy Reviews* 158: 112148.

Ballard, C. and Banks, G. (2003). Resource wars: the anthropology of mining. *Annual Review of Anthropology* 32 (1): 287–313.

Bartrem, C., von Lindern, I., von Braun, M., and Tirima, S. (2022). Climate change, conflict, and resource extraction: analyses of Nigerian artisanal mining communities and ominous global trends. *Annals of Global Health* 88 (1): 17.

Bazilian, M.D. (2018). The mineral foundation of the energy transition. *The Extractive Industries and Society* 5 (1): 93–97.

Blair, J.J.A., Vineyard, N., Mulvaney, D. et al. (2024a). Lithium and water: hydrosocial impacts across the life cycle of energy storage. *WIREs Water* 11: e1748. https://doi.org/10.1002/wat2.1748.

Blair, J.J., Balcázar, R.M., Barandiarán, J., and Maxwell, A. (2024b). Lithium mining and exhausted ecologies in the Atacama desert. *International Development Policy* 16: 257.

Blair, J.J.A., Hopkins, E., Koester, T.K. et al. (2025). *Community Benefits Agreements, Comprehensive Resources on How Communities Can Deploy CBAs to Shape Their Futures*. Climate and Community Institute https://climateandcommunity.org/research/cbas/.

Bos, V. and Forget, M. (2021). Global production networks and the lithium industry: a Bolivian perspective. *Geoforum* 125: 168–180.

Byemba, G.K. (2020). Formalization of artisanal and small-scale mining in eastern Democratic Republic of the Congo: an opportunity for women in the new tin, tantalum, tungsten and gold (3TG) supply chain? *The Extractive Industries and Society* 7 (2): 420–427.

Cobalt Institute (2000). Cobalt uses. https://www.cobaltinstitute.org/cobaltuses.html (accessed 14 November 2025).

Diami, S.M., Kusin, F.M., and Madzin, Z. (2016). Potential ecological and human health risks of heavy metals in surface soils associated with iron ore mining in Pahang, Malaysia. *Environmental Science and Pollution Research* 23: 21086–21097.

Farnham, A., Winkler, M.S., Zabre, H.R. et al. (2022). Spatial mobility and large-scale resource extraction: An analysis of community well-being and health in a copper mining area of Zambia. *The Extractive Industries and Society* 9: 101016.

Government of Canada (2023). Aluminum. https://natural-resources.canada.ca/our-natural-resources/minerals-mining/minerals-metals-facts/aluminum-facts/20510 (accessed 14 November 2025).

Henckens, M. and Worrell, E. (2020). Reviewing the availability of copper and nickel for future generations. The balance between production growth, sustainability and recycling rates. *Journal of Cleaner Production* 264: 121460.

Klinger, J.M. (2018). *Rare Earth Frontiers: From Terrestrial Subsoils to Lunar Landscapes*. Cornell University Press.

Kluger, J.R. (2016). *The Clifton-Morenci Strike: Labor Difficulty in Arizona, 1915–1916*. University of Arizona Press.

Li, H., Zhang, W., Sun, K. et al. (2021). Manganese-based materials for rechargeable batteries beyond lithium-ion. *Advanced Energy Materials* 11 (25): 2100867.

McKinsey & Company and Global Battery Alliance (2023). *Battery 2030: Resilient, Sustainable, and Circular*. Report.

Perreault, T. (2013). Dispossession by accumulation? Mining, water and the nature of enclosure on the Bolivian Altiplano. *Antipode* 45 (5): 1050–1069.

Radley, B. (2024). *Disrupted Development in the Congo. The fragile foundations of the African mining consensus*. Oxford University Press.

Radley, B. and Geenen, S. (2021). Struggles over value: corporate–state suppression of locally led mining mechanisation in the Democratic Republic of the Congo. *Review of African Political Economy* 48 (168): 161–177.

Responsible Minerals Initiative (2020). Company List. https://www.responsiblemineralsinitiative.org/about/members-and-collaborations/

Riofrancos, T. (2023). The security–sustainability nexus: lithium onshoring in the Global North. *Global Environmental Politics* 23 (1): 1. https://doi.org/10.1162/glep_a_00668.

Shade, L. (2015). Sustainable development or sacrifice zone? Politics below the surface in post-neoliberal Ecuador. *The Extractive Industries and Society* 2 (4): 775–784.

Sovacool, B.K., Ali, S.H., Bazilian, M. et al. (2020). Sustainable minerals and metals for a low-carbon future. *Science* 367 (6473): 30–33.

Spears, B.M., Brownlie, W.J., Cordell, D. et al. (2022). Concerns about global phosphorus demand for lithium-iron-phosphate batteries in the light electric vehicle sector. *Communications Materials* 3 (1): 14.

USGS (2022). Graphite deposits in the United States. http://www.usgs.gov/data/graphite-deposits-united-states (accessed 14 November 2025).

Valenta, R., Kemp, D., Owen, J. et al. (2019). Re-thinking complex orebodies: consequences for the future world supply of copper. *Journal of Cleaner Production* 220: 816–826.

Watt, R. (1956). History of Morenci, Arizona. Thesis, University of Arizona.

Willow, A.J. (2016). Indigenous ExtrACTIVISM in boreal Canada: colonial legacies, contemporary struggles and sovereign futures. *Humanities* 5 (3): 3. https://doi.org/10.3390/h5030055.

Young, B. (1965). Jamaica's bauxite and alumina industries. *Annals of the Association of American Geographers* 55 (3): 449–464.

Zabowski, D., Henry, C., Zheng, Z., and Zhang, X. (2001). Mining impacts on trace metal content of water, soil, and stream sediments in the Hei River basin, China. *Water, Air, and Soil Pollution* 131: 261–273.

Further Reading

Ballard, C. and Banks, G. (2003). Resource wars: the anthropology of mining. *Annual Review of Anthropology* 32 (1): 287–313.

Beiser, V. (2024). *Power Metal: The Race for the Resources that Will Shape the Future.* Penguin Group.

Chee, M.E.B. (2022). Ethical supply chains of metals and minerals. In: *Routledge Handbook of the Extractive Industries and Sustainable Development*, 480–502. Routledge.

Klinger, J.M. (2018). *Rare Earth Frontiers: From Terrestrial Subsoils to Lunar Landscapes*. Cornell University Press.

Riofrancos, T. (2022). The security–sustainability nexus: Lithium onshoring in the global north. *Global Environmental Politics* 1–22.

Sovacool, B.K. (2021). When subterranean slavery supports sustainability transitions? Power, patriarchy, and child labor in artisanal Congolese cobalt mining. *The Extractive Industries and Society* 8 (1): 271–293.

Part 2

Energy Futures

Part 2 explores the material, spatial, social, and political dimensions of energy transition through six interrelated chapters. It emphasizes the complexity of decarbonization beyond technology, focusing on how infrastructures, labor, landscapes, and justice must be reconceived to build a more sustainable and equitable energy future. Through the lenses of transportation, energy grids, buildings, industry, food systems, and environmental justice, these chapters draw attention to the vast, interconnected systems that power daily life—and the need to transform them.

We start in Chapter 10 with electricity grids and pipelines, which are the world's largest machines, stretching across continents to transport energy in its many forms. This chapter introduces key debates around the development and governance of electricity systems, from centralized bulk power networks to distributed smart and microgrids. As the energy transition scales up, grids must evolve rapidly to accommodate intermittent renewables and decentralized generation. Key policy and design issues include how to upgrade aging infrastructures, the politics of siting new transmission lines, and who should own and operate these grids. Public power, regional transmission operators, independent system operators, and merchant generators each play different roles, creating a patchwork of responsibilities and regulatory jurisdictions. Grid modernization is not only a technical project but also a political and social one, raising questions about cost, resilience, and access. The chapter also explores pipeline infrastructure for natural gas, oil, carbon dioxide, and hydrogen. While pipelines reduce some risks associated with transport by rail or truck, they also produce significant environmental and social externalities. Pipeline siting and operation have sparked major resistance movements, particularly from Indigenous communities. The future of these infrastructures remains contested, especially as new proposals for hydrogen and CO_2 pipelines emerge in net-zero planning. Ultimately, grid and pipeline design will shape the pace, equity, and environmental impact of the energy transition.

In Chapter 11, Mobility and Transportation, we learn that getting around is both a major source of greenhouse gas (GHG) emissions and a site of profound inequality and spatial injustice. This chapter reframes transportation as a question of mobility—how and for whom we design systems to move through space. It critiques car-centric planning and automobility as systems of energy dependence, racialized exclusion, and environmental degradation. The combustion engine remains one of the most polluting technologies in widespread use. Electrification is a dominant decarbonization strategy. Electric vehicles (EVs) offer energy efficiency and lower emissions, but widespread adoption still requires mining, materials, and energy infrastructure on a vast scale. Complementary policies such as feebates, fleet procurement, and shared mobility can help accelerate the transition, but challenges remain—particularly in addressing equity and public participation. The chapter also explores public transportation, aviation, and shipping. Public transit is framed not just as a climate solution but as a social good—essential for reducing mineral demand, creating more inclusive cities, and expanding mobility access. For aviation and shipping, decarbonization is complex. High-speed rail offers one alternative to short-haul flights, while sustainable aviation fuels, electric planes, and hybrid propulsion remain in development. In shipping, innovation focuses on wind-assisted vessels, electrified logistics, and the phaseout of dirty bunker fuels. Throughout, the chapter emphasizes the need to reimagine mobility as equitable, efficient, and human-centered.

Chapter 12 is about how the built environment is responsible for an estimated 40% of global resource use and a significant share of energy consumption and emissions. This chapter examines how energy flows through buildings, focusing on heating, cooling, lighting, and appliance use. It introduces concepts such as building envelopes, thermal efficiency, and the social construction of comfort. Electrification is a central strategy: heat pumps, induction stoves, and electric water heaters can dramatically reduce emissions—especially when powered by clean electricity. Policies in California and elsewhere promote net-zero homes that combine electrification with rooftop solar and battery storage. Yet barriers remain, particularly for renters, low-income households, and older buildings. The "split incentive" problem—where landlords have no reason to invest in energy efficiency—locks millions into energy-inefficient homes and high bills. Materials also matter. Concrete, glass, steel, and plastics carry significant embodied emissions. Wood and bio-based materials offer low-carbon alternatives, but recycling and circularity in construction remain limited. The chapter urges a deeper cultural and policy shift: designing buildings that are not only efficient, but healthy, adaptable, and equitable spaces to live and work.

Chapter 13 examines industrial sectors—including steel, cement, chemicals, and aluminum—that are responsible for roughly 25% of global GHG emissions and are among the most difficult to decarbonize. This chapter introduces industrial ecology as a framework for making industrial systems behave more like ecosystems—where waste from one process becomes input for another. Concepts like the circular economy, supply chain sustainability, and material valorization underpin efforts to redesign production for regenerative, not extractive, ends. Critical industry studies challenge traditional approaches by centering power, labor, and justice. Questions of who benefits, who is harmed, and how sustainability is defined must accompany technical innovation. From emissions in steel and cement to the toxic legacies of chemical production, industry's impacts ripple across geographies and generations. Clean energy manufacturing is not exempt. Solar panels, batteries, and wind turbines have their own intensive supply chains. The chapter calls for public policy, industrial planning, and community participation to ensure that cleaner technologies do not replicate the harms of fossil fuel systems.

Chapter 14 explores the interdependent flows of energy, water, and food—highlighting their deep entanglement in both ecological systems and modern infrastructure. Agriculture is energy-intensive, reliant on fossil fuels for fertilizers, irrigation, machinery, transport, and processing. Globally, agriculture contributes up to 25% of annual GHG emissions. Water, likewise, is both a critical input to food production and an energy consumer: pumping, transporting, and treating water all require energy. The nexus framework helps identify trade-offs and synergies. Solar irrigation and agrivoltaics—where crops and solar panels share land—are emerging approaches to co-optimizing food and energy. Food waste and wastewater treatment also intersect with energy, as methane from anaerobic digestion can generate heat or electricity. Water footprints—blue (consumptive use), green (rainwater), and gray (pollution)—offer useful metrics for assessing impacts. Equity issues abound, as low-income communities often lack access to clean water and bear the brunt of industrial agriculture's pollution. A nexus approach demands holistic planning across silos to manage planetary resources fairly and sustainably.

The final chapter of Part 2 centers on justice in energy transitions. Drawing on the environmental justice movement, it identifies distributional, procedural, and recognitional injustices embedded in energy systems—from fossil fuel sacrifice zones to the uneven benefits of solar adoption. Key topics include energy poverty, decolonizing energy, fenceline communities, and just transitions for labor. Energy justice is not just about equal access to power but also about correcting historical harm. The chapter discusses climate reparations, participatory research, and the role of grassroots organizing in shaping just outcomes. It examines the contradictions of clean energy development in marginalized regions and calls attention to how "green" projects can reproduce extraction without proper governance or consent. Ultimately, the chapter urges a justice-centered energy transition that integrates community participation, policy reform, labor rights, Indigenous sovereignty, and ecological integrity. Justice is not a side issue—it is foundational to durable, inclusive, and ethical climate action.

Chapter 10

Electricity Grids and Pipelines

The largest machines on Earth are the electricity and pipeline distribution networks that transport natural gas, petroleum, tar sands, hydrogen, ammonia, and carbon dioxide. These massive infrastructures are crucial to powering the world today. Every day, billions of gallons of petroleum are moved across continents. The North American electricity grid is considered by some to be the largest machine in the world, with the interstate pipeline system ranked as the second largest. Around the globe, similar systems are expanding and becoming increasingly interconnected. Many questions and debates surround the future of electricity grids. How can they be developed more rapidly? Are large-scale centralized grids preferable to decentralized ones? Likewise, there are ongoing discussions about emerging pipeline networks—such as those proposed for transporting natural gas, carbon dioxide, or hydrogen—and whether they might reduce risks associated with moving hazardous energy resources by road or rail, or whether they make sense at all. This chapter synthesizes current research and debates on the evolution, design, and politics of electricity and pipeline infrastructures.

Big Energy Infrastructures Move Energy Around

Energy is rarely used at the location where it is extracted. Sites of extraction are often spatially distant from where energy is transformed into usable fuels, and it is often moved yet again before being converted into heat, light, or motion. Historically, canals were built to transport coal by barge from rural mines to urban centers; later, railroads did the same. With the widespread adoption of electricity in the twentieth century, human civilization began constructing some of the largest machines ever built: electricity grids. These were soon followed by complex pipeline systems designed to transport oil, natural gas, coal slurries, and other energy sources—as well as water, wastewater, carbon dioxide, and chemicals. Electricity grids and pipelines now span thousands of miles, connecting sources of energy with the places where it is ultimately consumed. One of the earliest examples of an electricity grid was the Pearl Street Station in lower Manhattan, developed by the Edison Electric Company. Today, the construction and interconnection of electricity grids remains a vital concern—particularly in the context of scaling up renewable energy development (Box 10.1).

The development of pipelines is also a critical piece to contemporary narratives about energy, even the social construction of identity. Pipelines carry various energy resources from natural gas and oil to tar sands and coal slurries; they carry final fuels like gasoline and diesel fuels, and increasingly plans are suggesting growth in hydrogen and carbon dioxide. In addition to these seemingly innocuous infrastructures, pipelines are objects of social resistance for climate justice advocates and Indigenous water protectors, for example, with the Dakota Access and Keystone XL pipelines (Bosworth 2022) (Figure 10.4). Similar struggles over pipelines occurred a generation ago with the Trans Alaska pipeline, which would bring petroleum from the North Slope of Alaska and Prudhoe Bay to the port of Valdez (Berry 1975).

Box 10.1 The Electricity Grid

Electricity grids are complex systems composed of many interconnected parts. At their most basic, they include power generators, substations (which transform power voltages up or down), transmission and distribution lines, towers, poles, and in some cases, undersea or underwater routes. The transmission network moves electricity at high voltages across large distances. Higher voltages are used because they allow more power to be transmitted efficiently over thinner wires with less energy loss. In contrast, the distribution network is a much more extensive web of lower-voltage wires that deliver electricity from substations to homes, businesses, and other end users.

Efforts to modernize and decarbonize electricity systems are underway in many parts of the world. However, as political scientist Leah Stokes illustrates, transforming these systems often encounters significant political resistance (Stokes 2020). Entrenched interests—particularly fossil fuel companies and electric utilities—frequently shape energy choices in ways that prioritize profit over environmental and public interest. These political headwinds can slow or distort the transition toward cleaner, more equitable energy futures.

Electricity Grids

The first electric grids were not built to deliver bulk power to homes and businesses, but rather to enable communication. Early telegraph systems sent low-voltage electric signals that could be interpreted using Morse Code. Technically, these systems did not transmit power in the way we think of it today—they moved small amounts of electricity as telecommunications signals that activated specific devices. Nonetheless, they were remarkable innovations that stretched across vast distances. Early transcontinental and transoceanic telecommunication lines became celebrated feats of engineering.

It is particularly notable that during this era of early electrical transmission—when telegraph and telephone systems were being developed—engineers had not yet perfected the process of drawing copper wire strong enough to span long distances as we see today. Thus, the development of electricity grids is not merely a story of connecting wires and building transmission lines; it also required breakthroughs in metallurgy and other material sciences, as well as technological and social innovations.

The size and scope of modern electricity transmission systems are immense. In the western United States, for example, transmission systems span tens of thousands of circuit miles (Figure 10.1). The California Independent System Operator (CAISO) leads with approximately 26,000 circuit miles, followed by PacifiCorp at 17,100 miles, the Western Area Power Administration at 17,000 miles, the Bonneville Power Administration at 15,000 miles, and the Los Angeles Department of Water and Power at 3,500 miles. These are significant networks—but according to experts, they are still insufficient for the energy transition ahead. Research from Princeton University on U.S. national electrification efforts estimates that grid capacity will need to double or even triple by 2030 and 2040, respectively, in order to meet climate goals.

The question of how best to transmit electricity was already central at the end of the nineteenth century, when the first electric generators capable of powering neighborhoods and communities were being developed. Thomas Edison advocated for direct current (DC), while Nikola Tesla and others argued that alternating current (AC) was superior for long-distance transmission. Today, the debate is no longer AC versus DC—both have their applications—but rather how to build transmission infrastructure quickly enough to keep global warming below dangerous thresholds. Edison's system, developed by the Edison Electric Company, launched the first commercial electricity grid in the world in 1882 and operated on DC power. Meanwhile, Westinghouse championed the first AC grid, connecting hydropower from Niagara Falls to Buffalo, New York—an early demonstration of long-distance AC transmission.

Electrical grid development is often a leading example of state-led development efforts, as it enables industrialization and improvements in quality of life. In

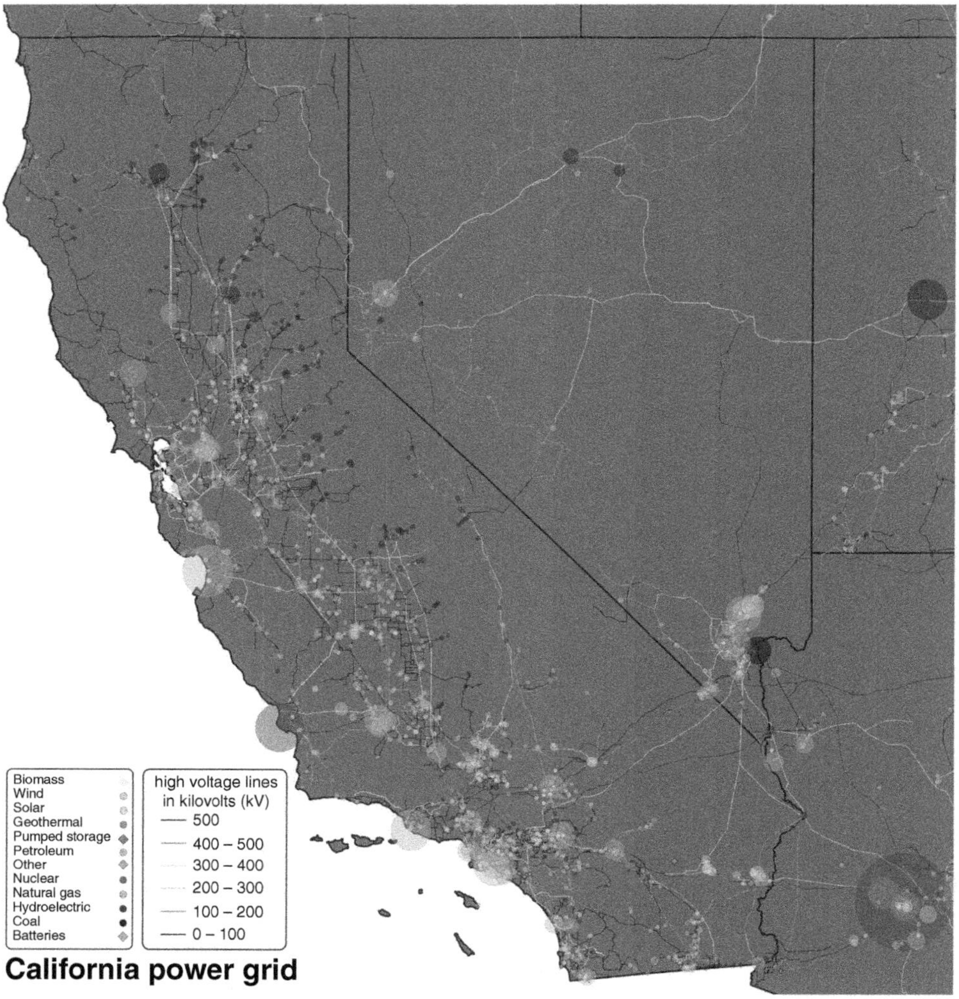

Figure 10.1 The California electricity grid. Source: Wikideas1/California power grid/ https://upload.wikimedia.org/wikipedia/commons/thumb/1/1d/California_power_grid.webp/4870px-California_power_grid.webp.png/CC0 1.0/last accessed on September 18, 2025.

Mozambique, geographers have shown how Frelimo—the Marxist-Leninist liberation movement—expanded electricity infrastructure into rural areas following national independence. This effort was not only about energy access; it was also a strategic move to strengthen territorial control and foster a sense of national unity (Power et al. 2016). In *Powerlines*, geographer Andrew Needham illustrates how the construction of major transmission lines into Arizona and the broader Southwest—delivering inexpensive hydropower and coal-fired electricity—enabled the rise of desert metropolises like Phoenix (Needham 2014). These cases show how energy infrastructure can be both a material and political force in shaping development trajectories (Box 10.2).

Box 10.2 RTOs, ISOs, Utilities, Merchant Generators. How Does This All Work?

The transmission and distribution of electricity is done by electric utilities. Some are owned by investors, others are public, while some are co-ops. Some utilities own and manage the power lines as well as generate electric power. In other grids, the utility and the power plants are owned by so-called "merchant" generators. Regional Transmission Operators (RTOs), or Independent System Operators (ISOs) are the typically non-profit organizations that ensure smooth operations of electricity flows by managing energy markets, ensuring reliability, dispatching power plants as needed, and planning for system upgrades, future power mixes, and other activities essential to delivering reliable power. These organizations maintain reliability standards and often coordinate utility governance and operations. They help maintain everything from capacity markets to make sure adequate power is available to manage grid operations during emergency situations (wildfires, rolling blackouts, etc.).

The regulation of electricity grids falls largely to states, but where power lines cross state or international boundaries, the Federal Energy Regulatory Commission may be involved. This agency is tasked with setting electricity rates and the market rules that shape what power is delivered when. Some have argued that the agency should have authority to help get transmission approvals that may be blocked by local jurisdictions or the state. Most research on this topic suggests that careful planning can help transmission project development outpace projects that are hastily fast-tracked.

Electricity grids have been historically centralized systems, often with only one owner, power plant operator, and planner. Some electricity grids have been partitioned into power line companies and power generation companies. The electricity generation companies are often called merchant power plants, and the goal of separating them from the utilities that deliver power and maintain wires was to encourage investment and competition in the power sector to drive down prices.

Since they are private profit-generating entities, investor-owned electric utilities often find tension between shareholder return and customer cost. In a distributed system, there's a tension between individual benefit and non-participant cost. Ideally, the operator is trying to maximize the value of both forms. But private companies are also competing to attract capital, so making companies look profitable can help attract the financing needed to grow or expand the electricity system (Box 10.3).

Sometimes electricity grids can be expanded not by building new transmission lines, but through "non-wires alternatives"—such as distributed energy resources (DERs) or advanced energy management systems. In the United States, DERs have been enabled in part by the Public Utility Regulatory Policies Act (PURPA) of 1978, a

Box 10.3 Should the Electric Power Grid Be Publicly Owned?

Many electricity grids in the world are privately owned by investors, but some are public owned. In the USA, the legacy of public power utilities is in part a legacy of uneven development in the country. Most transmission infrastructures are privately owned, but the federal government through the Department of Energy operates the Bonneville Power Administration, the Southeastern Power Administration, the Southwestern Power Administration, and the Western Area Power Administration. The first publicly owned transmission and electric generation facilities were built by the federal government by the Tennessee Valley Authority, who built hydropower dams and transmission infrastructure during the New Deal as an effort to bring electricity to parts of the country left behind by utility investors. Similarly, the Bonneville Power Authority had the goal of electrification of the Pacific Northwest, an effort popularized by Woodie Guthrie in folk songs about public power, including "Columbia roll-on," that were sung across the Pacific Northwest. A question facing our existing and future energy infrastructures is about ownership. Can and should the private sector or public, or some combination, run these important infrastructures?

foundational policy that, while often associated with the emergence of net metering, more simply required electricity grids to become open to competition and renewable energy. PURPA introduced the concept of the "qualifying facility" and required electric utilities to compete with third-party power producers, including small generators, cogeneration facilities, and distributed, net-metered solar systems. Utilities are mandated to purchase electricity from qualifying facilities at their avoided cost—the cost they would otherwise incur to generate or purchase the same power. In some states, policy prescribes utilities to procure electricity from net-metered solar from any utility customer. Thanks to advancements in communication and control technologies, net-metered solar systems can now be aggregated across many customers. With the appropriate regulatory authority and coordination, these systems can function as virtual power plants, providing reliable, distributed energy to the grid, avoiding costly transmission systems and power plants (Figure 10.2).

However, challenges go beyond policy and interconnection. In many parts of the world, electricity grid systems and their supporting infrastructures are aging, outdated, and in need of significant upgrades—both in transmission and distribution. These aging systems not only hinder the integration of modern power technologies but can also pose serious safety risks to communities. A tragic example occurred in 2019, when a 100-year-old transmission tower hook in PG&E territory failed in dry, high winds, sparking the devastating Camp Fire in Paradise, California, which destroyed entire towns and communities.

Figure 10.2 The Path 46 transmission corridor connects the Diablo Canyon nuclear power plant on the Central Coast of California to the Central Valley.

Another growing issue is social opposition to transmission infrastructure (Figure 10.3). Expanding transmission—seen by many as essential for decarbonizing the electricity system and integrating renewable energy—often faces resistance from communities along proposed corridors. One potential solution gaining traction is to site new transmission lines along existing infrastructure corridors—such as railroads or highways—which could reduce the potential for social and environmental disruption. In the case of highways, undergrounding the lines during roadwork may minimize surface impacts, aside from temporary construction-related traffic. Other suggestions for gaining community support for transmission infrastructure include reduced energy costs, co-ownership and revenue-sharing agreements, and community benefits agreements (Bristow et al. 2012).

Smart Grids and Microgrids

If the grid is a large machine for generating electricity, a smart grid or microgrid can be thought of similarly, except as smaller machines. DERs are types of devices or assets that can generate, consume, or store electricity, and that can respond to a signal remotely. Where there is grid access, DERs are connected to the larger system. But they would be able to function independently or as an island in situations where they wanted or needed to. Given the enormity and scale of large-scale transmission infrastructures, some point in contrast to the benefits of more distributed and local grid resilience in the form of smart grids and microgrids.

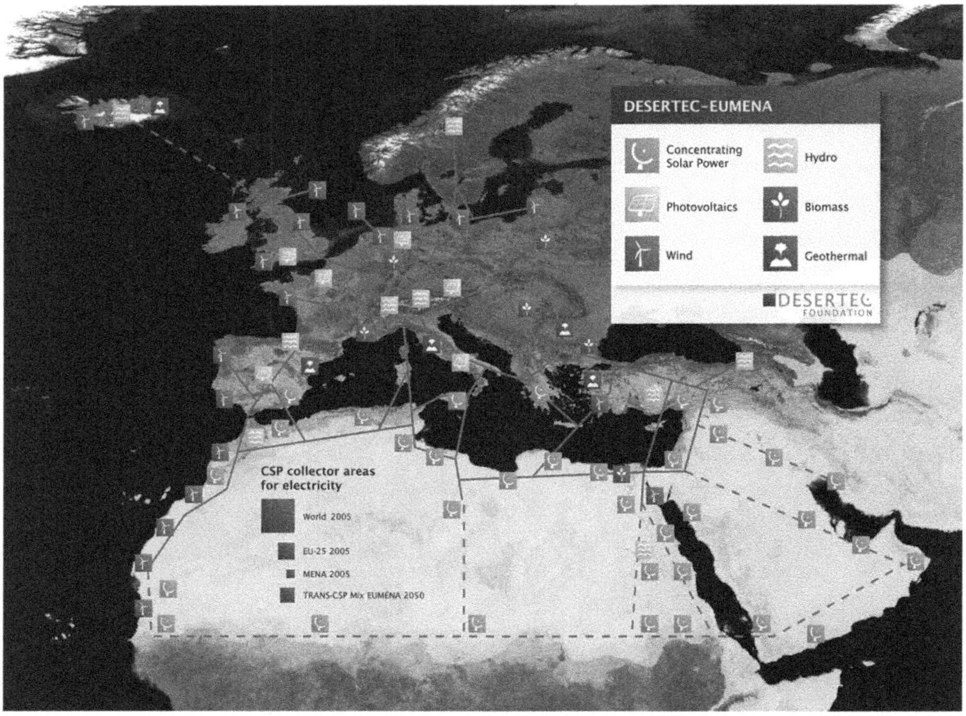

Figure 10.3 Desertec hopes to someday connect renewable energy resources across North Africa and Europe. Source: DESERTEC Foundation / Wikimedia Commons / CC BY-SA 2.5.

There are a handful of reasons one might pursue smart- and micro-grids. Having a reliable, uninterruptible supply of power might be one of these. Some simply want to get off the grid. Grid defection is the idea that if electricity prices rise high enough, it will spur customers to leave the electric grid in favor of household generation and storage. For a grid resiliency example in wildfire season in California, power lines can break or equipment can fail, and utilities are increasingly shutting off power to avoid power lines causing wind-driven wildfires. One might consider building microgrids for communities frequently shut down or exposed to storm outages, etc., or for emergency management facilities like firehouses or places that might serve as disaster locations (Box 10.4).

Critical energy geographers have taken up the study of microgrids in the United Kingdom, finding these technologies work to reshape human behavior. By getting people to understand and see more clearly their patterns of use, they got energy users to self-regulate their own behavior (Bulkeley et al. 2016). This shows how technologies can reshape social relations much in the way French Philosopher and social critic Michael Foucault described "governmentality," a way of reshaping our relation to the world and each other through knowledge, power, and practice.

Box 10.4 California's Over-the-Fence Rule and Smart Grids

Some great examples of political problems for microgrids can be found in some of the rules maintained by electric utility monopolies. One such example is California's "over the fence" rule, which prevents smaller enterprises from linking energy systems without triggering rules and regulations that treat their activities like they were a utility. This prevents parcels that might be adjacent from sharing energy or other resources like parking spaces from integrating across property lines. Whereas, cooperatively owned microgrids could serve a useful purpose in the integration of local energy resources into distribution grids and even lessen the need to build out transmission infrastructures. California Public Utilities Commission, which oversees such projects, is looking for ways to better incentivize smart microgrids in vulnerable communities or those susceptible to power outages.

Natural Gas, Oil, and Tar Sands Pipelines

Pipelines extend all over the world, moving not only familiar things like oil, natural gas, and water, but also rock slurries of coal or other minerals, and now there are new proposals to move around carbon dioxide for carbon capture schemes. Historian Christopher Jones describes how these large infrastructure projects that resulted in large energy transportation systems like coal barges and later oil pipelines and electricity lines helped establish the hegemony of the mineral or fossil fuel energy regime, locking society into the projects in a repeated way that makes it difficult to move toward other infrastructures (Jones 2014).

There are significant risks associated with the development and use of pipelines for transporting oil, gas, tar sands, and other fossil fuels. Historic accidents underscore these dangers. For example, a major pipeline owned by Texaco leaked over five million gallons of oil into the Gulf of Mexico off the coast of Louisiana in 1970—the largest oil spill in U.S. history until the Exxon Valdez supertanker oil spill disaster in 1989. Some types of liquid fossil fuels are especially difficult to clean up. In 2010, a tar sands pipeline spill in Kalamazoo, Michigan released viscous diluted bitumen (dilbit) that spread over 40 miles down the Kalamazoo River, requiring one of the most complex and expensive inland oil spill cleanups in U.S. history.

At the same time, many safety advocates point out that transporting fossil fuels by other means—such as trains—poses significant hazards. So-called "bomb trains", which carry highly volatile fuels like tar sands and crude oil, have been linked to catastrophic explosions and fires. Several communities have experienced devastating impacts from rail accidents involving fossil fuels. Some argue that pipeline construction should continue as long as these fuels are in use, because the alternative—transporting them by rail—may carry greater public safety risks, particularly for communities located near rail corridors. Railcars carrying volatile fuels, especially tar sands and crude oil, present serious risks of explosion, fire, and long-term environmental damage.

Pipelines are inherently spatial, and their spatial characteristics often manifest as economic value. For example, the value of natural gas can change as it moves through the pipeline network—some locations command higher prices than others. This variation is influenced by bottlenecks in the system that can prevent low-cost energy from reaching high-demand, high-price areas. The term "gas basis" refers to the difference in natural gas prices between two geographic locations, typically due to transportation constraints and supply–demand imbalances. A narrower basis spread indicates a smaller price difference, which can result from increased pipeline capacity that eases flow restrictions. Consequently, adding new pipelines is often seen as a strategy to reduce energy costs by relieving supply constraints in high-priced regions. However, the situation is not always straightforward. Some natural gas pipelines are built specifically to connect domestic supplies to export markets, such as LNG terminals. These export-oriented developments can put upward pressure on domestic gas prices, as they shift supply away from local markets to meet international demand (Figure 10.4).

Pipelines have become key objects of geopolitical struggle, as they shape not only energy markets but also international alliances and power dynamics. One of the most prominent examples is Nord Stream 2, a highly contested pipeline project designed to transport natural gas from Russia to Germany via the Baltic Sea. Intended to double

Figure 10.4 People hold signs opposing an Enbridge Tar Sands Pipeline expansions at Protect the Water, Revoke the Permits! Rally in St Paul, Minnesota. Source: Lorie Shaull / Flickr.

the capacity of the existing Nord Stream 1 pipeline, Nord Stream 2 bypasses traditional transit countries—most notably Ukraine—and has been at the center of geopolitical controversy for over a decade. The central issue involves the strategic relationship between Russia and Ukraine. Currently, a significant portion of Russian natural gas exports to Europe passes through Ukrainian pipeline networks, generating transit fees that contribute substantially to Ukraine's economy. Nord Stream 2 would reroute much of that gas, effectively undermining Ukraine's role as a transit country and reducing its political leverage over energy flows to the European Union.

Critics, including the United States, some EU member states, and Ukraine, have argued that the pipeline increases European dependency on Russian energy, potentially giving Russia greater geopolitical influence. Supporters, primarily Germany and some energy companies, have framed the project as a commercial endeavor aimed at ensuring stable gas supplies.

The situation escalated dramatically following Russia's invasion of Ukraine in 2022. In response, Germany suspended the certification of Nord Stream 2, and the pipeline has since become a symbol of the risks of energy entanglement with authoritarian regimes. The alleged sabotage of the Nord Stream pipelines in 2022 added a further layer of complexity, turning infrastructure itself into a potential target of hybrid conflict.

Oil Supertankers and Ports, LNG Infrastructures, Offshore Rigs

Critical nodes in fossil fuel infrastructures include supertankers, ports, complex drilling apparatus and means of transporting those fuels. These infrastructures comprise a vast and interconnected global system of maritime transport and offshore production that moves fossil fuels across continents and oceans. Three of the most important components of this network are oil supertankers, LNG infrastructure, and offshore drilling platforms. Together, they form a critical backbone of the global energy economy and have significant environmental, geopolitical, and engineering implications.

Oil supertankers are massive ships designed to transport crude oil or petroleum products across oceans. The largest of these, called Very Large Crude Carriers or Ultra Large Crude Carriers, can carry up to two million barrels of oil in a single journey. These tankers travel between export terminals near oil fields (such as in the Persian Gulf or West Africa) and import terminals in major energy-consuming regions, including the U.S., Europe, and East Asia. Ports that host these tankers—like Port of Houston, Rotterdam, or Singapore—require deepwater harbors, specialized unloading equipment, and spill prevention systems. These marine logistics systems are vital for maintaining global oil flows but also present major ecological risks, including oil spills and ballast water pollution.

Box 10.5 Fugitive Methane Emissions

Fugitive emissions from natural gas extraction activities, pipelines, and compressor stations are major sources of the greenhouse gas methane released into the atmosphere. The term *fugitive* refers to emissions that escape from a system that is intended to be sealed or controlled—leaking due to equipment damage, degradation, improper assembly, or during maintenance and transfer operations, rather than being released through designated exhaust or venting systems. These emissions are especially concerning because methane is a potent greenhouse gas, with a global warming potential significantly higher than carbon dioxide over a 20-year timescale. The diffuse and often unmonitored nature of fugitive emissions makes them difficult to detect, quantify, and regulate.

Looking to the future, natural gas pipelines may carry alternative energy gases, such as biomethane (biogas) or hydrogen-natural gas blends, as part of broader decarbonization strategies. This introduces new challenges: future fugitive emissions could be complex mixtures of methane, hydrogen, carbon dioxide, and other trace gases—each with different environmental and safety profiles. Hydrogen, for example, is highly flammable and can leak more easily than methane due to its smaller molecular size. As energy systems evolve, monitoring and managing fugitive emissions will require advanced leak detection technologies, updated regulatory frameworks, and a deeper understanding of the climate and safety risks posed by emerging gas mixtures.

To move natural gas requires enormous infrastructures too. LNG is natural gas cooled to reduce its volume by over 600 times, making it easier to store and transport. LNG is a critical component of the global energy trade, particularly for countries with limited pipeline access. The LNG system includes liquefaction plants (where natural gas is cooled into liquid form), cryogenic tankers (specialized ships that transport LNG), import terminals, and regasification facilities, where LNG is turned back into gas and injected into local pipeline networks. Countries like Qatar, Australia, and the U.S. are major LNG exporters, while Japan, South Korea, and parts of Europe are major importers. LNG has been promoted as a “bridge fuel” in energy transitions, but concerns persist around methane leakage, lifecycle emissions, and the long-term carbon lock-in from building such large-scale infrastructure (Box 10.5).

Offshore rigs are platforms used to explore and extract hydrocarbons from beneath the ocean floor. These can range from shallow-water jack-up rigs to deepwater semisubmersibles and floating production units. Offshore fields are often located in geopolitically sensitive and environmentally fragile regions, such as the Gulf of Mexico, the North Sea, the Arctic, and the South China Sea. These platforms are engineering feats, operating in extreme weather and deep-sea environments. But they also come with substantial risks: blowouts, oil spills, and seismic impacts can have catastrophic consequences, as seen in disasters like the Deepwater Horizon spill in 2010. The infrastructure also has long-term consequences for marine ecosystems and coastal communities.

Piping Carbon Dioxide and Hydrogen Around for the Net Zero Economy

New pipelines may be necessary to move new cleaner energy fuels, biproducts, or wastes. For example, captured carbon dioxide may need to be moved from sites of capture to sites of injection and permanent storage, while hydrogen may need to be moved from sites where electrolysis is built to where hydrogen will be stored or used. Carbon dioxide is already moved around in some energy systems. Enhanced oil recovery separates CO_2 from oil processing and is used to pressurize oil fields to "enhance" their productivity, squeezing the remaining hydrocarbons out of the oil or gas field like a sponge (Box 10.6).

Hydrogen pipeline superhighways will have to connect sites of electrolysis where cheap renewable energy is available in rural areas or offshore. These would connect hydrogen pipelines to where it might be used as blended into heating fuels or in fuel cells or whatever is determined to be the best use of the fuel. Many industries may require hydrogen as a source of fuel to make high-temperature heat.

Box 10.6 Enhanced Oil Recovery and Carbon Dioxide Pipelines

In 2023, the carbon dioxide pipeline network was over 5,000 miles in the United States (U.S. Department of Transportation 2023). Much of this network is moving carbon dioxide for enhanced oil recovery operations (EOR). The use of EOR is described as one pathway to capturing some of the greenhouse gas emissions from oil production and use. The way it works is that most oil field reservoirs are readily accessed by conventional approaches. As the oil field is depleted, the remaining oil can be difficult to get out of the reservoir. Think of the reservoir as a sponge that you need to "wring" tightly to get the last water out. In many cases you need to change the chemical properties of the remaining oil to get it out. EOR can use heat, but also gas pressure can be applied to the oil field to liberate the oils. Today in the United States, this is most common in places like the Permian Basin, where CO_2 is piped in from various natural and industrial sources. While CO_2 is captured it does spur more oil production, so the amount of emissions this process captures compared to what it allows is not always straightforward to assess.

Green Infrastructures, Siting Big Infrastructures, and the Just Transition

Just transitions governance in addition to improving access to energy, lessening burdens on communities from fossil fuel pollution, prompts us to look for contradictions related to environmental equity that might flow from efforts to pursue green infrastructures (Newell and Mulvaney 2013). Greater procedural equity on the siting of infrastructures is complicated by the typical ways that these projects are proposed, permitted, and developed. Typical projects advance through a "decide-announce-defend" decision-making process, where projects are decided without much input, announced to the public, and the public consultation process is mostly spent defending the proposal instead of a process that is more participatory and collaborative (Walesh 1999). This consultation process in US federal policy-making has particularly failed Native American Tribes, where meaningful consultation is meant to bring input to protect sacred sites and cultural resources on public lands, off-reservation, from development (Bathke 2014). Collaboration with representatives of disadvantaged stakeholders and fenceline communities are opportunities to improve natural resource governance (Dobbin and Lubbell 2021) and more informed ecological decisions (Box 10.7).

Owing to its history of colonization and dispossession, in the United States, nearly 80% of all known lithium deposits sit within 35 miles of tribal reservations. The US Department of Interior's Bureau of Land Management claims to use best practices when it comes to Tribal engagement and consultation (Bureau of Land Management 2023).

To lessen the frictions of renewable energy development, some researchers suggest addressing some of the procedural inequities in the development process. For example, developers can offer something to non-compensated land owners in a region, instead of just working with those landholders who stand to benefit. They can offer

Box 10.7 Section 7 of the U.S. Endangered Species Act

The Endangered Species Act (ESA) was enacted in 1972 and immediately became a landmark public policy for the protection of species threatened and endangered against development. The jeopardy provision of the ESA is in Section 7 bringing agency expertise to weigh in on the implication for a project to jeopardize a listed species. In the environmental review process coordinated under the National Environmental Policy Act, the Section 7 consultation is in the Biological Opinion, written by staff at the US Fish and Wildlife Service or National Marine Fisheries Service depending on what species are under consideration. A study led by researchers and legal staff at the Defenders of Wildlife found that 88,290 consultations from 2008 to 2015 led to no projects being substantially modified as a result of the Section 7 review. This suggests that far from the received wisdom that the ESA is causing major delays to infrastructures, in fact the policy has had little effect in shaping project directions.

more knowledge-sharing and public-input-sharing opportunities. It is also important they suggest to understand a community's history with power generation and their connections to the land (Elmallah and Rand 2022).

Another area where consultation and prior feedback are important is regarding environmental justice communities. Formal environmental justice policy in the USA can be traced back to a working group from the U.S. EPA in 1990, which eventually informed the text of the Executive Order signed into law by the President in 1994. This law required that federal permitting agencies examine the implications of their decisions in the environmental review process regarding environmental justice and environmental racism (Holifield 2001). While the policy has been criticized for an inability to meaningfully shape outcomes, it served as a model for other states. California in 2001 assembled an environmental justice board at the California EPA, and has integrated it into stronger policies since culminating in the establishment of a Bureau of Environmental Justice and specific powers to reject infrastructure permits on the grounds of environmental inequity at some state agencies.

Whether environmental justice policy is effectively incorporating otherwise marginalized views and driving environmentally equitable outcomes. Much of environmental justice activism starts from the idea that environmental policy-making is contentious and adversarial; change is a result of protest, litigation, and other efforts to stop or block new infrastructures or locally unwanted land uses. The question of whether effective tactics for social and environmental change can be collaborative instead of adversarial, given the appropriate context and capable governance, is a key question for scholars in the social science of energy and water.

For us to understand the ways that environmental injustices caused by infrastructures are created, sustained, and reproduced, it is helpful for us to understand how marginalized voices are included in the decision-making and permitting process (Box 10.8). While environmental injustices are spatially distributed inequities, they

Box 10.8 "Decide, Announce, Defend" Planning Has Left a Legacy of Gridlock

A legacy of gridlock has widely discredited the "decide, announce, defend" approach to infrastructure planning and development and environmental decision-making, in which agencies confront the public only after determining where projects go (Beierle 1999). This is referred to as the social gap in siting infrastructures the public finds favorable like renewable energy systems, but has difficulty in building them. Much of the popular discourse on social resistance to some kinds of energy infrastructures focuses on not-in-my-backyard (NIMBY) motivations, but researchers on the social gap find great nuance in these explanations with notable takeaways for developers and policymakers.

Source: Phadke (2013) / Taylor and Francis.

are often a result of distributive inequity (Walker 2009). Procedural inequity during the public participation process for infrastructures can be a site of inequality formation for fenceline communities.

Highlights/Takeaways

- The development of transmission systems is a major undertaking in developing as well as developed economies, for different reasons.
- Extensive pipeline networks and infrastructures move all sorts of materials around today and will likely continue to into the future, which makes understanding social interactions with these projects important topics of research and inquiry.
- Major infrastructures are built for the long haul. It is critical that society agree with planned projects in a way that reflects the long-term commitments to those projects and their associated impacts.
- Giving communities a sense of ownership over infrastructures or even shareholding, collective ownership, or even public ownership could make it easier to site proposed energy projects.

Questions for Review

1. Who are the key organizations that manage the distribution of electric power generation? What are the key roles that they play?
2. What would your twenty-year plan for transmission development look like? How would you start? What key information would you seek out?
3. What are the implications for a community for building infrastructures? What are the implications of retiring infrastructure after its useful life?
4. What are the main components of a smart grid, and what is the case for them from a resilience perspective?

Exercise 10.1

Conduct a mock Environmental Impact Statement (EIS) process for a proposed high-voltage transmission line crossing a wilderness preserve that is also recognized as sacred by an Indigenous community.

Classroom Instructions: Form small groups representing different stakeholders: Indigenous community leaders, Transmission company engineers, Environmental advocacy groups, Labor unions, Federal or state energy regulators, Wildlife and land management agencies, local citizens, businesses, and property owners. Each group will research their stakeholder perspective, develop a position statement regarding the project, present their environmental, cultural, or economic concerns. Prepare an EIS outline, including project purpose and need, alternatives considered (e.g., underground routing, alternative siting), description of the affected environment, analysis of potential environmental, cultural, and social impacts, mitigation measures, and recommendations.

Deliverables: One-page position paper per stakeholder group, Group EIS presentation (7–10 minutes) with visuals, Reflective essay (500–700 words): What tensions or trade-offs emerged? What did the process reveal about infrastructure planning and cultural respect?

Exercise 10.2

Choose a controversial pipeline project (e.g., Dakota Access, Keystone XL, Mountain Valley Pipeline) and reconstruct the timeline of development, opposition, and regulatory decisions.

Research and develop a **chronological timeline** highlighting the project proposal date and company involved, major milestones: permits, approvals, protests, court rulings; Moments of escalation or public attention (e.g., protests, spills, legal action). Create a figure or concept map identifying key stakeholder groups (e.g., Indigenous communities, environmental groups, state/federal agencies, landowners, fossil fuel companies), their interests and tactics (e.g., legal filings, direct action, lobbying, media campaigns), and points of convergence or negotiation.

Include a **short analysis** (750–1,000 words) addressing the following questions: What were the main arguments for and against the pipeline? How were concerns about land use, climate, water, or sovereignty addressed or ignored? What does this case reveal about energy justice and infrastructure decision-making?

References

Bathke, J.P. (2014). Ocotillo wind: a case study of how tribal-federal governmental consultation is failing tribal governments and their spiritual landscapes through renewable energy development. *Human Geography* 7 (2): 46–59.

Beierle, T.C. (1999). Using social goals to evaluate public participation in environmental decisions. *Review of Policy Research* 16 (3–4): 75–103.

Berry, M.C. (1975). *Alaska Pipeline. The Politics of Oil and Native Land Claims*. Indiana University Press.

Bosworth, K. (2022). *Pipeline Populism: Grassroots Environmentalism in the Twenty-First Century*. University of Minnesota Press.

Bristow, G., Cowell, R., and Munday, M. (2012). Windfalls for whom? The evolving notion of 'community' in community benefit provisions from wind farms. *Geoforum* 43 (6): 1108–1120.

Bulkeley, H., Powells, G., and Bell, S. (2016). Smart grids and the constitution of solar electricity conduct. *Environment and Planning A* 48 (1): 7–23.

Bureau of Land Management. 2023. "Tribal consultation." https://www.blm.gov/services/tribal-consultation.

Dobbin, K.B. and Lubell, M. (2021). Collaborative governance and environmental justice: Disadvantaged community representation in California sustainable groundwater management. *Policy Studies Journal* 49 (2): 562–590.

Elmallah, S. and Rand, J. (2022). "After the leases are signed, it's a done deal": exploring procedural injustices for utility-scale wind energy planning in the United States. *Energy Research & Social Science* 89: 102549.

Holifield, R. (2001). Defining environmental justice and environmental racism. *Urban Geography* 22 (1): 78–90.

Jones, C.F. (2014). *Routes of Power: Energy and Modern America*. Harvard University Press.

Needham, A. (2014). *Power Lines: Phoenix and the Making of the Modern Southwest*. Princeton University Press.

Newell, P. and Mulvaney, D. (2013). The political economy of the 'just transition'. *The Geographical Journal* 179 (2): 132–140.

Phadke, R. (2013). Public deliberation and the geographies of wind justice. *Science as Culture* 22 (2): 247–255.

Power, M., Newell, P., Baker, L. et al. (2016). The political economy of energy transitions in Mozambique and South Africa: the role of the rising powers. *Energy Research & Social Science* 17: 10–19.

Stokes, L.C. (2020). *Short Circuiting Policy: Interest Groups and the Battle over Clean Energy and Climate Policy in the American States*. USA: Oxford University Press.

U.S. Department of Transportation. 2023. "Annual report mileage for hazardous liquid or carbon dioxide systems." https://www.phmsa.dot.gov/data-and-statistics/pipeline/annual-report-mileage-hazardous-liquid-or-carbon-dioxide-systems.

Walesh, S.G. (1999). DAD IS OUT, POP IS IN. *Journal of the American Water Resources Association* 35 (3): 535–544.

Walker, G. (2009). Beyond distribution and proximity: exploring the multiple spatialities of environmental justice. *Antipode* 41 (4): 614–636.

Further Reading

Hughes, T.P. (1993). *Networks of Power: Electrification in Western Society, 1880–1930*. JHU Press.

Jones, C.F. (2014). *Routes of Power: Energy and Modern America*. Harvard University Press.

Needham, A. (2014). *Power Lines: Phoenix and the Making of the Modern Southwest*. Princeton University Press.

Chapter 11

Mobility and Transportation

"Beyond a certain speed, motorized vehicles create remoteness which they alone can shrink. They create distances for all and shrink them for only a few."

—Ivan Illich, Energy & Equity

Most of global transportation remains powered by petroleum products, with some increases in electric vehicles (EVs) and natural gas over the past decade. The combustion engine is one of human civilization's most polluting technologies. From criteria air pollution to greenhouse gas emissions and nitrogen pollution from corn ethanol, the fuels we burn in automobiles and other vehicles are a major environmental issue. EVs passed another milestone with 10 million cars manufactured in 2022. However, there are two billion cars that need to be retired or decarbonized, and that's assuming the same proportion of people today are in vehicles. That challenge is daunting. This chapter gets at the core issues shaping mobility, and resists framing the challenge as a simple substitution of internal combustion vehicles with a cleaner one.

It's About Mobility, Not Transportation

Mobility refers to the ways we move ourselves through space—while transportation is the broader system that moves both people and goods. In some areas, the priority may be to improve mobility options for individuals, and in others, it may be to optimize

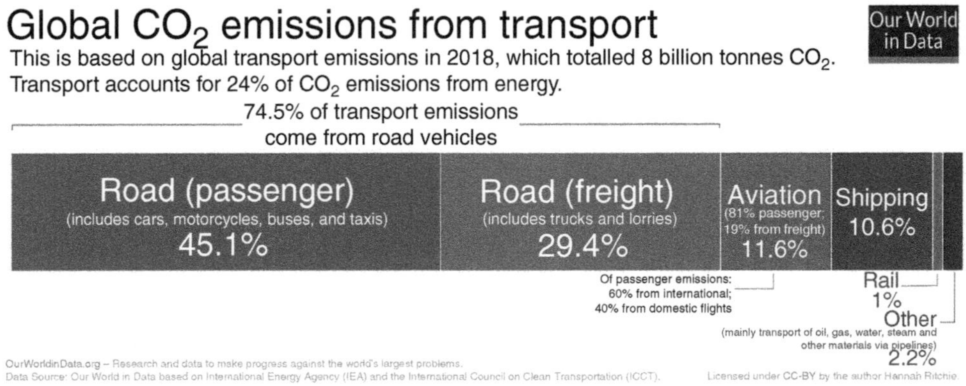

Figure 11.1 Global emissions from transportation. Source: Our World in Data/Cars, planes, trains: where do CO_2 emissions from transport come from/https://ourworldindata.org/co2-emissions-from-transport/CC BY 4.0/last accessed on September 18, 2025.

transportation systems for efficiency and sustainability. For the past half-century, however, mobility—particularly in the United States—has been largely equated with private car use. While the car has offered convenience and independence, it has also brought significant social, environmental, and public health consequences (Figure 11.1).

Transitioning to low-carbon transportation will require us to rethink the systems we've built. Promoting human-scale mobility, such as walking, biking, and micromobility options like scooters and e-bikes, will mean redesigning the physical spaces of our cities and suburbs. This includes investments in protected bike lanes, secure bike storage, accessible sidewalks, and clear signage and road markings—all of which make it safer and more appealing to leave the car behind. At the same time, digital innovations are reshaping how people move. The rise of ride-sharing platforms such as Uber and Lyft, along with the development of autonomous vehicles, raises new questions about urban form, infrastructure, equity, and environmental impact. Will these technologies decrease car ownership and emissions, or will they reinforce car dependence and congestion in new ways? How we shape these systems will determine whether cities become more connected and livable—or more fragmented and polluting.

Finally, it's important to confront the reality that car culture is costing lives. Traffic fatalities, especially among pedestrians and cyclists, have risen in many countries. Streets are too often designed with cars as the priority, leading to dangerous environments for other road users. Reversing these trends will require cultural, political, and design changes to put people—not vehicles—at the center of transportation planning. This chapter explores how low-carbon transportation can reduce emissions, enhance quality of life, and create more just and sustainable communities—while challenging the dominance of car-centric planning (Figure 11.2).

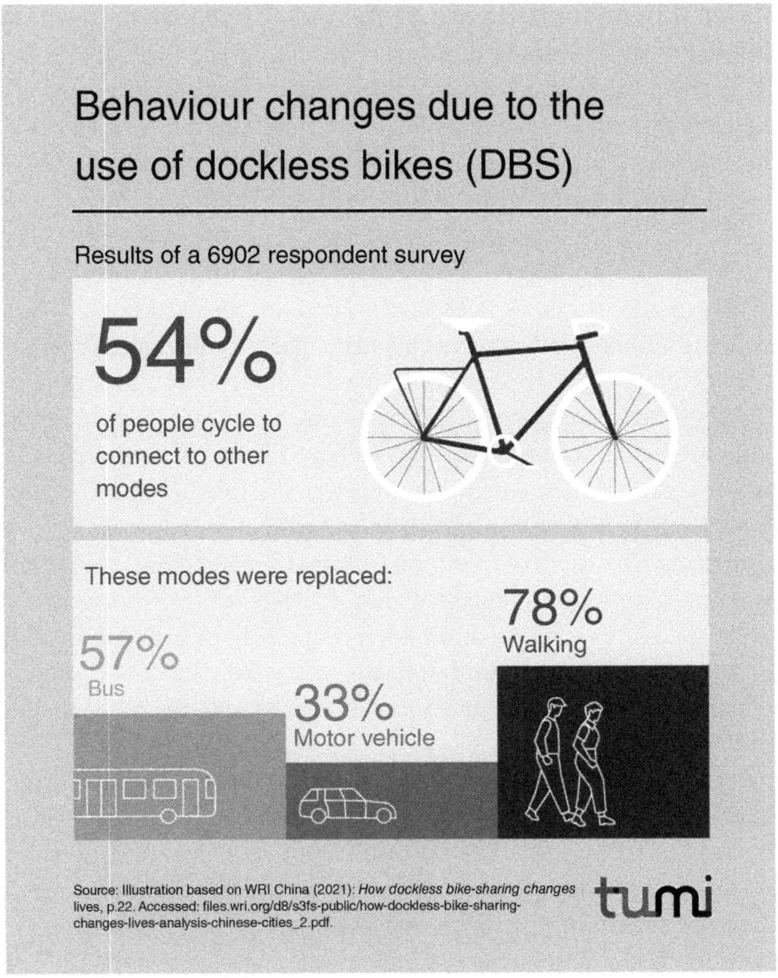

Figure 11.2 How bicycles lead to transportation mode shifts. Source: TUMInitiative/ https://upload.wikimedia.org/wikipedia/commons/0/07/Behaviour_change_due_to_use_of_dockless_bikes.png/CC BY SA 4.0/last accessed on September 18, 2025.

Automobility and the Combustion Engine

It would be hard to imagine the evolution of our contemporary transportation systems simply by looking at the earliest automobiles of the late nineteenth century. These first motor vehicles were experiments in propulsion, powered by a variety of technologies, including steam engines, early combustion engines, battery-powered electric motors, and even hybrid designs. The first internal combustion engines (ICEs) emerged in the mid-1800s and were refined over the following decades to become more efficient and compact. By the early twentieth century, gasoline combustion engines had become dominant, aided by mass production innovations such as Henry Ford's assembly line, which made cars more affordable and accessible to middle-class consumers.

At the dawn of the automobile era, there was still a degree of uncertainty about which technology would become dominant. In fact, in the early 1900s, battery EVs were widely used in urban areas due to their quiet operation and ease of use, while steam-powered cars and diesel engines had their own niche applications. However, the growing infrastructure for gasoline distribution, combined with the rise of oil economies and road-building programs, helped ICEs outpace their competitors by the 1920s.

The concept of automobility—developed within cultural studies, geography, and science and technology studies—refers not only to the car itself but also to the socio-technical systems and cultural imaginaries that support and sustain car-based transportation (Beckmann 2001). Automobility is shaped by institutions such as automakers, petroleum companies, governments, insurance industries, and urban planners, which together construct a world where the car is normalized, celebrated, and, in many contexts, seen as essential to modern life (Sheller and Urry 2000).

In many parts of the world, automobility has been elevated to the level of cultural identity, reinforcing narratives of freedom, personal achievement, safety, and national pride. At the same time, automobility has marginalized non-car modes of transport by eroding public spaces for walking and biking, especially in urban and suburban settings. While some cities have begun to reclaim space from cars—through pedestrian zones, bike lanes, or car-free infrastructure—others remain deeply embedded in car-centric planning.

Automobility also shapes material economies. The design and production of vehicles are influenced by global supply chains, many of which are carbon-intensive and resource-extractive, relying on metals, rare earths, plastics, and petroleum-based products. In this sense, automobility extends beyond personal mobility to encompass a global industrial system of extraction, logistics, and consumption.

Critically, automobility intersects with race, class, and spatial politics. In the United States, for example, the automobile played a central role in enabling suburbanization and white flight, while facilitating the disinvestment of urban cores through racist planning practices like redlining (Box 11.2), highway construction, and zoning laws that prioritized vehicular access over equitable development (Pulido 2017). As a result, automobility can be understood not just as a neutral mode of transport, but as a politically charged system that has shaped segregation, inequality, and patterns of environmental injustice. As we confront the challenge of decarbonizing transportation, it's important to recognize that replacing gasoline cars with EVs alone will not undo the deeper structural issues of automobility. Transforming transportation in a truly sustainable and equitable way will require rethinking land use, rebalancing infrastructure investments, and prioritizing public and active transportation alternatives (Henderson 2006).

It was the development of the automobile of course that led to the passage of the original Clean Air Act in California and eventually the federal policy. It turns out it's very difficult to monitor air pollution from millions of cars, each with their own separate pollution control device (Box 11.1). So the early years of air pollution regulation had an important role to play in the development of automobiles. Part

Box 11.1 Legacy Pollution from Fuel Additives to Limit Air Pollution from Combustion

Combustion engines take oxygen from the air and combust liquid hydrocarbon fuels. It was discovered long ago that adding extra oxygen through chemical additives can help automobile engines run more smoothly. Tetraethyllead is an additive initially used as an anti-knocking agent, initially developed by Delco, a subsidiary of General Motors. Its development coincided with the prohibition on the production of alcohol or ethanol, so it was used in gasoline until the 1980s. It was phased out starting in 1975 when the Clean Air Act required gasoline retailers to offer unleaded gasoline, which contained additives like methyl tertiary butyl ether (MTBE) or ethanol added instead (Nriagu 1990). Leaded gasoline fouled the recently installed air pollution equipment the Clean Air Act recently required on all new automobiles, called catalytic converters. MTBE would later be found to be a difficult water pollutant that led to the widespread reintroduction of ethanol in place of a fuel oxygenate. In fact, the ethanol mandate in the Energy Policy Act of 2005 that required the production of 36,000,000 gal of ethanol by 2022, described in the bioenergy chapter, eliminated the fuel oxygenate requirement for gasoline blenders and retail sales.

Box 11.2 How Redlining Shaped Housing and Transportation

Redlining is a racial discrimination practice that resulted in the exclusion of Blacks from housing in predominantly or completely white neighborhoods in the USA (Nardone et al. 2020). The practice was most widespread in the 1930s, particularly with lending institutions but also realtors. It still happens today in various forms, but many places have laws explicitly prohibiting the practice. Either way, it left a legacy that lives with communities today. Redlining resulted in minority communities living disproportionately in polluted neighborhoods, near industrial fencelines, or highways. It also often left Blacks and Latinos and other minorities and people with low incomes with long commutes from where they lived to the neighborhoods they might work in, the same neighborhoods that they were excluded from. Research on redlining in the East Bay, California, suggests that the process left behind the community with inferior public transportation, while whiter suburban communities had far more expensive road and highways paid for with taxpayer dollars (Golub et al. 2013). There are some questions about the cost of gasoline and a few other energy items from a redlining perspective.

of the rules and regulations for automobiles that shaped the types of cars we built were also the corporate average fuel economy standards, commonly called the CAFE standards. Later regulators developed renewable fuel standards to extensively lessen the imports of petroleum as well as reduce air pollution, and reduce the carbon intensity of liquid fuels sold. It is widely criticized for failing to deliver particularly on greenhouse gas emissions goals, but also had negative implications for habitat and water quality, and even caused significant price increases for foodstuff (Lark et al. 2022).

Electric Vehicles

Vehicle electrification is a central strategy for achieving low-carbon transportation systems. The main advantage of EVs lies in the superior energy efficiency of electric motors, which convert approximately 85–90% of input energy into motion, compared to ICEs that typically operate at 20–30% efficiency. In addition, EVs emit zero tailpipe emissions, which improves local air quality, especially in urban areas.

Global EV adoption has accelerated significantly in recent years. By 2026, there will be more than 100 million EVs on the road worldwide, with China leading in both sales and infrastructure development. In Norway, over 80% of new car sales are fully electric, supported by aggressive policy incentives and infrastructure investments. The United States has seen steady growth as well, with EVs comprising approximately 10% of new vehicle sales in 2025, and projections showing rapid scaling due to improved models, charging infrastructure, and supportive policies. Globally in 2025, one in four vehicles sold will be EVs.

A variety of policy mechanisms have been proposed and implemented to encourage EV adoption:

- Cash-for-clunkers programs offer financial incentives to trade in older, high-emission vehicles for newer, cleaner alternatives—potentially accelerating the retirement of the most polluting cars from the fleet.
- Feebate programs tax higher-emission vehicles (e.g., SUVs or sports cars) and use the revenue to subsidize low-emission or zero-emission vehicles, creating market pressure to shift demand toward cleaner options.
- Federal and municipal fleet procurement can drive demand by leveraging public purchasing power to create economies of scale for EV production. The push to electrify the U.S. Postal Service fleet—spurred by public pressure and legal advocacy—is a recent example of how federal fleet electrification can serve both symbolic and material roles in market transformation.

The rise of the sharing economy has also influenced the adoption of EVs. Carsharing services, fleet-based rideshare companies, and subscription vehicle services are increasingly incorporating electric models. In these contexts, centralized fleet management can make EVs more cost-effective, and shared mobility options offer lower barriers to EV access for consumers (Lim et al. 2022).

Public Transportation

Improving public transportation in urban and suburban areas is not only essential for reducing climate pollution and airborne toxins, but also critical for alleviating traffic congestion, enhancing mobility, and improving overall quality of life. By offering efficient alternatives to car travel, public transit systems can help cities reclaim space from automobiles and reinvest in walkable, bikeable, and more socially inclusive infrastructure. One of the most powerful arguments for public transit lies in its potential to reduce the material footprint of decarbonization. While electrifying cars is a necessary step toward a low-carbon future, mass vehicle electrification requires vast quantities of critical minerals and metals—especially lithium, cobalt, nickel, and graphite—for battery manufacturing. Expanding bus and rail systems can lower per capita vehicle use, which in turn decreases the number of EVs needed. A report by the Climate + Community Project found that in a best-case scenario—where vehicle sizes are reduced through lightweighting and public transit uptake is maximized—lithium demand could be cut by as much as 92% compared to a business-as-usual EV rollout. This dramatic reduction illustrates how a focus on systemic transportation reform could help avoid replicating the extractive burdens of the fossil fuel era.

Beyond emissions and materials, public transit is also a social equity solution. Transit systems provide essential mobility for those who cannot afford a car, are unable to drive, or live in communities underserved by road infrastructure. Expanding and improving public transit helps ensure that the benefits of clean energy transitions are shared more broadly, rather than accruing only to those who can afford personal EVs. At the same time, investing in transit-oriented development, bus rapid transit (BRT), and electrified and high speed rail (Box 11.3) can reshape urban land use patterns to be more sustainable, inclusive, and connected. Ultimately, public transportation is not a side note in energy policy—it is a core strategy for achieving climate goals, reducing resource demand, and creating livable, low-carbon cities. Prioritizing transit alongside electrification ensures a more just, material-efficient, and people-centered energy transition (Riofrancos et al. 2023).

Significant federal investments in highways have helped bolster transportation and subsidize personal transportation. Big investments really began with the 1956 Highway Investment Act in the United States. Similar road build-outs have happened elsewhere in the world. Public policies could invest more in reconnecting communities divided by highway infrastructures (Freemark et al. 2023). A major concern in redevelopment is the idea that rebuilding places would lead to green gentrification, pushing those with lower incomes out of the places just revitalized. The idea of green gentrification acknowledges that efforts to revitalize or redevelop around green initiatives tend to

Box 11.3 California's High-Speed Rail

The California HSR project represents one of the most ambitious efforts in the United States to provide a low-carbon, electrified alternative to air travel, particularly for trips between California's major cities. The vision is to connect San Francisco and Los Angeles with a 220 mph (354 km/h) electric train, reducing travel time to under three hours and significantly cutting emissions compared to flying or driving. Initiated by voter approval of Proposition 1A in 2008, the HSR system was proposed as a key part of California's broader climate mitigation strategy, helping to reduce emissions from the state's largest-emitting sector: transportation. The train will be powered by 100% renewable and low carbon energy, offering a model for sustainable intercity travel.

lead to replacement of the inhabitants of those neighborhoods as they are outbid for those new green urban amenities (Gould and Lewis 2016).

Aviation and Rail

Greenhouse gas (GHG) emissions from aviation are projected to increase significantly in the coming decades, driven by rising demand for global passenger travel. By some estimates, the volume of air travel in 2050 could be more than double that of 2010 levels. The aviation sector already emits nearly one gigaton of CO_2 annually, making it one of the fastest-growing contributors to climate change. Decarbonizing aviation is uniquely difficult for several reasons. Jet fuel has exceptionally high energy density—providing a lot of energy per kilogram—which is essential for flight, where both fuel weight and volume are critical constraints. This makes it hard to substitute conventional fuels with low-carbon alternatives. Moreover, the aircraft fleet turns over slowly: commercial jets typically remain in service for 20–30 years, meaning that even with rapid technological advancements, it would take decades to significantly reduce emissions from the global fleet. Airports, fueling infrastructure, and maintenance equipment are also built around the existing fossil fuel supply chain, further complicating the transition.

One potential near-term solution is to replace short-haul flights with high-speed rail (HSR), which offers a lower-carbon and often faster alternative for city pairs within 500–800 km. Expanding rail infrastructure—especially electrified, high-speed lines powered by renewable energy—can reduce the need for air travel where alternative options exist, while also offering benefits such as greater passenger comfort, station accessibility, and lower lifecycle emissions.

In the air, electric planes are being developed for short-haul, low-passenger routes, with several companies testing small aircraft for regional use. However, the current limitations of battery weight, energy density, and the need for rigorous safety testing and certification make electric propulsion viable only for very specific use cases in the near term.

The most viable solution for long-distance, commercial aviation in the short to medium term is the use of sustainable aviation fuels (SAFs). These "drop-in" fuels can be blended with conventional jet fuel and used in existing aircraft without requiring new engines or infrastructure. SAFs can be derived from a range of feedstocks, including used cooking oils, tallow, and other waste oils (limited in supply) oilseed crops like camelina and jatropha, sugarcane and corn ethanol converted through advanced processing, electrofuels, synthesized from captured CO_2 and green hydrogen. While SAFs can significantly reduce life-cycle emissions, challenges remain around feedstock availability, production costs, and scalability. Governments and airlines have begun investing in SAF development, but without strong mandates or incentives, adoption is likely to remain limited in the near future.

To reduce aviation emissions meaningfully, a multi-pronged strategy will be needed, combining greater investment in SAF development and deployment, expanding rail alternatives to replace regional flights, continuing research into electric and hybrid aircraft, supporting behavioral shifts such as reduced flying or modal shifts (Figure 11.3).

Progress has been uneven. As of 2024, construction is underway on a 171-mile stretch in California's Central Valley, between Merced and Bakersfield, with an estimated completion date in the early 2030s. The project has faced cost overruns, political pushback, and logistical delays, and its future beyond the Central Valley remains uncertain. Nonetheless, proponents argue that this initial segment is critical to demonstrating feasibility, building institutional capacity, and laying the groundwork for future extensions to the Bay Area and Southern California. If fully realized, the California HSR system could divert millions of trips from regional aviation and freeway traffic, support compact, transit-oriented urban development, significantly reduce per-passenger emissions, serve as a blueprint for other HSR efforts in the US. Critics point to the challenges of funding and project governance, but supporters see it as a long-term investment in a resilient, low-carbon transportation future—a key component of any strategy to reduce air travel demand within high-traffic corridors.

Shipping Industries

Global shipping industries are responsible for a few percent of global greenhouse gas emissions, and flows of international trade are destined to increase this into the future. That said, it is estimated that nearly 50% of global shipping is moving energy around. So if petroleum use diminishes, some of this will go away, but there could also be

Figure 11.3 High-speed rail in Europe connects much of the continent and makes train travel more popular. Source: Bernese media/High Speed Railroad Map of Europe/https://commons.wikimedia.org/wiki/File:High_Speed_Railroad_Map_of_Europe.svg/CC BY SA 3.0/last accessed on September 18, 2025.

increases in global liquefied natural gas (LNG) flows and of course many manufactured commodities like solar panels, wind turbines, and batteries and their supply chains are global cargo in shipping containers.

The key fuel powering global shipping in 2022 is heavy fuel oil, the fraction of hydrocarbon remaining after the gasoline and other light fractions are removed. It is a particularly dirty fuel. Research near the Ports of Los Angeles and Long Beach, California, shows environmental justice communities receive the brunt of cargo ship pollution, and more importantly the pollution from diesel trucks that take cargo away from ports to logistics centers for further sorting before moving to final customers.

Innovations in global shipping will include renewable liquid fuels, electrification, and the incorporation of sails to capture the wind where appropriate. Imagine a situation where instead of diesel trucks idling at warehouses and polluting communities, they were charging electric batteries from photovoltaics installed on the roof of the warehouses and logistics centers without burning a thing.

Highlights/Takeaways

- The challenges with mobility and transportation can be reduced to three main themes: (1) reduce the carbon intensity of fuels/energy, (2) improve the efficiency of converting energy to mileage (fuel economy, energy per distance), and (3) expand modes of mobility that encourage walking, biking, smaller vehicles, and telemobility.
- Electrification is largely seen as the solution to much of the mobility challenges for much of transportation. But aviation and some kinds of cargo transit may require more drop-in fuels.
- We still have to move stuff around! Transportation emissions from human products and services are on the rise and will also increase total transportation in the future.

Questions for Review

1. Think of the mobility and transportation problem as resting on three foundations: energy efficiency, decreased vehicle travel, and low-carbon fuels. What key policies or technological innovations would you pursue in each of these domains to achieve the lowest possible greenhouse gas emissions from the transportation sector? What are the main challenges in implementing these solutions across urban and rural contexts?
2. How does automobility shape the built environment, social inequality, and environmental outcomes? Reflect on how car-centric planning affects urban form, access to public space, and the distribution of benefits and harms across different populations. What alternatives to automobility are being pursued, and what barriers prevent their broader adoption?
3. What are the benefits and trade-offs of replacing short-haul flights with HSR? Consider climate impacts, infrastructure investment, regional equity, and energy use. What conditions need to be in place—geographic, political, and financial—for HSR to succeed as a viable low-carbon alternative to aviation?

Exercise 11.1

Supply Chains, Sustainability, and Environmental Justice

Biofuels and clean synthetic fuels used in ICEs, EVs, and hydrogen fuel cells (HFCs) vehicles could provide cleaner, lower carbon options for vehicle transportation in California than gasoline and diesel from petroleum. Yet, there are legitimate critiques regarding the sustainability and environmental justice issues in the supply chain for several of these emerging low carbon prime mover and fuel combinations. Biofuels researcher Dr. John Sheehan in 2009 wrote, we need to "avoid the never ending cycle of solving problems with new technology solutions from which new problems arise that need new technological solutions." So lets think ahead when it comes to this next technological fix and plan to make our future cars better.

Each student will be assigned one of these supply chains:

- **Electric vehicle batteries:** cobalt, graphite, lithium, manganese, nickel, bauxite
- **Fuel cells and electrolyzers:** hydrogen, platinum, iridium
- **Ethanol:** sugarcane, corn, cellulosic biomass
- **Biodiesel or renewable diesel:** soybean oil, palm oil, canola oil, jatropha, algae

In this assignment you have been hired as a sustainability advisor to the California Governor's office tasked with developing criteria to evaluate the sustainability and social justice dimensions of transportation options. The criteria you propose for sustainability and social justice will be used to benchmark and select which kinds of policies can be pursued to maximize ecological and social benefits of sustainable energy strategies. How can we pursue the prime-mover and fuel options above without reproducing the social and environmental impacts they currently cause (or may cause if they are not widely used)?

800–1,000-word written summary, 1.5 line spacing, 1 inch margins, 12 pt. Times font, word count listed, name/title/date at the top, with at least 5 peer reviewed research articles or high-quality white papers and reports. Your summary should (1) introduce the reader to the issue that low-carbon technologies will require new materials to be produced to meet 2050 climate action targets, (2) highlight the major socio-ecological impacts of the assigned batteries/fuel cells/fuel sources, and (3) propose a set of criteria, regulations, and/or best practices to improve, mitigate, or minimize these impacts so the state can meet the highest criteria for sustainable and socially just transportation fuels.

Address your letter to the Governor's Office and the Chair of the California Air Resources Board. Focus on what you deem to be the most important issues to screen for based on your understandings of the readings. Your description of impacts need not be exhaustive, but it should contain the major considerations and emphasize the ones most important. Describe qualitative criteria, certifications, and quantitative metrics (use tables if it is easier to present metrics this way), and make sure you consider people and the environment.

References

Beckmann, J. (2001). Automobility—a social problem and theoretical concept. *Environment and Planning D: Society and Space* 19 (5): 593–607.

Freemark, Y., Rajninger, T., and Hermans, A. (2023). Federal support could help reconnect communities divided by highways—but more funding Is needed. The Urban Institute. https://www.urban.org/urban-wire/federal-support-could-help-reconnect-communities-divided-highways-more-funding-needed (accessed January 4, 2025).

Golub, A., Marcantonio, R.A., and Sanchez, T.W. (2013). Race, space, and struggles for mobility: transportation impacts on African Americans in Oakland and the East Bay. *Urban Geography* 34 (5): 699–728.

Gould, K.A. and Lewis, T.L. (2016). *Green Gentrification: Urban Sustainability and the Struggle for Environmental Justice*. Routledge.

Henderson, J. (2006). Secessionist automobility: racism, anti-urbanism, and the politics of automobility in Atlanta, Georgia. *International Journal of Urban and Regional Research* 30 (2): 293–307.

Lark, T.J., Hendricks, N.P., Smith, A. et al. (2022). Environmental outcomes of the US renewable fuel standard. *Proceedings of the National Academy of Sciences* 119 (9): e2101084119. https://doi.org/10.1073/pnas.2101084119.

Lim, S., Dolsak, N., Prakash, A., and Tanaka, S. (2022). Distributional concerns and public opinion: EV subsidies in the US and Japan. *Energy Policy* 164: 112883.

Nardone, A., Chiang, J., and Corburn, J. (2020). Historic redlining and urban health today in US cities. *Environmental Justice* 13 (4): 109–119.

Nriagu, J.O. (1990). The rise and fall of leaded gasoline. *Science of the Total Environment* 92: 13–28.

Pulido, L. (2017). Rethinking environmental racism: white privilege and urban development in Southern California. In: *Environment* (ed. K. Anderson), 379–410. Routledge.

Riofrancos, T., Kendell, A., Dayemo, K. K, et al. (2023). "Achieving zero emissions with more mobility and less mining." Climate + Community Project, January. https://www.climateandcommunity.org/more-mobility-less-mining (accessed January 5, 2025).

Sheller, M. and Urry, J. (2000). The City and the car. *International Journal of Urban and Regional Research* 24 (4): 737–757.

Further Reading

Beckmann, J. (2001). Automobility—a social problem and theoretical concept. *Environment and Planning D: Society and Space* 19 (5): 593–607.

Chapter 12

Homes, Buildings, and Occupants

"There is an extraordinary irony in government efforts to promote more rational and sustainable forms of behavior in individuals in economies which are dominated by the imperative to make profit through capital accumulation, and which are inevitably addicted to unsustainable growth. Too much of the practice literature ignores political economic matters mediating practices."

—(Shove and Spurling 2013)

This chapter explores the built environments in which people live, work, and play, with a focus on how energy flows through buildings and how our infrastructure choices shape both local and global resource consumption. The built environment—including homes, offices, schools, factories, and recreational spaces—represents one of the most significant investments of natural resources by human societies. Some estimates suggest that up to 40% of all natural resources extracted by humans are ultimately embodied in buildings, from the concrete and steel used in construction to the energy required for heating, cooling, lighting, and powering appliances. We review where most energy in buildings is used, with a particular emphasis on thermal conditioning—the heating and cooling of indoor air. In many regions, especially those with variable climates, space heating and cooling account for the largest share of building energy consumption. We also explore how heat escapes from or enters buildings, examining the roles of insulation, ventilation, building orientation, window performance, and other design factors. Understanding these dynamics allows us to identify opportunities for improving energy efficiency, thereby reducing overall demand and lowering operating costs. This is particularly relevant when integrating renewable energy systems, such as rooftop photovoltaics, where energy efficiency can help better align supply and demand on-site. Buildings are not just individual energy users—they are nodes in larger urban systems. Cities concentrate population, capital, infrastructure, and innovation, and they also serve as engines of economic growth. Yet they are also major consumers of energy and materials, and often depend on extractive activities carried out in distant, rural, or global supply chain regions. The spatial organization of cities, the zoning of neighborhoods, the design of transportation networks, and the patterns of housing development all influence energy use and emissions at a systemic level. We also consider the social and environmental justice dimensions of building energy use. Who gets to live in energy-efficient homes? Who bears the cost of leaky, inefficient buildings and the burden of high utility bills? How do building codes, subsidies, and retrofitting programs address—or perpetuate—energy inequities? By examining policies, technologies, and cultural patterns shaping building energy systems, we can better understand how to design and retrofit spaces that are not only climate-responsive, but also equitable, resilient, and regenerative.

Building Resources and Envelopes

Since the Neolithic Revolution, humans have constructed protective shelters to shield themselves from the elements. Early dwellings—made from locally available materials like mud, thatch, stone, and wood—offered both physical protection and a means of maintaining thermal comfort in the face of changing weather. Over time, these structures evolved into complex buildings with increasingly specialized materials, mechanical systems, and thermal control strategies. In today's built environment, building envelopes—the physical barrier between indoor and outdoor spaces, including walls, roofs, windows, and insulation layers—play a central role in determining a structure's energy performance and comfort levels. Modern building envelopes are essential to understanding how energy is retained or lost in residential and commercial spaces. In cold climates, heat escapes through poorly insulated walls and single-pane windows. In hot climates, solar heat gain through the roof or unshaded windows increases the need for mechanical cooling. Because of this, the thermal performance of the envelope can significantly influence the heating, ventilation, and air conditioning (HVAC) loads, which account for a substantial share of building energy use globally.

But beyond their operational impacts, buildings also represent massive material investments. Construction consumes enormous amounts of natural resources—from metals and minerals like copper, aluminum, iron, and sand, to high-embodied-energy materials like cement, steel, and glass. The global building sector is responsible for nearly 40% of energy-related carbon dioxide emissions, when considering both operational and embodied carbon. As such, buildings are not only energy users—they are also material-intensive infrastructures whose design, construction, and maintenance shape global patterns of resource extraction.

Architectural historian Daniel A. Barber, in his book *Modern Architecture and Climate*, explores how earlier generations of architects addressed thermal challenges before the rise of mechanical air conditioning. Drawing on case studies from the early to mid-twentieth century, Barber shows that passive design strategies—such as natural ventilation, building orientation, shading devices, thermal mass, and cross-ventilation—were once widely used to create cooler and more comfortable indoor environments, even in hot and humid regions. These strategies demonstrate that architecture itself can function as a climate technology, and that design choices deeply influence energy demand. Today, many of these lessons are being revisited in the push for low-energy, climate-adaptive buildings. As energy systems decarbonize and global temperatures rise, the building envelope is becoming a critical site of intervention—not just for energy efficiency, but for comfort, resilience, and sustainability (Box 12.1). By investing in better building materials, smarter envelope designs, and locally responsive architecture, we can dramatically reduce energy consumption while improving the quality of life for occupants across diverse climate zones (Barber 2023).

Box 12.1 Heat Transfer Flows in Buildings

To understand how buildings retain or lose heat, it's essential to grasp the three primary mechanisms of heat transfer: convection, conduction, and radiation. **Convection** involves the movement of heat through fluids—most often air—and plays a major role in how warm or cool air circulates within a space. For example, warm air rising and cool air sinking sets up convective currents inside a room or between interior and exterior zones. **Conduction** refers to the transfer of heat through solid materials. Heat flows through walls, windows, roofs, and floors from warmer to cooler areas. Poorly insulated buildings allow heat to escape in winter and enter in summer through this process.

Radiation is the emission of energy in the form of infrared waves, which travel through space and are absorbed by surfaces. Sunlight passing through a window and warming the floor is a familiar example of radiant heat transfer. While our scientific understanding of thermal dynamics is relatively recent, the use of fuel-based heating in interior spaces dates back thousands of years. Early civilizations relied on the combustion of wood, dung, or other biomass to generate warmth, often in open hearths or pit fires. Archaeological evidence suggests that even Neolithic dwellings had designated spaces for indoor fires, reflecting the central role of heat in human settlement.

The use of coal as a domestic heating fuel can be traced back nearly 1,000 years, with more widespread adoption in Europe during the 13th and 14th centuries, especially in regions where forests were depleted or where coal was abundant and accessible. As the technology of home heating evolved, so too did the architecture of buildings. The development of hearths, fireplaces, and cast-iron stoves allowed more efficient use of fuel, while the invention and integration of chimneys dramatically improved indoor air quality by venting smoke away from living spaces and out of homes and villages.

These historical shifts in heating technology not only changed how people lived but also influenced urban form, building design, and energy infrastructure—laying the groundwork for the modern building systems we rely on today. (Smil 2021).

Electrify Everything, Solarize it

The call to "electrify everything" has emerged as a powerful guiding principle for decarbonizing the residential sector. This movement, now gaining momentum in cities, states, and utilities across the globe—particularly in parts of the United States and Europe—seeks to phase out all forms of indoor fossil fuel combustion, replacing them with efficient electric technologies powered increasingly by clean energy sources (Box 12.2)(Sterchele et al. 2018). At the core of this transition is the removal of appliances and systems that run on distillate fuels, natural gas, or propane, which are widely used for space heating, water heating, cooking, and even clothes drying. Mounting scientific evidence has shown that indoor combustion releases harmful pollutants, including nitrogen dioxide (NO_2), carbon monoxide (CO), and fine particulate matter (PM2.5)—all of which contribute to asthma, cardiovascular disease, and other serious health risks. In response, several jurisdictions, including Berkeley, New York City, and parts of Europe, have enacted policies to ban gas hookups in new residential construction, citing public health and environmental protection.

However, electrifying homes presents several technical, financial, and logistical challenges. One of the most difficult hurdles is the existing stock of residential appliances and infrastructure. Many households still rely on gas-powered furnaces, water heaters, and stoves, and swapping these out can be expensive and time-consuming. For example, heat pump water heaters (HPWHs)—a clean and highly efficient alternative—may require electrical upgrades, such as new wiring or dedicated circuits. These upgrades often fall outside the scope of a typical repair service, meaning that homeowners may feel locked into replacing broken gas appliances with more of the same, especially during emergencies. Additionally, this transition places new demands on the electrical grid, which must expand its capacity, flexibility, and reliability to support a growing number of electric homes. Significant investments in grid infrastructure, demand-side management, and distributed energy resources (like rooftop solar) will be necessary to ensure that the benefits of electrification can be realized equitably and affordably.

Box 12.2 Electrification Coupled with Solarization is the Zero Net Energy Pathway for California

California has long been a leader in building energy efficiency, with stringent codes and standards for insulation, windows, HVAC systems, and appliance performance. But as the state pushes deeper into decarbonization, its new frontier is pairing household electrification with solar power to achieve net-zero energy homes. Under this model, California aims to eliminate residential natural gas use by replacing combustion-based heating and cooking systems with electric alternatives—especially air-source heat pumps (ASHPs), induction cooktops, and HPWHs (Box 12.3). These appliances are not only more efficient but also compatible with rooftop solar systems, enabling households to self-generate a significant share of their electricity needs.

The goal is to reduce operating costs, lower emissions, and create a more resilient, grid-integrated housing stock. For new construction, the Title 24 Building Energy Efficiency Standards now require that most new homes be built solar-ready or include rooftop solar systems by default. For existing buildings, incentives and financing mechanisms are being expanded to support retrofits and appliance replacement. The state envisions a future where households are all-electric, solar-powered, and eventually equipped with on-site battery storage and electric vehicle chargers, turning homes into active participants in a decentralized energy system. In this way, electrification isn't just a technology shift—it's a pathway to redefine the relationship between homes, energy systems, and the climate.

Box 12.3 How a Heat Pump Works

Heat pumps operate using the same thermodynamic principles that govern air conditioners and refrigerators, specifically the processes of adiabatic compression and expansion. Rather than generating heat through combustion, heat pumps work by moving heat from one place to another using a refrigerant loop and a small amount of electricity.

Here's the basic cycle.

Evaporation & expansion (cooling phase): A liquid refrigerant absorbs heat from its surroundings—either outdoor air, ground, or indoor air, depending on the system. As it absorbs heat, the refrigerant evaporates and expands, turning into a gas. This process pulls in heat energy, even from cold air.

Compression (heating phase): The refrigerant gas is then compressed by a compressor, which raises its pressure and temperature significantly. This hot, high-pressure gas is now ready to release heat.

Condensation (heat delivery): The hot gas passes through a heat exchanger (also called a condenser), where it releases the stored heat into the indoor air or water system. As it releases heat, the refrigerant condenses back into a liquid.

Expansion valve: The liquid refrigerant then passes through an expansion valve, lowering its pressure and temperature, and the cycle begins again.

This process allows a heat pump to extract heat from cold outdoor air (or the ground) and use it to warm indoor spaces—even in temperatures well below freezing. Conversely, in cooling mode, the process runs in reverse: the pump extracts heat from inside the building and releases it outdoors, functioning like an air conditioner. Because heat pumps move heat rather than generate it, they can be much more efficient than traditional furnaces or boilers. For every unit of electricity consumed, a modern heat pump can deliver 2–4 units of heat energy—sometimes more—making them 200–400% efficient, depending on the conditions. Even when powered by fossil-fuel-generated electricity, heat pumps can reduce greenhouse gas emissions compared to directly burning gas or oil for heat.

Beyond the technical considerations, there are behavioral and equity dimensions to the electrification effort (Bednar et al. 2017). Some communities—particularly renters, low-income households, and residents of older homes—face higher barriers to adoption due to cost, ownership structures, or lack of access to rebates and programs. Overcoming these barriers will require targeted policies, incentive structures, and community-based solutions that prioritize the most vulnerable and underserved populations in the energy transition.

The electrify everything movement confronts several challenges, especially in households where combustion is still used. These typically include space heating, but fossil fuel combustion is also commonly used for heating water, cooking, and even some clothes dryers use natural gas or propane. This shift to electricity will require a significant expansion of our electricity grid as discussed in Chapter 10. But it also will require the replacement of many older heating and hot water systems that rely on combustion. These can take a long time to swap out at a home. Hot water heaters also provide a service—warm showers—that people want, and when they break is when people usually decide to upgrade them. The challenge is that some of these devices might need home-side upgrades, for example, running electricity to them, and those projects are not typically the type of project that can be done by the water heater repair service. So this is one way customers get locked into the same equipment.

True to our critical energy studies approach, this is an area ripe for investigation, and questions about how electrification is framed, contested, and resisted are all topics that critical approaches can help us understand.

Social Construction of Comfort and the Role of Behavior

The question of household comfort is far more complex than it might first appear. It is shaped by a combination of lived experience, climate, cultural norms, and the availability and affordability of energy. In colder climates, a lack of heating often signals energy poverty—a condition experienced by households living in cold, damp, poorly insulated, and under-ventilated buildings. These physical conditions not only affect comfort but are also closely linked to respiratory illness, mental health challenges, and social isolation.

In hotter climates, the issue is often a lack of access to cooling, which can lead to dangerous indoor temperatures, especially during heatwaves. In such contexts, the air conditioner becomes not only a symbol of thermal comfort but also a marker of social status, reflecting a household's ability to control its environment and signal participation in middle-class consumer norms.

Sociologist Dr. Elizabeth Shove and colleagues have critically examined the cultural construction of comfort, emphasizing that comfort is not purely physiological, nor is it simply "in our heads." Rather, it is a socially and historically produced expectation, shaped by technologies, architectural design, routines, and media representations. For example, the modern expectation that indoor spaces should maintain a constant temperature year-round—usually around 21 °C (70 °F)—is relatively recent and deeply tied to air conditioning, heating systems, and the design of modern buildings. Shove's work encourages a shift from thinking of energy demand as a fixed need to seeing it as emergent from social practices. This has important implications: if energy use patterns are shaped by habit and infrastructure, then households offer real opportunities for transforming energy systems—not just through technology, but through behavioral change and reimagined norms of comfort. While this argument may challenge libertarian or techno-centric perspectives, it opens the door to subtle, socially aware interventions that respect diversity in household routines and values.

This sociological lens aligns with broader research on behavioral responses to energy conservation. For instance, evidence shows that voluntary environmental actions by firms can induce positive responses among citizens, who are then more likely to support favorable policies—such as public procurement contracts, tax breaks, or lenient regulatory enforcement—for those firms (Kolcava et al. 2021). In other words, good environmental behavior can be socially contagious.

In the residential sector, demand-side behavior will play a vital role in future energy systems. Some existing tools, such as time-of-use pricing, are already designed to influence when people consume electricity. Similarly, the success of demand response programs often hinges on economic incentives—rewards for reducing demand during peak periods. Yet economics alone may not be enough. Sociologists and behavioral scientists have shown that altruism, fairness, and community identity can also be powerful motivators for participation in energy efficiency programs (Dietz 2015). Understanding household comfort as a social and cultural phenomenon, rather than merely a technical or economic one, opens up new possibilities for inclusive, flexible, and community-centered approaches to energy transition. It challenges us to think not just about what technologies we adopt, but about how we live, and how those living patterns shape and are shaped by energy use.

Heat Pumps

Heat pumps are widely considered one of the most effective technologies for electrifying space and water heating—two of the largest sources of energy consumption and greenhouse gas emissions in the residential and commercial building sectors. Their importance in climate mitigation stems from both their high efficiency and their compatibility with decarbonized electricity grids. A heat pump is essentially an air conditioner in reverse. While air conditioners remove heat from indoor air and release it outside, heat pumps can reverse this process to extract heat from outdoor air (even when it's cold) and transfer it indoors. Heat pumps rely on the principles of thermodynamics, particularly adiabatic compression and expansion, to move heat rather than generate it through combustion.

To illustrate, if you've ever used a can of spray paint or hairspray, you may have noticed that the can gets cold when sprayed. That's due to adiabatic expansion—when a compressed fluid rapidly expands, it absorbs heat from its surroundings, causing the can to cool. The reverse process—adiabatic compression—is what makes a heat pump work: when a refrigerant is compressed, it increases in temperature, and this heat can then be transferred to air or water inside the building. This same principle explains natural phenomena like the Santa Ana or Diablo winds in California, and the Chinook winds of the Pacific Northwest and Rockies. These downslope winds are compressed and heated as they descend in elevation, resulting in warm, dry conditions that can exacerbate wildfire risks.

Unlike traditional heating systems that burn fuel to generate heat, heat pumps move heat from one place to another. This makes them much more efficient than combustion-based systems. Typical ASHPs can achieve coefficient of performance (COP) values between 2.0 and 4.0, meaning they can deliver 200–400% more heat energy than the electrical energy they consume. In contrast, even the most efficient natural gas furnaces max out around 95–98% efficiency, as combustion devices are limited by the laws of thermodynamics to less than 100% efficiency.

Because of this efficiency, heat pumps are not just good for reducing household energy use—they're also a critical climate solution. Even when powered by natural gas-generated electricity, a heat pump can still emit fewer greenhouse gases than directly burning natural gas for heat. And as the electricity grid becomes cleaner, the emissions savings grow even greater. This is especially true in regions powered by renewables like wind, solar, or hydropower, where heat pumps can operate with near-zero carbon emissions.

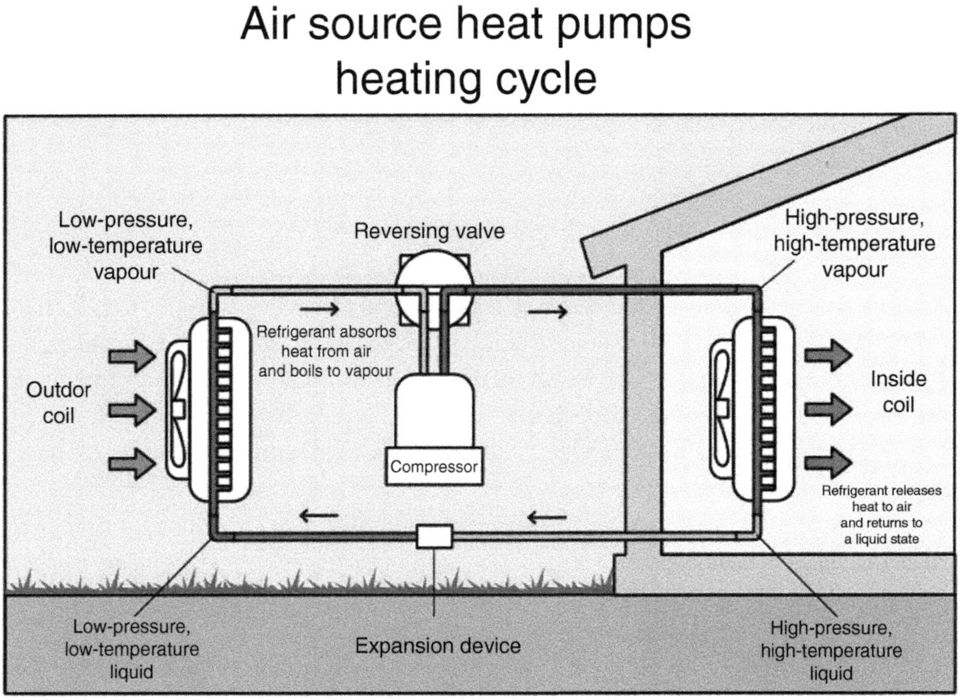

Figure 12.1 Air-sourced heat pumps typically have an outdoor compressor and an indoor air handling unit. Source: U.S. Department of Energy /Air-Source Heat Pumps/https://www.energy.gov/energysaver/air-source-heat-pumps /Public Domain/last accessed on September 18, 2025.

Heat pumps are available in several configurations:

- ASHPs extract heat from outdoor air and are the most common and affordable.
- Ground-source or geothermal heat pumps (GSHPs) draw heat from the more stable underground temperature and can be even more efficient but require higher upfront installation costs.
- HPWHs work similarly but are optimized for heating domestic hot water.

As building codes shift toward electrification, many regions and utilities are offering rebates and incentives to accelerate heat pump adoption. Retrofitting homes with heat pumps can not only lower emissions but also reduce long-term energy bills, especially when combined with energy-efficient building envelopes and on-site solar energy generation (Figure 12.1).

Materials: Wood, Concrete, Glass

The materials we use to construct our buildings—whether wood, concrete, glass, steel, or composites—play a central role in shaping the environmental footprint of the built environment. Globally, the extraction and use of materials for construction is one of the most resource-intensive sectors of the economy. In 2019, approximately 30 gigatons of nonmetallic minerals were extracted worldwide, with a staggering 95% destined for construction purposes (Hertwich et al. 2019). The bulk of this is gravel, sand, and limestone, used for roads, concrete, and cement, as well as massive infrastructure projects like bridges, tunnels, dams, and urban expansion.

Concrete, and more specifically cement (its key binding ingredient), is one of the most used construction materials on Earth—and one of the most carbon-intensive. Cement production alone accounts for roughly 7–8% of global CO_2 emissions, primarily due to the calcination of limestone and the high temperatures (around 1,450 °C) required in kilns, which are often powered by fossil fuels. While concrete is strong, durable, and relatively inexpensive, its environmental impacts are significant. That said, concrete also offers thermal mass benefits, which can reduce operational energy use in well-designed buildings.

Researchers and engineers are exploring low-carbon alternatives such as carbon-cured concrete, cement substitutes (like fly ash or slag), and recycled aggregate materials to reduce the material's climate impact. Still, reducing emissions from concrete will require both technical innovation and a rethinking of how and when we use it.

In contrast to high-emission materials like concrete and steel, wood is gaining renewed attention as a sustainable construction material. When sourced from sustainably managed forests, wood stores atmospheric carbon in its cellular structure—essentially locking carbon away for the lifetime of a building. Modern engineered wood products such as cross-laminated timber (CLT) and glulam beams are increasingly used in mid-rise and even high-rise construction, offering lightweight, strong, and low-carbon alternatives to traditional concrete and steel frames. Architects, engineers, and environmental advocates have embraced wood as part of a broader strategy for decarbonizing buildings, but its benefits are contingent upon responsible forestry practices, long-term building life, and protection from degradation or fire. If burned or left to decay at the end of life, stored carbon is released back into the atmosphere.

Glass is another ubiquitous material in modern architecture—used for windows, curtain walls, skylights, and facades. Glass production is also energy-intensive, involving high-temperature furnaces to melt silica sand and other materials. While clear, open facades are valued for aesthetics and daylighting, poorly insulated or non-coated glass can lead to significant energy losses due to heat gain or loss through windows. Technological advances in high-performance glazing, low-emissivity (low-e) coatings, and vacuum-insulated glass have helped reduce these impacts, but widespread adoption varies by region, cost, and building type.

As buildings age and are demolished, they generate enormous quantities of construction and demolition (C&D) waste. Materials like steel, aluminum, and concrete are commonly recycled or reused due to their high economic value and established recycling infrastructure. In contrast, materials like wood, glass, insulation foam, drywall, and plastic composites are often landfilled or incinerated, representing lost resources and embedded emissions.

Efforts to improve circularity in the building sector include deconstruction practices that preserve reusable components, design for disassembly (DfD), enabling future reuse or easier recycling, and urban mining, which treats buildings as resource banks of recoverable materials.

These approaches not only reduce waste and resource extraction but also align with broader goals of sustainable construction and material efficiency. As building codes evolve to incorporate life cycle carbon accounting, and as cities confront material scarcity and climate constraints, choosing the right materials—and planning for their reuse—will become central to the future of climate-resilient, resource-smart building design.

Building stocks take forever to turn over. Look around any major city or metropolitan area and note how many buildings are 80 or 100 or more years older. The importance of adopting building codes that are implemented sooner rather than later cannot be underestimated. The lifetime of buildings is such that there may never be a meaningful renovation in a cost-effective way for some buildings.

Many area utilities have weatherization assistance programs available to the poor, but programs often go undersubscribed for a variety of reasons. Energy justice scholar Tony Reames and colleagues have found energy efficiency programs can be more expensive in environmental justice communities (Reames et al. 2018).

Renters and Low-Income Buildings

Rising electricity and heating costs have become a growing concern in the transition to clean energy—especially for low-income households, who are often the most vulnerable to fluctuations in utility prices. These households frequently live in older or poorly maintained buildings, often as renters, and are at the mercy of landlords who lack incentives to invest in modern energy efficiency upgrades. This situation creates a well-documented issue known as the "split incentive" problem, where the landlord owns the property, but the tenant pays the energy bills. Because landlords do not directly benefit from lower energy costs, they often neglect upgrades such as improved insulation, air sealing, energy-efficient appliances, or modern heating systems. As a result, renters end up paying disproportionately high energy bills for homes that waste energy—while remaining locked out of the cost-saving and comfort benefits of energy efficiency improvements. One solution to this problem could be to help renters and low-income households become "prosumers" of energy, where they produce and consume electricity, reducing their electricity burden while generating cleaner energy (Figure 12.2).

Research from the Climate + Community Project highlights that low-income renters face some of the highest energy burdens in the country—defined as the percentage of household income spent on energy bills. In some cases, this burden exceeds 20% of monthly income, especially in under-insulated buildings, in extreme climates, or where electric baseboard or fuel-oil heating is used. Compounding this, these households are less likely to access rebates, incentives, or solar programs, which often prioritize homeowners or rely on credit and capital that renters lack. The consequences are wide-ranging. Families in energy-inefficient rental housing may be forced to choose

Figure 12.2 Prosumers produce and consume electricity, instead of just consuming it from the utility. Source: U.S. Department of Energy/Consumer vs Prosumer: What's the Difference?/https://www.energy.gov/eere/articles/consumer-vs-prosumer-whats-difference / Public Domain/last accessed on September 18, 2025.

between heating and other essentials, rely on dangerous coping strategies (like using ovens for heat), or suffer from chronic under-heating or over-heating of living spaces. These conditions pose health risks, particularly for children, the elderly, and individuals with pre-existing conditions. Furthermore, the persistence of energy-inefficient rental housing slows national decarbonization efforts and deepens energy inequality.

Highlights/Takeaways

- Buildings are both energy systems and material systems. They consume enormous amounts of energy for heating, cooling, and lighting, while also embodying vast quantities of materials—such as concrete, steel, glass, and wood—that carry their own carbon and resource impacts. Nearly 40% of global energy-related CO_2 emissions are tied to the construction and operation of buildings.
- Thermal comfort is culturally shaped, not just technically achieved. Comfort is a social expectation, formed through historical routines, infrastructure, and media. Re-examining cultural norms—such as expecting constant indoor temperatures—can reveal new opportunities for energy conservation and social resilience (Shove and Spurling 2013).
- Electrification of buildings is essential but uneven. Replacing gas-powered appliances with electric heat pumps, stoves, and water heaters is key to decarbonization. However, this shift faces barriers including outdated infrastructure, cost, and inequitable access—especially for low-income renters, who often live in inefficient buildings and face the "split incentive" problem.

- Policy and design must confront systemic inequalities in housing and energy access. Energy justice requires more than better technology; it demands inclusive programs, targeted retrofits, and attention to who benefits from building upgrades. Without interventions, the clean energy transition risks deepening housing and utility inequalities for vulnerable populations.

Questions for Review

1. What are the three ways that heat leaves a building?
2. What are key challenges facing energy efficiency and electrification upgrades in low-income buildings? What is the split incentive problem? And how can we solve it?
3. How do building materials and design choices influence both operational and embodied energy use? Consider how materials like concrete, wood, and glass contribute to a building's carbon footprint during construction and over its lifetime. How might design strategies like passive heating/cooling or high-performance insulation reduce energy demand?
4. What are the social and structural barriers to energy-efficient housing, particularly for renters and low-income households? Explore how the "split incentive" problem and housing inequality intersect with energy justice. What kinds of policy interventions could address these barriers?
5. **In what ways are comfort and energy use socially constructed?** Drawing from Elizabeth Shove's work, how do norms, habits, and technologies shape our expectations for indoor temperatures and energy services? How might shifting these norms reduce energy demand?
6. **Why are heat pumps considered a transformative technology for residential decarbonization?** Discuss how heat pumps work, their efficiency compared to combustion-based systems, and the infrastructural or behavioral challenges that may affect their adoption.

Exercise 12.1

Reflect on the role of behavior, social norms, and cultural expectations in shaping how energy is used in buildings, especially in the residential sector. Describe a few common household behaviors or expectations (e.g., preferred indoor temperatures, appliance usage patterns, reliance on air conditioning or heating).

1. Identify examples of how **social or cultural norms** support or inhibit energy-saving practices in homes or workplaces. Think about:
   - Media portrayals of comfort (e.g., thermostats, home design)

- Peer pressure, status, or routines
- Barriers to behavior change (e.g., convenience, aesthetics, housing stock)

Drawing from the chapter or external research, describe **two behavioral interventions** that have succeeded in reducing household energy use (e.g., time-of-use pricing, feedback apps, thermostat nudges, energy use competitions). Reflect: Where are the **opportunities** for change? What social dynamics could be **leveraged or shifted** to support lower-carbon behaviors? Deliverable: 500–700 word reflection or 5-minute recorded presentation.

Exercise 12.2

Use a basic spreadsheet tool (e.g., Excel or Google Sheets) to track and calculate your household's annual electricity use and greenhouse gas (GHG) emissions from plug-in devices.

1. Instructions:
   - Create a spreadsheet with the following **columns**:
   - Device/Appliance
   - Power (watts or kW)
   - Daily Usage (hours)
   - Annual Usage (hours)
   - Annual Energy Use (kWh)
   - Emission Factor (kg CO_2e/kWh)
   - Annual GHG Emissions (kg CO_2e)
2. Choose at least 5–10 devices from your home (e.g., TV, fridge, computer, microwave, heat pump, lights).
3. Use appliance labels, manuals, or online databases (e.g., EnergyGuide, ENERGY STAR, manufacturer sites) to fill in missing power values. For emissions, assume an emission factor of 0.4 kg CO_2e/kWh, or use a regional grid factor if you want to personalize it. Summarize your total household electricity use and emissions from the devices you tracked. Identify three ways to reduce usage, swap devices, or shift behaviors to lower your emissions.

Deliverable: Completed spreadsheet and a short (300-word) summary of your findings and proposed changes.

Exercise 12.3

Analyze how housing type, tenure, and income level affect access to energy efficiency, and identify policy interventions that could address energy injustice. Choose two different housing types in your region or a city of your choice: One should be owner-occupied, the other rental or low-income housing. Look up or estimate the average energy efficiency, monthly energy bills, and any available rebate or assistance programs. Do quick research. What are the barriers to energy upgrades in rental or low-income housing? (Consider the "split incentive" problem, structural issues, or financing limitations.). Are there utility, city, or federal programs (e.g., weatherization assistance, heat pump incentives) that serve low-income renters? Reflect on what the equity implications are of failing to address building energy inequality? Propose two policy or design interventions that would make electrification and energy efficiency more accessible and fair. Deliverable: 750–1,000 word case comparison and policy proposal or video.

References

Barber, D.A. (2023). *Modern Architecture and Climate: Design before Air Conditioning*. Princeton University Press.

Bednar, D.J., Reames, T.G., and Keoleian, G.A. (2017). The intersection of energy and justice: Modeling the spatial, racial/ethnic and socioeconomic patterns of urban residential heating consumption and efficiency in Detroit, Michigan. *Energy and Buildings* 143: 25–34.

Dietz, T. (2015). Altruism, self-interest, and energy consumption. *Proceedings of the National Academy of Sciences* 112 (6): 1654–1655.

Hertwich, E.G., Ali, S., Ciacci, L. et al. (2019). Material efficiency strategies to reducing greenhouse gas emissions associated with buildings, vehicles, and electronics—a review. *Environmental Research Letters* 14 (4): 043004.

Kolcava, D., Scholderer, J., and Bernauer, T. (2021). Do citizens provide political rewards to firms engaging in voluntary environmental action? *Journal of Cleaner Production* 279: 123564.

Reames, T.G., Reiner, M.A., and Stacey, M.B. (2018). An incandescent truth: disparities in energy-efficient lighting availability and prices in an urban U.S. county. *Applied Energy* 218: 95–103. https://doi.org/10.1016/j.apenergy.2018.02.143.

Shove, E. and Spurling, N. (2013). *Sustainable Practices: Social Theory and Climate Change*. Routledge.

Smil, V. (2021). *Grand Transitions: How the Modern World was Made*. Oxford University Press.

Sterchele, P., Palzer, A., and Henning, H.-M. (2018). Electrify Everything?: Exploring the Role of the Electric Sector in a Nearly CO2<\/sub>-Neutral National Energy System. *IEEE Power and Energy Magazine* 16 (4): 24–33. https://doi.org/10.1109/MPE.2018.2824100.

Further Reading

Shove, E. and Walker, G. (2010). Governing transitions in the sustainability of everyday life. *Research policy* 39 (4): 471–476.

Shove, E., Pantzar, M., and Watson, M. (2012). *The Dynamics of Social Practice: Everyday Life and how it Changes*. Sage.

Chapter 13

Industrial Ecology and Manufacturing

Take. Make. Waste. This simple phrase captures the foundational material flow of nearly all of the global economy—an extractive, linear model in which we remove resources from the Earth, transform them into goods, and ultimately discard them as waste. According to one study, "[h]umanity has deposited 2,500 gigatons of waste and emissions into the environment since 1900," illustrating the immense scale of material throughput and its consequences.

Industry and manufacturing leave a particularly large footprint because they not only rely on vast quantities of material feedstocks, but also require intensive energy inputs to produce the goods that underpin modern life—from concrete and steel to fertilizers, plastics, and electronics. The industrial sector accounts for approximately 25% of global greenhouse gas emissions annually, making it one of the highest-priority areas for climate mitigation.

Some industrial processes, like those used to produce iron and steel, are considered among the hardest to decarbonize due to the high temperatures and chemical reactions involved. Others, like aluminum production, may be more amenable to electrification or low-carbon alternatives.

Yet beyond decarbonization lies a deeper and more transformative question: How can we reduce material use altogether? Concepts like dematerialization, the circular economy, and industrial ecology push us to rethink production systems not just in terms of efficiency, but in terms of regeneration—designing systems that reuse materials, reduce extraction, and prevent waste. These approaches envision an economy that loops rather than leaks, where waste becomes a resource, and manufacturing aligns more closely with ecological principles rather than depleting them.

Rethinking Waste: Industrial Ecology and Circular Economies

Industrial sectors are some of the most material- and energy-intensive components of the global economy. "Industry" typically refers to the manufacturing and processing activities that support modern life—including petroleum refining, chemical production, construction, and heavy manufacturing. The most emissions- and energy-intensive industries are often grouped under the category of "hard-to-abate" sectors, which include iron and steel, cement, chemicals and plastics, aluminum, and petroleum refining. These sectors together account for a significant share of global greenhouse gas emissions, not only because of their fuel use and process heat demands, but also because of the inherent emissions released during chemical transformations, such as the conversion of limestone into clinker in cement production or coke in steelmaking.

Addressing the environmental impacts of these sectors requires more than just cleaner energy. It calls for rethinking how materials flow through industrial systems, shifting from linear "take-make-waste" models to more circular and regenerative production systems (Box 13.1). This is the central challenge of industrial ecology—a field that examines how we can make industrial systems behave more like ecosystems, where waste from one process becomes the nutrient or feedstock for another.

A key concept in this shift is valorization—the process of giving value to waste by converting it into a useful product. For example, mine tailings (the waste material left after ore processing) can sometimes be processed into materials for cement, bricks, or road base, reducing the need for virgin raw materials and avoiding the environmental impacts of waste disposal. Similarly, industrial byproducts like fly ash, slag, or CO_2 emissions are increasingly being used in secondary industries such as green concrete production, chemical synthesis, or algae-based fuels.

The circular economy offers a conceptual framework for operationalizing these shifts, emphasizing the need to eliminate waste, reduce material intensity, reuse products and components, and recycle materials—in that order of priority. This approach is often illustrated through the waste hierarchy, which guides both industry and policy by placing the greatest value on prevention and reuse before considering recycling and disposal. In contrast to conventional efficiency frameworks that focus narrowly on minimizing harm, the circular economy aims to redesign systems from the ground up, transforming waste into a design flaw rather than an inevitability.

A number of foundational thinkers have shaped the principles that underpin circularity. Michael Braungart and William McDonough's *Cradle to Cradle* (McDonough and Braungart 2010) book and design philosophy emphasizes creating closed-loop material cycles, where materials are either biological nutrients (that safely return to nature) or technical nutrients (that circulate indefinitely through industrial systems). Their certification system encourages product design that supports disassembly,

Box 13.1 Rethinking Waste: Circular Economy Principles and Practice

The circular economy is an alternative to the traditional linear model of "take-make-waste," where resources are extracted, manufactured into products, and eventually discarded. Instead, the circular economy seeks to design waste out of the system entirely, inspired by biological cycles in nature where materials are continually reused and regenerated. According to the Ellen MacArthur Foundation, a leading voice in circular economy thinking, the model rests on three core principles: (1) Design out waste and pollution, (2) Keep products and materials in use, and (3) Regenerate natural systems. This model challenges the dominant logic of planned obsolescence and disposable culture. Rather than simply recycling at the end of a product's life, circular design encourages rethinking the entire product lifecycle—from modular phones that can be upgraded rather than discarded to leasing models that incentivize durability and repair to biodegradable packaging made from agricultural waste. A notable example is the Dutch company Fairphone, which produces smartphones designed to be modular, repairable, and ethically sourced. Users can easily replace a cracked screen or aging battery with simple tools, extending the phone's lifespan and reducing e-waste. Fairphone also prioritizes conflict-free minerals and transparency in its supply chain, aligning its operations with circular and ethical production principles. Circular economy thinking pushes beyond efficiency to ask: How can economic systems work more like ecosystems? How do we eliminate the very concept of waste? As climate change, material scarcity, and global waste crises converge, the circular economy offers a compelling blueprint—not just for making things better, but for making better things. At the same time, some of these ideas are vulnerable to greenwashing, and claims should be substantiated. They are often also led by private voluntary governance, which brings its own questions about efficacy.

repair, and recovery. Janine Benyus's *Biomimicry* (Benyus 1997) encourages innovation that learns from and emulates natural systems, where everything produced is functionally integrated, multi-purpose, and waste-free. Nature, in this framing, becomes a mentor for sustainable design. Walter Stahel's *Performance Economy* (Stahel 2019) shifts attention from products to services and longevity, suggesting that we can meet human needs more efficiently by leasing or servicing products (like tools, cars, or appliances) rather than selling them outright. This incentivizes durability, repairability, and component reuse.

Industrial ecology, as an academic and practical field, envisions networks of interconnected industries that exchange energy and materials in ways analogous to food webs in ecosystems—minimizing net resource extraction and maximizing material circulation. These frameworks are increasingly important as we transition to cleaner

Table 13.1 Selected materials used to produce renewable and low energy devices and infrastructures.

Technology	Inventory of materials and commodities
Photovoltaics	Silicon, glass, aluminum, silver, copper, steel, polymers, electronics, indium, gallium, tellurium, germanium, cadmium, and zinc
Wind turbines	Steel, copper, aluminum, iron, cement, glass-reinforced plastics, plastic resins, dysprosium, neodymium, praseodymium, and electronics
Electric vehicles	Steel, aluminum, glass, copper, metal composites, fiberglass, rubber, ceramics and magnets, polymers, lead-acid batteries, lithium-ion batteries, electronics
Lithium-ion batteries	Steel, iron, nickel, copper, manganese, lithium, cobalt, graphite, zinc, polymers, and electronics

energy systems. As wind turbines, solar panels, batteries, and electric vehicles replace combustion-based technologies, we will require huge volumes of metals, minerals, glass, and rare earths. Without circular approaches, this transition risks replicating the extractive logic of fossil fuels with a new wave of mining and waste generation. The path forward is not just about reducing emissions—it's about designing an economy that regenerates ecosystems, respects planetary boundaries, and creates long-term material resilience. Circular economy principles, when applied rigorously, offer one of the most promising routes to achieving that vision (Table 13.1).

Critical Industry Studies

For much of this book we have been talking about critical energy studies, and in the same spirit critical industry studies would take a similar approach. If industrial ecology helps us understand how material and energy flows move through production systems, then critical industry studies ask why those systems are built the way they are, who benefits, and who is harmed. Rooted in traditions of critical political ecology, political economy, and socio-technical systems analysis, this emerging perspective moves beyond questions of efficiency or innovation to confront the power relations, historical legacies, and socio-environmental injustices embedded in the "hidden abodes" of industrial production (Huber 2017).

Where traditional industrial ecology might model how one factory's waste can become another's input, a critical industry study might ask whose land was used for the factory or the mine? What labor conditions shape the supply chain? How are environmental burdens distributed across race, class, and geography? Which industries get labeled as "green," and who decides?

Drawing from critical political ecology, scholars have long emphasized that environmental change is inseparable from social and political processes. In this view, industrial systems do not simply emerge from technological progress—they are actively shaped by colonial histories, state subsidies, trade regimes, and corporate lobbying. The global concentration of heavy industry in the Global South, for instance, is not a natural outcome of efficiency but often the product of uneven development, labor exploitation, and environmental deregulation.

Critical industry studies also build on critiques of sustainability discourse, which can obscure or depoliticize underlying drivers of ecological harm. As scholars like Melissa Leach and James Fairhead have shown, even well-intentioned green reforms can become vehicles for dispossession, green grabbing, or technocratic control when implemented without democratic engagement or attention to context (Fairhead et al. 2012). Similarly, in the manufacturing and energy sectors, efforts to "green" production often prioritize efficiency and innovation over justice, reparations, or structural change. From a critical industrial ecology standpoint, then, it's not enough to design systems that recycle or reduce waste (Box 13.3). We must also interrogate the political economy of industrial land use, the militarization of extractive frontiers, the racialized division of labor, and the institutional logics that normalize overproduction and planned obsolescence.

This perspective aligns with the work of Joshua Lepawsky (Lepawsky 2018), who examines how narratives of circularity in global e-waste systems often obscure the displacement of harm to informal laborers in the Global South. It also echoes Max Liboiron's work on pollution and harm (Liboiron 2021), which highlights how industrial systems rely on slow violence, making harm appear invisible, gradual, or acceptable. Ultimately, critical industry studies push us to ask not just how to make industry more sustainable, but how to democratize industrial decisions, revalue labor and care, and reimagine economies that are organized around sufficiency, equity, and ecological reciprocity—not just profit and throughput.

Circular Economy and Sustainable Materials Use in Global Supply Chains

Supply chains are the sequential linkages of firms and processes that connect natural resources to finished products. While some disciplines use synonymous terms like commodity chains, production networks, or the French term "filière," the idea is the same: industries are not isolated—they are embedded in global systems of extraction, transformation, logistics, labor, and consumption (Crang et al. 2013). In the context of sustainability, supply chains are not just logistical—they are ecological and political spaces. They begin at sites of extraction, such as mines, forests, and oil fields—where some of the most visible environmental impacts and community conflicts occur. These impacts include deforestation, tailings ponds, water depletion, and land dispossession. From there, materials move through refineries and manufacturing plants, where energy is consumed, and pollutants are emitted. They eventually become part of consumer goods and infrastructure and, after use, become waste or enter secondary markets through reuse or recycling.

For renewable energy, understanding the supply chain is crucial. These systems—whether solar photovoltaics, wind turbines, or electric vehicles (EVs)—rely on metals and materials sourced from all over the world, often from countries with weak environmental protections or histories of labor and human rights violations. For example, large portions of cobalt, lithium, and rare earth metals come from regions with artisanal or informal mining sectors, raising concerns about child labor, unsafe conditions, and inequitable economic benefits. As Sovacool et al. (2020) argue, sustainable energy transitions must be critically examined across the full length of the supply chain, especially to ensure that new forms of exploitation and extractivism do not replace old ones (Sovacool, Ali, et al. 2020).

Steel and coal-derived coke

Steel is a pillar of industrial society and a major contributor to climate change. It is produced largely in blast furnaces fueled by coal-derived coke, which acts both as a fuel and a chemical reducing agent. These are among the most energy-intensive facilities on the planet, often using more than three million tons of coal equivalent per year (Smil 2002). Steel and iron today are still largely made in blast furnaces. We are already seeing announcements by steel manufacturers to become zero carbon by directly purchasing renewable energy. A more ultimate objective would be the replacement of processes that directly cause greenhouse gas emissions. The challenge is not just energy use, but also direct process emissions from the reduction of iron ore. While some steelmakers are pursuing renewable electricity and green hydrogen, widespread transformation will require deep structural changes (Figure 13.1).

Figure 13.1 A blast furnace uses high temperatures to make steel. Source: Goodwin Steel Casting / Flickr.

Aluminum

Aluminum is also highly energy-intensive, though it relies mostly on electricity rather than combustion. That makes it more readily decarbonizable, provided low-carbon electricity is available. Aluminum was the fourth-largest industrial emitter globally in 2023. Circular economy practices like recycling aluminum scrap can reduce emissions by up to 95% compared to primary aluminum production. However, the increasing use of composite materials and plastic coatings is making it harder to recover clean scrap. Electrifying aluminum smelting and ensuring clean electricity inputs will be crucial.

Copper

Copper is the circulatory system of the energy transition, essential for power transmission, motors, and electronics. Some forecasts estimate that decarbonization could require 2–3 times more copper than is currently mined. Yet copper mining comes with significant environmental costs, especially in terms of land use, water pollution, and toxic tailings. Moreover, copper ore often contains small amounts of other metals—indium, gallium, cobalt, tellurium, silver, and molybdenum—that are extracted as co-products. This makes copper mines a linchpin for multiple critical materials, though it also complicates waste and pollution management. China dominates the refining and processing of many of these metals, while 60% of copper reserves lie in Chile, Peru, Australia, Mexico, and the United States.

Concrete and cement

Cement is the second most used substance on Earth after water and a cornerstone of modern construction. Its main component, clinker, is made by heating limestone ($CaCO_3$) to very high temperatures, releasing CO_2 both from fuel combustion and chemical reaction: The main input for cement is limestone, but also calcium and silicon are also needed. So the supply chains for cement reach back to limestone quarries, but also to sand, another key input. Cement is critical to construction industries for everything from residential homes to bridges, roads, sidewalks, and commercial buildings. The technique most commonly used to produce cement in kilns is not only energy intensive, but also emits CO_2 by chemical reaction. Between energy consumption and chemical reaction, energy inputs, and each ton of cement results in about a ton of CO_2. In fact, estimates of global CO_2 emissions put cement at around 6% of overall greenhouse gas (GHG) emissions.

The stoichiometry of cement processing that produces CO_2 emissions is

$$CaCO_3 + \text{heat} \rightarrow CaO + CO_2$$

There are some more short-term changes to cement production that could reduce emissions. Fuel switching to less carbon-intensive fuels could reduce emissions.

Some efforts are pursuing improved energy efficiency for this very heat-intensive energy through technological change. Wet kilns are less efficient than dry kilns, so some changes may be needed to cement-making equipment. The standard clinker technologies for Portland cement—the most common kind—are nearing the limits of their ability to get more efficient. Because cement plants are capital-intensive and long-lived, turnover is slow. But retrofits, process innovation, and alternative binders like calcium sulfoaluminate cements could help reduce emissions. Occupational health also remains a concern, particularly cement kiln dust exposure.

Other air pollutants from cement production include persistent organic pollutants, dioxins, heavy metals, sulfur dioxide, and particulate matter. Several cement companies across the globe are experimenting with means to capture the carbon emissions from cement production. Some companies are even attracting venture capital. Without better carbon pricing or voluntary efforts to invest in carbon capture, it is unlikely that cement industries will become low carbon on their own. Growth of cement is still slated to continue to rise in coming decades. We are not near peak cement, so it is critical to produce with fewer emissions in this industry as we transition forward, especially given the degree of infrastructure, homes, and buildings that are candidates for retrofits. However, cement plants are very capital-intensive, so turnover can take a long time. Other considerations with cement industries are related to occupational safety. Cement kiln dust is a widely accepted occupational hazard (Figure 13.2).

Figure 13.2 Cement making begins with quarrying and crushing, before other processes, including heating and clinker production. This plant in Norway is equipped with carbon capture technology, the first such facility in the world. Source: Eldart / Wikimedia Commons / CC BY-SA 3.0.

Ammonia

Ammonia production, primarily through the Haber-Bosch process, is heavily dependent on natural gas as both fuel and feedstock. The industry emits both CO_2 and N_2O (nitrous oxide)—the latter being a potent greenhouse gas. Ammonia is essential for fertilizer but poses safety risks, including catastrophic explosions. Globally, production is concentrated in about 300 large facilities. Decarbonizing ammonia requires electrification of hydrogen production via electrolysis (green hydrogen) and possibly capturing CO_2 from the process. Ammonia may also serve as a future energy carrier or shipping fuel.

Glass

Glass production involves high-temperature processing of quartz, soda ash, and lime, resulting in both energy-related and process-related CO_2 emissions. Different types of glass—like bottle glass and flat glass—have varying material and energy intensities. Flat glass is particularly important in green buildings, electric vehicles, and solar panels. Demand is expected to surge, yet no clear pathway to net zero has emerged for the glass sector (Del Rio et al. 2022). Innovations in electric melting furnaces, waste glass (cullet) recycling, and chemical substitutes are being explored.

Specialty chemicals and major alloys

One of the fastest-growing and most profitable areas for petrochemical industries is specialty chemicals. Since the production of these requires heat and steam, it is another place where green hydrogen or some other energy-dense fuels would help these industries decarbonize.

An important area of decarbonization is the chemical industry, which is responsible for 10% of global GHG emissions, as well as numerous other materials (McKinsey and Company 2018). About 2/3rds of overall energy use in industry is used by manufacturers using high-temperature heat, such as in making plastics and ammonia. Many chemical industries like these rely on steam and super-heated steam for heating reactor vessels and other tasks. This includes plastics manufacturers, paper and pulp producers, and others that take natural resource feedstocks and turn them into products. Electrification of many of these processes could also reduce emissions in this space. But others will need significant innovations. Some chemical processes could be expanded to become sources of energy. Ammonia could be used as a fuel directly or cracked for its hydrogen (Davis et al. 2018).

Chemical industries are intimately interconnected with energy companies. Many of the feedstocks used in chemical industries are purchased from the oil and gas industries. The chemical industries are often customers of petroleum companies who sell them raw materials. One common example is ethane, which is separated by a cracker. They are also critical suppliers of materials to the photovoltaic industries, making everything from metallization pastes and solders to backsheets and encapsulants.

Moving away from fossil fuel energy sources will impact the plastics industry. Finding greener, compostable, recyclable, biodegradable substitutes for plastics will require major initiatives in green chemistry, particularly with plastics. Green chemistry was made popular by Dr. Paul Anastas, who directed the US Environmental Protection Agency (EPA)'s green chemistry program, and John Warner of a company called Polaroid. It will be important for discoveries to be made in bioplastics to replace plastics made from natural gas or ethane. There are many questions still about the quality and durability of plastics that will make it challenging to replace. But transitions away from fossil fuels will imply transitions away from plastics, and this move will help with issues related to plastics in water and oceans, the Pacific garbage patch, plastic bag pollution, and micro-plastics, all of which can be problematic for wildlife and natural resources, like salt, with the increasing density of plastic pollution in the water column.

Pulp and paper

The pulp and paper industry is energy-intensive but also one of the few sectors with potential for negative emissions via bioenergy with carbon capture and storage (BECCS). Since paper mills often already burn biomass residues, they could be well-positioned to integrate carbon capture and storage (CCS) infrastructure (Sagues et al. 2020). However, environmental concerns persist around water use, deforestation, chemical discharge, and air pollution—particularly from chlorine bleaching and volatile organic compound emissions. Pulp and paper production is one industry where BECCS (see chapter 5) might be more viable.

This broader framing of sustainable industry and supply chains emphasizes that a just transition is not only about decarbonization, but also about transforming extractive systems, redistributing value, and reimagining material economies. Future efforts must combine technical innovation, policy reform, and social justice frameworks to ensure that climate solutions do not replicate the harms of the fossil-fueled industrial era.

Industry Powered by Clean Energy?

The decarbonization of industry is one of the most difficult and urgent challenges for global climate policy. Industrial facilities require enormous quantities of high-temperature heat, often for reactors, kilns, boilers, or steam generation. In fact, about two-thirds of all energy used in industry is dedicated to producing heat. Many of these processes require temperatures above 400 °C, which are difficult to achieve with conventional renewable electricity alone. This makes sectors like steel, cement, glass, and chemical production some of the hardest to electrify and thus hardest to decarbonize.

One promising pathway is green hydrogen–hydrogen produced using renewable electricity through electrolysis. Because hydrogen has a very high energy density, it is considered a key energy carrier for industries that cannot easily be electrified. For example, green hydrogen could replace coal in steelmaking, fuel high-heat chemical reactors, or be blended into ammonia production. However, the scale of green hydrogen required to meaningfully replace fossil fuels in industry is vast and would require massive renewable energy deployment, new infrastructure, and market coordination.

Another emerging opportunity lies in the decarbonization of industrial parks—zones where multiple manufacturing facilities share energy systems, such as combined heat and power plants or steam networks. One research team (Guo et al. 2020) found that China's industrial parks accounted for 11% of the country's GHG emissions in 2015, rising to 18% by 2018. Notably, these zones are often more carbon-intensive than the average electricity grid. This underscores the need to target industrial parks for electrification, efficiency upgrades, and heat recovery innovations, such as using waste heat to supply nearby cities. Similar zones are common in South Korea, Japan, Malaysia, and elsewhere, suggesting opportunities for international knowledge sharing and policy development.

Manufacturing clean energy technologies also poses its own challenges. For example, polysilicon production—a key step in making solar photovoltaic modules—is one of the most energy-intensive steps in the clean energy supply chain (see Box 13.2). Emissions from producing solar-grade polysilicon depend significantly on the carbon intensity of local electricity. Many facilities are located in industrial parks with coal-heavy power sources, so using grid-average emissions factors can drastically understate the true carbon footprint of solar panel production.

Box 13.2 Energy Star manufacturing equipment

Most people associate ENERGY STAR with household appliances or office electronics, but the label also applies to industrial manufacturing equipment. The US EPA and Department of Energy certify high-efficiency industrial boilers, process heaters, compressed air systems, and motor-driven equipment under the ENERGY STAR label. These certifications help manufacturers reduce operating costs, fuel consumption, and greenhouse gas emissions—while offering a clear benchmark for procurement decisions.

ENERGY STAR also supports facilities in benchmarking their performance using tools like the Energy Performance Indicator, which enables companies to compare their energy efficiency against industry peers. As industrial emissions come under increasing regulatory scrutiny, ENERGY STAR tools offer a cost-effective entry point for continuous improvement and emissions reductions.

Many of the environmental problems tied to manufacturing originate not just from waste or emissions—but from how products are designed in the first place. The production and use of persistent, bioaccumulative, and toxic chemicals—especially in plastics, coatings, solvents, and electronics—have long-term impacts that extend far beyond the factory floor. As products deteriorate, break down, or are discarded, they can leach pollutants into ecosystems, bioaccumulate in wildlife and humans, or persist in soils and sediments for decades. Sustainability experts often argue that focusing on the design phase of products offers the greatest leverage for change. This is sometimes referred to as "designing out pollution." In the United States, this idea is supported through voluntary tools like the EPA's Green Chemistry Expert System, while Europe has taken a stronger regulatory stance through programs like REACH. These initiatives aim to phase out hazardous chemicals, promote safer alternatives, and require extended producer responsibility for the materials and waste generated by industrial production.

However, as sociologist David Hess and others have shown, many sustainable production policies lack accountability mechanisms and are rarely co-designed with public input. Industry often uses sustainability language as compliance branding—claiming "sustainability" simply because they meet baseline regulatory requirements, even as supply chains remain opaque and social harms persist. Researchers who study the issue argue that without public participation and structural change, environmental gains will remain incremental and fragile.

More broadly, we must question how production decisions are made, who has power in design processes, and how alternative technologies or materials are marginalized. Current systems overwhelmingly treat sustainability as a technical challenge for engineers or chemists, while excluding the values and knowledge of affected communities, consumers, or workers. But there is growing evidence—from science and technology studies, anthropology, and design research—that lay actors can and do influence production practices when given access to tools, information, and agency. In sum, the future of sustainable manufacturing must go beyond decarbonization. It must integrate design justice, green chemistry, worker safety, and public accountability into the heart of industrial policy. Whether we're producing solar panels, hydrogen turbines, or recycled concrete, the question remains: not just how efficiently we manufacture, but who gets to decide how and for whom things are made.

Production and manufacturing of things are associated with "unsustainable" environmental impacts. Many products are recognized as containing toxic, persistent, or bioaccumulative chemicals, things often not learned until they are out in the public for some time. Their use, storage, transport, and deterioration over time, not just manufacturing or disposal, can lead to dissipation into the environment. By targeting how products are chosen and made, a troublesome source of contamination can be addressed at the outset rather than at the tailpipe. This view seems to prevail in the United States to date. Second, focusing on production is said by some industry and environmentalist actors to promise a fundamental "solution" to environmental problems. If industry is to attain greater "sustainability," it needs to reconsider how it makes its production choices. By targeting the materials, water, energy, ecological, and social

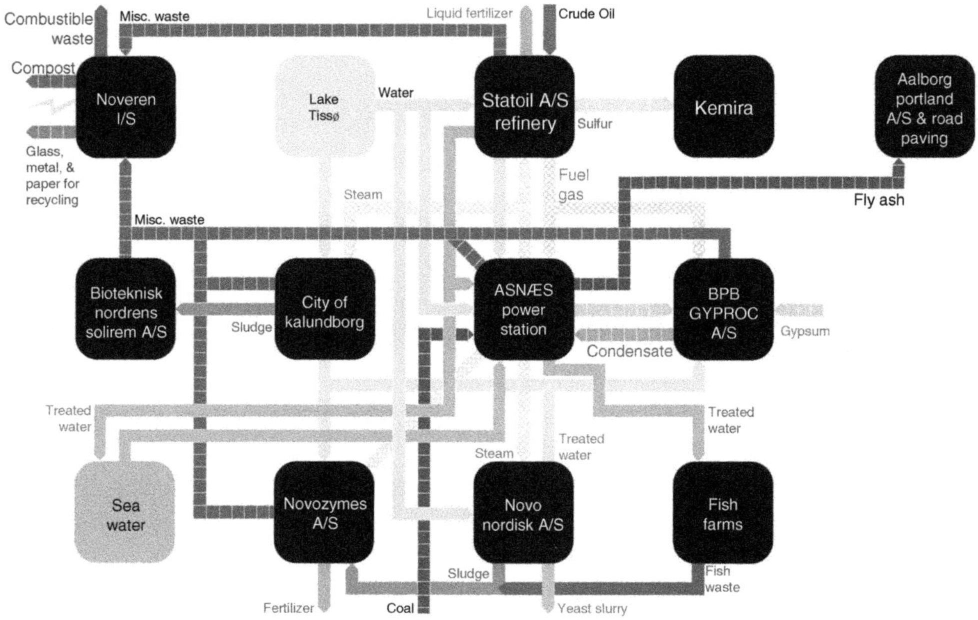

Figure 13.3 The Kalundborg industrial symbiosis example. Source: Rosenfeld Media / Flickr.

impacts of production at the stage of making design choices, it may be feasible to "design these out" to some degree. This view seems more prevalent within in Europe so far.

There is still much to do to probe what "sustainability" might mean and how to make production choices transparent, traceable, and accountable to minimal impacts. Chemical and manufacturing industries claim that production is "sustainable" when they may just mean they are not breaking any environmental laws (Figure 13.3).

Box 13.3 Is the famous industrial ecology case at Kalundborg as circular as it claims?

While the Kalundborg Symbiosis is often celebrated as a model of circular industrial design, some researchers and critics have raised important questions about how truly sustainable or regenerative it really is. At first glance, the exchange of waste heat, steam, gypsum, and biosolids among companies appears to mirror the closed-loop processes of natural ecosystems. However, deeper analysis reveals that many of the primary industrial processes at Kalundborg are still highly extractive and carbon-intensive—especially the coal- and gas-fired power generation that anchors many of the symbiotic relationships. Rather than eliminating environmental harm, the symbiosis might instead be making conventional industrial activities more efficient without fundamentally transforming their environmental impact. In this way, Kalundborg could be seen more as a case of "eco-efficiency" than true "circularity" or ecological regeneration. Moreover, the continued reliance on fossil fuels and high-temperature industrial processes means that many material flows are not indefinitely recyclable; for example ash and sludge may have secondary uses, but eventually accumulate as waste or pollution. Critics argue that the system does not fully account for the limits of material reuse, the embodied emissions of inputs, or the social dimensions of sustainability, such as labor rights and global supply chain impacts. Thus, while Kalundborg remains an inspiring and instructive case of localized industrial collaboration, it may represent an intermediate step toward sustainability—useful, but insufficient—if not accompanied by broader efforts to decarbonize, dematerialize, and restructure production systems more fundamentally.

Highlights/Takeaways

- Traditional industrial production follows a "take-make-waste" model, driving vast material and energy use along with significant emissions and waste. Industrial ecology and circular economy approaches challenge this linearity by promoting material reuse, waste valorization, and system-level redesign—transforming waste into resources and reimagining industry more like ecosystems.
- Clean energy alone is not enough. While decarbonization of industry is crucial, many of the hardest-to-abate sectors—like steel, cement, and chemicals—require innovations beyond just switching to clean electricity. Solutions like green hydrogen, process redesign, material substitution, and carbon capture must be integrated with deeper questions about how and why we produce in the first place.
- Supply chains and justice are inseparable. Critical minerals and industrial goods travel through complex, often opaque global supply chains that embed environmental and social injustices—from child labor in cobalt mines to toxic waste near refineries. A just transition must extend beyond the factory gate to consider extraction, labor, and the communities along the entire chain of production.
- Redesigning production means rethinking power. Sustainability in manufacturing isn't only about efficiency or pollution control—it's also about who has power over design, production, and environmental decisions. Critical industry studies call for greater transparency, worker safety, and public engagement in shaping how things are made, ensuring that climate solutions don't simply replicate the harms of the fossil-fueled past.

Questions for Review

1. How does the circular economy challenge traditional industrial models, and what limitations or contradictions arise when trying to apply circularity to carbon-intensive sectors like steel, cement, or chemicals? Consider both the material and systemic barriers to achieving closed-loop production, and whether concepts like industrial symbiosis go far enough to meet ecological goals.
2. In what ways do global supply chains complicate the pursuit of just and sustainable manufacturing? Reflect on how power, labor, geography, and environmental regulation shape where and how goods are made and who bears the costs or benefits of energy transitions in industries like mining, refining, and heavy manufacturing.
3. What does a critical approach to industrial sustainability look like, and how does it differ from mainstream narratives of innovation and efficiency? Explore the roles of public participation, design justice, and structural reform in shaping more equitable production systems—and consider how these might shift responsibility from individual firms to broader socio-political institutions.

Exercise 13.1

Imagine you are the Corporate Social Responsibility or Sustainability officer or manager for a manufacturing company. You realize that the company you work for is very far from meeting the goals and principles of a circular economy in manufacturing. What are some early first steps a company might take to begin this process?

Exercise 13.2

What policies could a government develop to strengthen sustainability and low-carbon efforts in industry, particularly in sectors that are hard to decarbonize. What kinds of incentives or penalties could help foster innovation in this space?

References

Benyus, J.M. (1997). *Biomimicry: Innovation Inspired by Nature*, vol. 688136915. New York: Morrow.

Crang, M., Hughes, A., Gregson, N. et al. (2013). Rethinking governance and value in commodity chains through global recycling networks. *Transactions of the Institute of British Geographers* 38 (1): 12–24.

Del Rio, D.D.F., Sovacool, B.K., Foley, A.M. et al. (2022). Decarbonizing the glass industry: a critical and systematic review of developments, sociotechnical systems and policy options. *Renewable and Sustainable Energy Reviews* 155: 111885.

Fairhead, J., Leach, M., and Scoones, I. (2012). Green grabbing: a new appropriation of nature? *Journal of Peasant Studies* 39 (2): 237–261.

Guo, Y., Tian, J., and Chen, L. (2020). Managing energy infrastructure to decarbonize industrial parks in China. *Nature Communications* 11 (1): 981. https://doi.org/10.1038/s41467-020-14805-z.

Huber, M.T. (2017). Hidden abodes: industrializing political ecology. *Annals of the American Association of Geographers* 107 (1): 151–166.

Lepawsky, J. (2018). *Reassembling Rubbish: Worlding Electronic Waste*. MIT Press.

Liboiron, M. (2021). *Pollution Is Colonialism*. Duke University Press.

McDonough, W. and Braungart, M. (2010). *Cradle to Cradle: Remaking the Way we Make Things*. North point press.

Sagues, W.J., Jameel, H., Sanchez, D.L., and Park, S. (2020). Prospects for bioenergy with carbon capture & storage (BECCS) in the United States pulp and paper industry. *Energy & Environmental Science* 13 (8): 2243–2261. https://doi.org/10.1039/D0EE01107J.

Stahel, W.R. (2019). Innovation in the circular and the performance economy. In: *Handbook of Sustainable Innovation* (ed. F. Boons and A. McMeekin), 38–58. Edward Elgar Publishing.

Further Reading

Conway, E. (2023). *Material World: The Six Raw Materials that Shape Modern Civilization*. Vintage.

Smil, V. (2016). *Still the Iron Age: Iron and Steel in the Modern World*. Butterworth-Heinemann.

Chapter 14

Food-Energy-Water

"*Yields in US agricultural production have increased rapidly since the 1940s ... specifically, these inputs include fertilizers, pesticides, and irrigation, all of which depend primarily upon fossil energy.*"

—David Pimentel (Pimentel 2019)

The production of food represents one of the most enduring and transformative human interactions with Earth's systems. Since the dawn of agriculture, activities like deforestation, irrigation, and the domestication of animals have reshaped landscapes and biogeochemical cycles. Land use change—especially forest clearing for cropland and grazing—is responsible for an estimated one-quarter of historical greenhouse gas (GHG) emissions since 1750. Today, the global food system—including crop cultivation, livestock production, transportation, processing, and refrigeration—accounts for nearly 25% of annual GHG emissions, making it a critical component of climate and sustainability discussions. Yet food production does not operate in isolation. It is deeply entangled with energy and water systems in ways that are both resource-intensive and vulnerable to disruption. Modern food production relies heavily on fossil fuels—for everything from fertilizers (e.g., ammonia synthesized from natural gas) to mechanized farming to long-distance cold-chain logistics. At the same time, agriculture is the largest user of freshwater globally, with roughly 70% of global freshwater withdrawals going to irrigation. Human-engineered water delivery systems—from dams, canals, and aqueducts to deep groundwater pumps—all require varying levels of energy input, meaning there is embedded energy in water (or "energy-for-water"), particularly in water-stressed regions that rely on desalination, recycled water, or inter-basin transfers. Water is also deeply affected by the food system. Excess fertilizers, manure, and pesticide runoff degrade water quality, leading to nutrient pollution, dead zones, and contaminated drinking water supplies. Livestock operations and confined animal feeding operations (CAFOs) are particularly impactful, contributing significantly to both methane and nitrous oxide emissions as well as pollution of surface and groundwater through waste mismanagement. In turn, the treatment of polluted water and municipal wastewater—essential to protect human and ecosystem health—can be highly energy-intensive, especially in high-tech or decentralized water recycling facilities. As demands on land, food, water, and energy grow, the food-energy-water (FEW) nexus approach offers an integrated framework for understanding the interdependencies and trade-offs among these systems. It recognizes that interventions in one domain—such as solar-powered irrigation or water reuse—can have cascading effects on others. This chapter explores how agrivoltaics (the dual use of land for solar generation and crop cultivation) and solar irrigation technologies are emerging as promising tools to optimize land use, reduce water stress, and lower the carbon intensity of farming. From the industrialization of agriculture to the rise of desalination plants in drought-prone regions, and from energy use in wastewater treatment to climate-smart food production, this chapter emphasizes the importance of thinking holistically about infrastructure, resources, and sustainability. By focusing on system interactions, rather than siloed sectors, we can better evaluate feedback loops, vulnerabilities, and opportunities for synergy, especially as we prepare for a future shaped by climate uncertainty, population growth, and resource constraints.

The Industrialization of Food Production

The story of agriculture is one of increasing industrial production, mechanization, and escalating energy use and GHG emissions. Over time, fossil fuels and electricity have steadily replaced human and animal labor with mechanical force, enabling greater productivity—but also introducing immense environmental costs. In 1700, it took approximately 300 labor-hours to work a hectare of land; by 1900, that number had dropped to 100; and by the year 2000, with fossil fuel-powered tractors and machinery, it had fallen to just one labor-hour (Smil 2000). This profound transformation—sometimes described as the second agricultural revolution—reflects not only a substitution of labor with energy but also a shift in how we conceptualize agricultural efficiency and productivity.

Industrial agriculture expanded far beyond traditional cropping and livestock techniques to encompass energy-intensive methods across the entire food system, from crop production, aquaculture, and fisheries to food processing, preservation, distribution, and retail. Where earlier food systems relied heavily on nutrient cycling, manure reuse, and solar drying, modern systems are built on synthetic fertilizers, chemical pesticides, motorized harvesting, and cold storage chains that span continents. One estimate suggests the American food system consumes the equivalent of 528 gal of oil per person annually (Pimentel and Pimentel 2007). That includes roughly 12 cal of fossil energy used for every one calorie of food consumed, a ratio that points to the systemic inefficiencies and externalized costs embedded in the industrial food model (Figure 14.1).

Some of the most significant shifts toward energy dependency in agriculture stem from the development and mass deployment of petrochemical-based fertilizers and

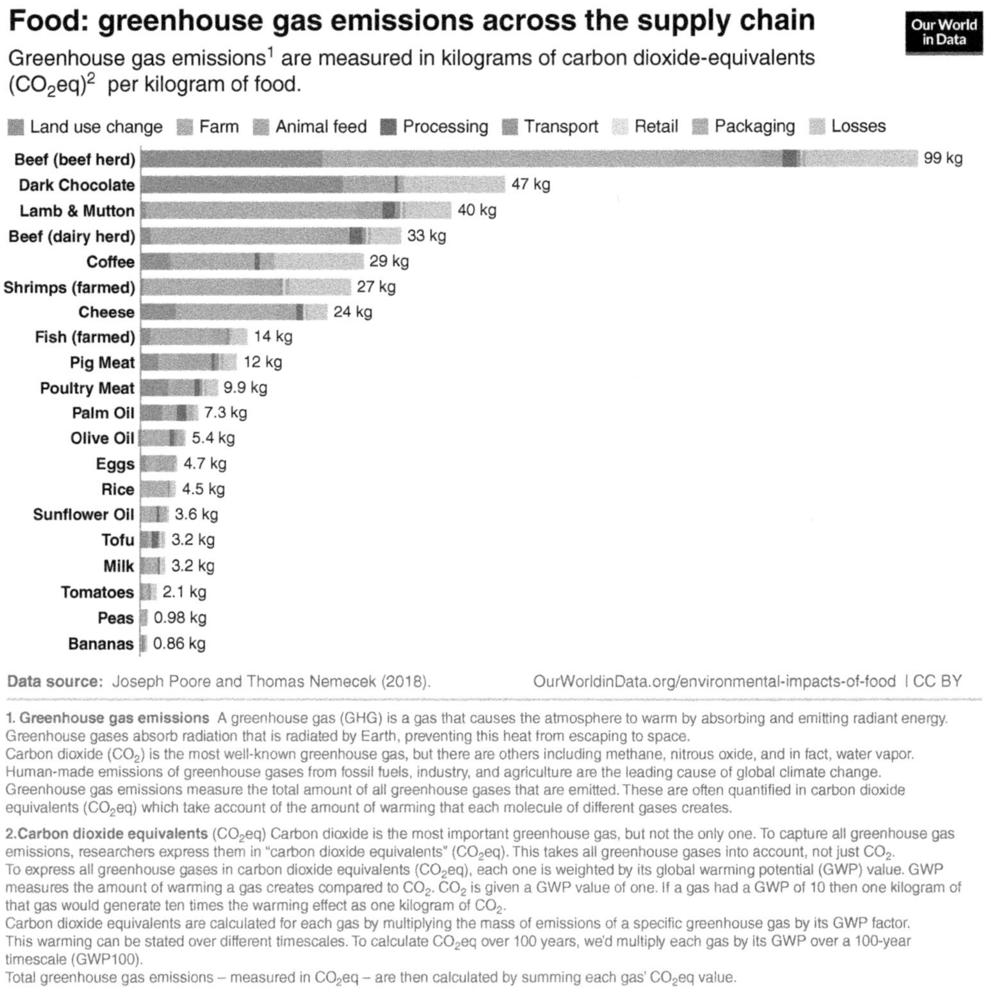

Figure 14.1 Greenhouse gas emissions from food across the supply chain. Source: Our World in Data/Food: greenhouse gas emissions across the supply chain/https://ourworldindata.org/grapher/food-emissions-supply-chain/CC BY 4.0/last accessed on September 18, 2025.

pesticides, the mechanization of tillage and harvesting, and the rise of energy-intensive irrigation systems. Combustion- or electric-powered crop drying replaced sun-drying, while food storage and refrigeration chains now require continuous electricity input. Meanwhile, food transportation—including by airplane—has introduced new layers of energy use, allowing for unprecedented food miles and the year-round availability of out-of-season crops. Globalization of trade has amplified these patterns: fresh berries flown from Chile to Boston, lamb shipped from New Zealand to London, and avocados harvested in Mexico and consumed in Norway.

In terms of energy inputs, direct uses such as diesel fuel and electricity dominate, particularly in mechanized agricultural systems. Indirect energy use includes the embedded energy in fertilizers, pesticides, plastics, packaging, and processing equipment. In many low-income regions of the world, these inputs remain prohibitively expensive or inaccessible. As a result, smallholders often rely on practices that more closely resemble early twentieth-century agriculture, using animal power, compost, and hand tools. While such practices can achieve respectable yields—particularly with appropriate ecological management—they remain vulnerable to labor shortages, climate variability, and limited access to markets or subsidies.

Layered onto this is the tremendous energy and emissions footprint of animal agriculture. Livestock production—especially of ruminants like cattle—contributes significantly to methane emissions, a potent GHG with over 80 times the warming potential of carbon dioxide over a 20-year horizon. Coupled with the land use change associated with pasture expansion and feed crop cultivation, animal agriculture is a leading contributor to deforestation, biodiversity loss, and nitrogen and phosphorus pollution (Fargione et al. 2008). When taken together, crop production, animal agriculture, food processing, transport, storage, and disposal—the global food system—are responsible for roughly one-quarter of global GHG emissions, according to the Intergovernmental Panel on Climate Change (IPCC 2022). Ultimately, the industrialization of food production has enabled vast increases in yield and availability, but at significant ecological and social cost. Understanding agriculture as a food–energy–water system, rather than a linear food supply chain, helps illuminate the complexity of these interactions and the opportunities for more integrated, resilient, and equitable food systems moving forward.

Food–Energy–Water Interactions

The food, energy, and water systems are deeply intertwined, forming a dynamic nexus at the heart of modern civilization. This relationship has been evident since the earliest human settlements, which emerged in fertile river valleys and floodplains such as the Nile, the Tigris-Euphrates, and the Indus. These regions offered not just rich soils, but predictable seasonal water flows essential for irrigation and food production. The ancient principle that "food grows where water flows" is still true today, especially in water-scarce regions where agricultural productivity depends heavily on engineered water delivery systems. In California, this saying has taken on political resonance as agriculture, cities, and ecosystems compete for increasingly scarce water resources.

Globally, over half of all agricultural land now relies on some form of irrigation. This marks a shift away from rainfed farming and toward more water-intensive systems that depend on canals, pumps, groundwater wells, and dams. As irrigation infrastructure expands, it demands not only water but also energy—to pump, transport, filter, and pressurize that water. In fact, agricultural irrigation is one of the largest consumers of freshwater globally, but also a growing consumer of electricity and liquid fuels. The energy required to manage water for agriculture is a major, often overlooked contributor to farm-level GHG emissions.

Energy production too requires water—at nearly every stage of the life cycle. From extraction (e.g., hydraulic fracturing for oil and gas or mining for coal and lithium) to refining and processing to cooling during power generation and even in end-of-life disposal, water is a critical input. This makes the energy sector highly water-dependent and vulnerable to droughts, water contamination, and conflict over water rights. Energy-water trade-offs can ripple into food systems: for example, water allocated for power plant cooling may not be available for irrigation, or a drought-induced hydropower shortfall could limit electricity used to pump water for crops. These feedback loops reveal just how connected these systems really are.

The most water-intensive component of the energy system is thermal power plant cooling—especially in fossil fuel and nuclear facilities. These power plants often employ once-through cooling systems, where water is drawn from a nearby lake, river, or ocean to absorb waste heat before being discharged. While most of this water is technically returned to the source—hence counted as a withdrawal rather than consumption—it is returned at a significantly higher temperature, which can disrupt aquatic ecosystems and reduce dissolved oxygen levels, affecting fish and other aquatic life (Figure 14.2).

Understanding the difference between water withdrawal and water consumption is key to evaluating energy and food system sustainability. Water withdrawals refer to the total volume of water removed from a source for use. This includes water used for irrigation or cooling that may be returned later. Water consumption, in contrast, is the portion of water that is no longer available for reuse because it has been evaporated, transpired by plants, incorporated into products, or otherwise lost to the immediate watershed (Figure 14.3).

In this context, bioenergy production—including corn ethanol, sugarcane ethanol, and soy-based biodiesel—represents a particularly water- and land-intensive energy source (Buck 2019). These crops require substantial irrigation and fertilizer inputs, and when used for fuel rather than food, they introduce competition between energy and food security. Water pollution from nutrient runoff (e.g., nitrogen and phosphorus) used in these biofuel crops can also create dead zones in downstream ecosystems such as the Gulf of Mexico (Pimentel and Pimentel 2007).

Food, energy, and water systems—the FEWS nexus—are not just co-dependent, but also co-evolving. Technological developments in one domain can have profound

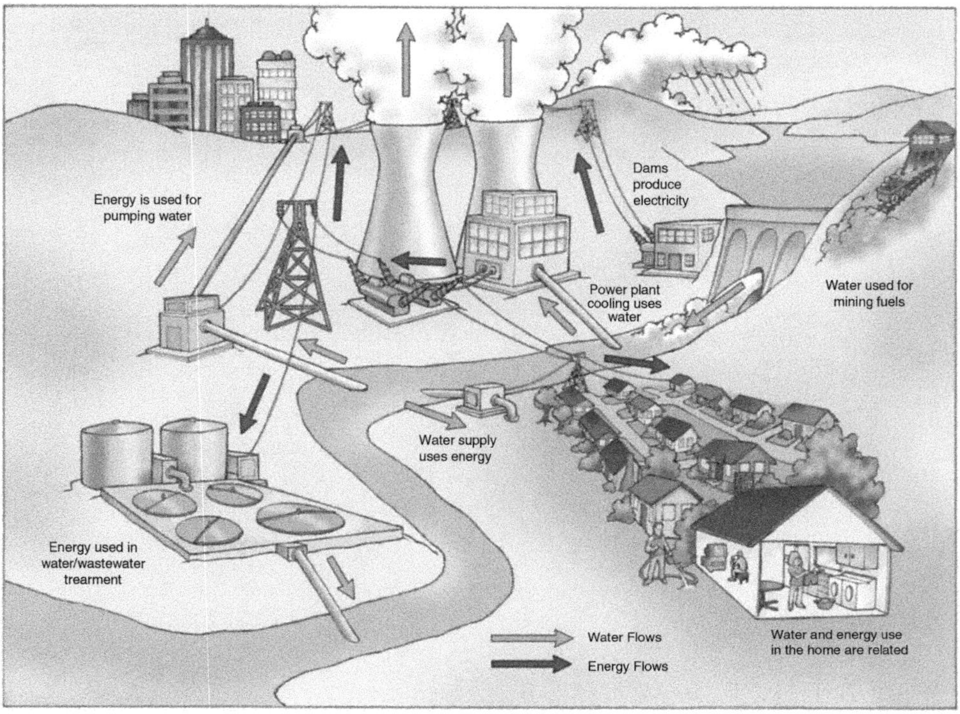

Figure 14.2 Energy demand on water resources. Source: U.S. Department of Energy/ Energy Demands on Water Resources. Report to Congress on the Interdependency of Energy and Water/https://ntrl.ntis.gov/NTRL/dashboard/searchResults/titleDetail/PB2011103657. xhtml /Public Domain/last accessed on September 18, 2025.

consequences for the others. As climate change increases the variability of water supplies and stresses food production systems, and as the energy transition reshapes global energy demand, adopting a nexus perspective becomes essential. Understanding and managing the interdependencies between food, energy, and water is not only key to sustainability—but to resilience and equity as well.

Water Footprints

Water footprinting is a valuable method for quantifying the volume and quality impacts of water used in the production of goods and services. In the context of agriculture, energy, and industrial production, water footprints help researchers, regulators, and firms better assess the full water impacts of their supply chains—not just how much water is withdrawn, but also how much is consumed, polluted, or lost. Water footprints are typically divided into three components: blue, grey, and green, each capturing a different dimension of water use and impact.

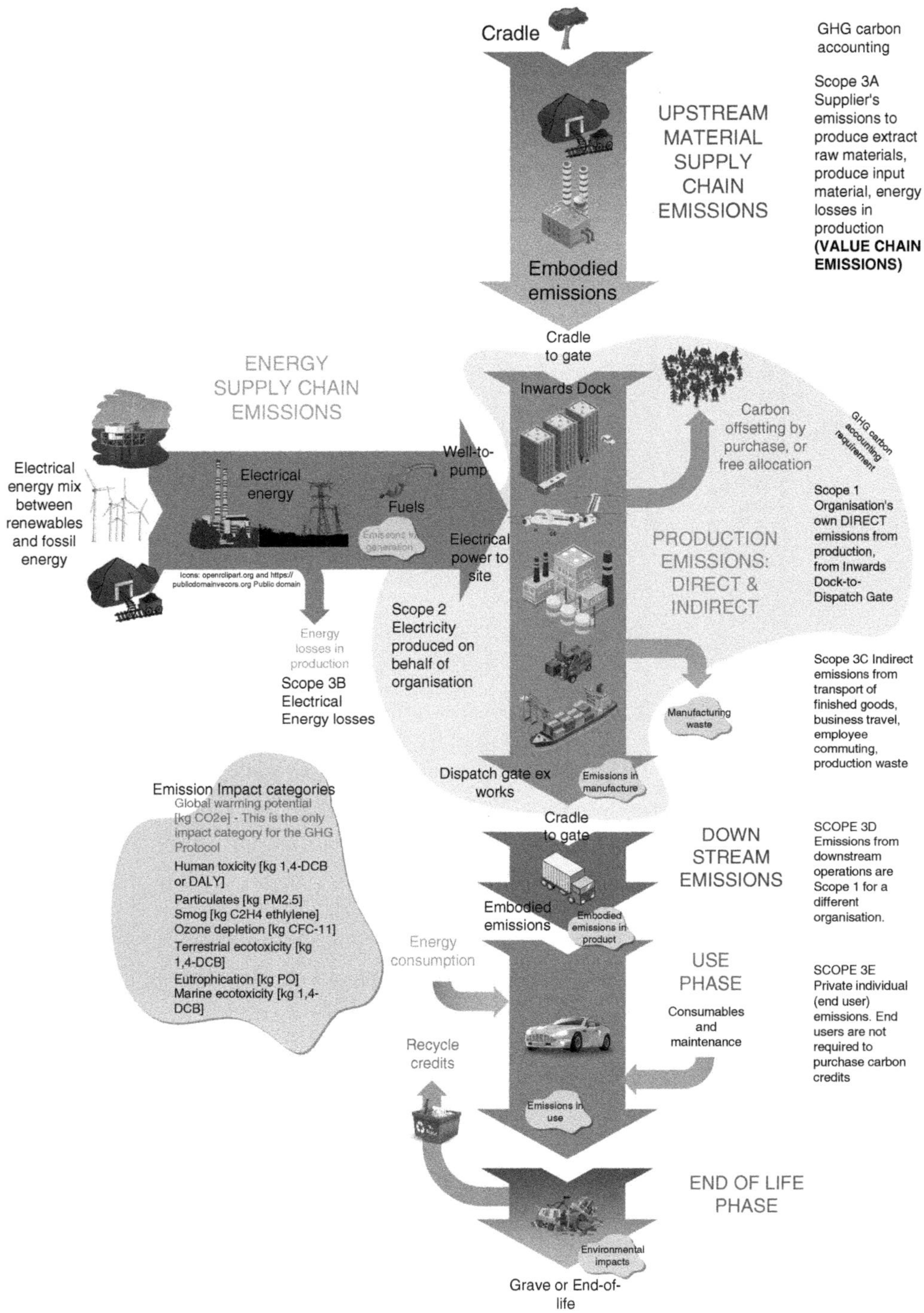

Figure 14.3 Life cycle assessment is used in greenhouse gas accounting to characterize the cradle to grave impacts of energy and materials. Source: John Pons/https://upload.wikimedia.org/wikipedia/commons/f/f2/Life_cycle_analysis_and_GHG_carbon_accounting.jpg/CC0 1.0/last accessed on September 18, 2025.

Blue water footprint—withdrawn water

The blue water footprint refers to the volume of surface or groundwater withdrawn and consumed during production processes. This includes water used for irrigation, cooling, or industrial processing that does not return to the same watershed. Because freshwater is often scarce or over-allocated, the blue water footprint is especially critical for understanding pressure on regional water supplies.

Blue water footprint = Volume of freshwater withdrawals/Units of energy or product output

This measure provides a direct estimate of **how much water is taken** from the environment to support human activity.

Grey water footprint—polluted water

The grey water footprint measures the volume of freshwater required to dilute pollutants discharged from a production process to meet safe water quality standards. In essence, it estimates how much water is needed to assimilate pollution without exceeding environmental thresholds.

This footprint is calculated using a two-step process:

Dilution factor = Contaminant concentration ÷ Water quality standard This measures how much the pollution exceeds acceptable limits.

Grey water footprint = (Volume of wastewater discharged × dilution factor)/Units of production

This indicator helps translate pollution (e.g., nitrogen runoff, chemical effluent) into a comparable volume metric, making it easier to benchmark water quality performance across sectors.

Green water footprint—evaporated or transpired water

The green water footprint represents evaporative water loss from soil and vegetation, primarily through transpiration in rainfed agriculture and forestry systems. It accounts for the portion of rainfall used by crops and plants that never returns to surface or groundwater systems.

Green water is crucial in agriculture and forestry and is especially important in discussions of rainfed systems, ecosystem services, and land-use planning.

Green water footprint = Evapotranspired water volume/Crop or biomass yield

To get a complete picture of water-related impacts, water quality indicators are also essential. These include nitrate and phosphorus linked to fertilizer use; these nutrients can trigger eutrophication, hypoxia, and algal blooms in freshwater bodies. Nitrate contamination is also linked to blue-baby syndrome, affecting rural communities dependent on groundwater. Total suspended solids are elevated solids that result from soil erosion and poor land management (e.g., tillage, no cover crops). Suspended solids reduce water clarity, smother aquatic life, and contribute to siltation in rivers and streams. Pesticide and herbicide residues can include carcinogenic or endocrine-disrupting compounds, and their concentrations reflect the degree of chemical input used in crop systems. Runoff can cause widespread ecosystem damage and contaminate drinking water supplies. Each of these metrics is influenced by crop type, land use practices, fertilizer and pesticide application rates, and conservation efforts (Searchinger et al. 2008).

Water footprint color schemes in energy and agriculture

Color-coding water footprints (e.g., blue, grey, green) provides a simple visual and conceptual framework for understanding the distinct ways water is used and impacted in production systems. Energy systems researchers like Emily Grubert (2022) have emphasized how color-coded water-energy footprinting helps track both quantity and quality dimensions of water impacts across energy and agricultural supply chains—allowing for better integration of water resource planning with climate and energy policy (Grubert 2022). Ultimately, water footprints offer a powerful lens to evaluate the sustainability, equity, and efficiency of resource use across sectors—and help illuminate where reforms or innovations are needed most.

Water Delivery Systems, Wastewater, and Food Waste

One of the most energy-intensive and GHG-emitting components of municipal infrastructure is the provision of drinking water and the treatment of wastewater. Across the globe, extensive water conveyance systems have been developed to move water from distant aquifers, rivers, and reservoirs to areas where natural water availability would otherwise be insufficient to support large populations. These systems—especially in arid or semi-arid regions—have enabled urban development in places where access to local water was historically too limited. However, this has come with trade-offs. The further water must be transported, and the more it must be treated, the higher its embodied energy—the total energy embedded in extraction, pumping, purification, and delivery.

In the United States, municipal wastewater treatment facilities (WWTFs) are typically the single most energy-intensive operation within a community. They are also one of the most promising opportunities for energy and cost savings. It is estimated that US WWTFs consume more than 30 terawatt-hours (TWh) of electricity per year, costing over $3 billion annually, and accounting for approximately 4% of total US electricity consumption for water-related services (US Department of Energy, n.d.). Electricity alone can represent 25–40% of a wastewater treatment plant's operating budget and may comprise 15–30% of a municipality's total energy bill. As urban populations grow and water quality standards become stricter, energy demand for wastewater treatment is expected to increase unless major efficiency upgrades are implemented.

Despite these challenges, opportunities for emissions reduction and resource recovery abound. Wastewater contains roughly five times more energy than is required to treat it, mostly in the form of chemical energy in organic matter and thermal energy from warm water flows. Increasingly, utilities are installing anaerobic digesters, biogas recovery systems, and onsite combined heat and power (CHP) units to tap into this energy, turning treatment facilities into net energy producers or even energy-neutral operations in some cases. These systems not only cut energy costs but also reduce methane emissions from uncontrolled decomposition.

Speaking of emissions, the GHG footprint from global wastewater treatment is estimated to be equivalent to the combined emissions of the aviation and container shipping sectors—about 1% of global GHG emissions. This highlights the significance of even incremental improvements in wastewater management as part of broader climate strategies.

The energy required to provide drinking water varies widely by geography and infrastructure. In regions like California, where water is conveyed over long distances and steep terrain, water delivery can be extremely energy-intensive. For example, Los Angeles imports much of its water from the Sierra Nevada Mountains, hundreds of miles away, using an elaborate system of aqueducts and pumps. In contrast, cities closer to water sources or with gravity-fed systems—like San Francisco—require far less energy. In rural areas, where households often rely on private groundwater wells, energy costs for pumping can be a major burden, particularly when powered by diesel or inefficient electric systems.

Groundwater pumping for agriculture is another major energy consumer, especially in regions where water tables are dropping due to over-extraction. In response, several regions—including India and parts of California—are experimenting with solar-powered irrigation pumps as a strategy to displace diesel and reduce both costs and emissions. A small but innovative example of energy recovery from drinking water systems is found in Portland, Oregon, where a section of the city's gravity-fed water pipeline has been fitted with in-pipe hydroelectric generators, harnessing the energy of moving water to generate electricity. Such niche solutions, when replicated at scale, could meaningfully offset municipal electricity use.

Globally, access to clean drinking water and sanitation remains an acute equity issue. According to the World Health Organization, billions of people still lack access to basic sanitation services or safe, reliable water for daily needs. In rural and peri-urban communities, wastewater is often managed through decentralized systems such as septic tanks, leach fields, or open drains, many of which pose health and environmental hazards when not properly maintained.

Methane emissions from wastewater are especially important to address. Anaerobic conditions in sewers and treatment plants—where oxygen is limited—favor methanogenesis, the microbial process that releases methane (CH_4), a potent GHG. Capturing and combusting this methane (e.g., through anaerobic digestion with biogas recovery) can not only reduce GHG emissions but also generate renewable energy for onsite heating or electricity.

A related issue is food waste, which represents both a resource and a problem. In many municipalities, food scraps and organic waste are being diverted from landfills and routed into anaerobic digestion facilities. These facilities break down organic matter to produce biogas—a methane-rich fuel—and digestate, a nutrient-rich slurry that can be used as a soil amendment. In San José, California, for example, the world's largest anaerobic dry fermentation digester opened in 2013, turning urban food waste into usable energy and compost, demonstrating the potential of circular resource recovery in urban waste systems. Together, investments in water-energy-waste integration—such as co-located food waste and wastewater treatment, solarized pumps, and smart water reuse—represent a powerful avenue for reducing GHGs, saving money, and improving public health. As pressures on water and energy systems intensify, these innovations will be critical to building resilient, equitable, and sustainable cities.

Desalination and Recycled Water

One critically important question facing particularly drought-prone areas is where they will secure future water supplies. Freshwater loss from lost snowpack from climate change, over-drafted reservoirs, overallocated river systems, and competition from other human uses are evidence that new approaches to water management and provisioning may be needed (Box 14.1).

Questions around desalination can be complicated by specific local contexts, so it is difficult to generalize about the implications of such projects. Recycled water is likely to be an increasing part of drinking water supplies into the future. Using recycled water for irrigation is a practice that has been in place for several decades; San José, California, has had a purple pipe system for reused water. Recycled water is shown to be more energy efficient in terms of use than say, desalination. Yet, there remain questions around micropollutants and public acceptance.

Agrivoltaics, Floatovoltaics, and Over-Canal Solar

As the global demand for both renewable energy and agricultural land increases, new strategies are emerging to maximize land use efficiency and minimize land-use conflicts (Pascaris et al. 2021b). One such strategy is agrivoltaics—the dual use of land for both photovoltaic (PV) energy generation and agricultural production. Rather than viewing land as suitable for either farming or solar development, agrivoltaics explores

Box 14.1 Desalination and Environmental Justice

Energy use and brine disposal are typically the main considerations with desalination, but other issues might be more salient in some cases. A desalination project near Monterey, California, failed to get approval because of environmental justice issues. One of the issues is that the project relied on a saline aquifer, which was at risk of affecting the seawater intrusion into the local groundwater that supported agriculture in a disadvantaged community. In that case the California Coastal Commission invoked its Environmental Justice policy, which noted that the desalinated water would be delivered to a wealthier community on the Monterey Peninsula.

California, like other arid places around the world, suffers from seasonal and chronic water shortages. Many of these regions have turned to desalinization to augment water supplies. This case looks at the hydrosocial imaginaries constructed around a proposed desalination project in Marina, California. The project developer, an investor-owned water utility that would use slant wells to extract saline groundwater and seawater for the plant. The permits were denied by the California Coastal Commission, which justified the decision based on their new environmental justice policy, arguably the strongest such policy by an infrastructure permitting agency in the United States. The commission argued that the underrepresented community, which has long been a fenceline community with sand mines, a regional wastewater treatment facility, and a landfill, would not receive the benefits. Instead, that water would be sent to a better-off community on the Monterey Peninsula. Agency staff found there were more equitable alternatives to the water supply problem.

how food, fiber, and energy can be co-produced in the same space. Agrivoltaic systems involve strategically spacing or elevating solar panels to allow enough sunlight to reach the crops below while still optimizing electricity generation. These configurations often use tracking systems, semi-transparent panels, or elevated mounting structures to balance shading and light transmission. Several studies have shown that under certain conditions, growing crops under partial shade provided by PV panels can actually improve yields by reducing heat stress, conserving soil moisture, and moderating microclimates. Research from the US Department of Energy and universities like the University of Arizona suggests that land productivity can increase by up to 30% when food and energy are co-produced, compared to their independent production on separate parcels of land.

In addition to crops, agrivoltaics has been tested with grazing animals (sometimes called "ranch-voltaics"), aquaponics, and even pollinator habitat restoration. Livestock like sheep are often used to maintain vegetation under solar arrays, providing a cost-effective and low-impact form of site maintenance while preserving agricultural livelihoods. Furthermore, studies suggest that agrivoltaics can improve public perception and social acceptance of solar farms, particularly in rural areas where residents are concerned about losing productive farmland. In regions where agricultural land is protected or where zoning regulations prohibit utility-scale PV, agrivoltaics may provide a promising policy compromise—allowing energy development without displacing food production (Pascaris et al. 2021b).

Beyond land, floatovoltaics represent a similar approach to dual-use in water environments. These systems install PV arrays on floating platforms placed on irrigation canals, reservoirs, ponds, or wastewater treatment basins. Floatovoltaics have multiple co-benefits: they reduce water evaporation, help prevent algal blooms by shading the water surface, and in some cases, even improve PV efficiency due to the cooling effects of water. Countries such as India, Japan, and the Netherlands are already developing floatovoltaics at scale, and the potential in arid or land-constrained regions is immense.

Over-canal solar—a newer innovation gaining attention—extends this concept by placing solar panels directly over irrigation canals. This method, piloted in places like Gujarat, India, and now being tested in California's Central Valley, provides similar advantages: reducing water evaporation, protecting water quality, and avoiding land-use competition entirely. By co-locating solar generation with critical water infrastructure, over-canal solar may deliver multiple sustainability wins at once, while reducing the need for new transmission infrastructure in remote areas.

However, integrating food and energy production on the same land is not without challenges. Agrivoltaic and floatovoltaic systems often involve higher upfront costs, customized engineering, and interdisciplinary planning between energy developers, farmers, and local governments. Incentive structures in many countries continue to favor either agriculture or solar, rather than both. Unlocking the full potential of dual-use landscapes may require new policies, financing mechanisms, and land-use reforms that reward multifunctionality.

In addition to technical and economic dimensions, the intangible and cultural values of landscapes—such as food sovereignty, ecological stewardship, or community identity—must be part of the planning process. Agrivoltaics and its analogs offer an invitation to rethink how we produce and share resources in an increasingly climate-constrained world (Box 14.2) (Figure 14.4).

Solar Irrigation

Solar irrigation is an increasingly viable strategy for powering water delivery systems in agriculture, particularly in remote, off-grid, or energy-insecure areas (Durga et al. 2024). By harnessing solar PV panels to power pumps, solar irrigation systems reduce reliance on diesel generators or grid electricity—both of which can be expensive, unreliable, or unavailable in many rural regions. These systems can be installed on underutilized land or existing agricultural infrastructure such as barn rooftops, the

Box 14.2 Agrivoltaics for Multi-Functionality and Techno-Ecological Synergies

Agrivoltaics represents a promising frontier in sustainable land use, offering a model where agriculture and renewable energy generation work together rather than compete (Hernandez et al. 2019). This concept embodies techno-ecological synergy—a design principle where technological systems are deliberately integrated with ecological systems to produce co-benefits. For example, in arid or semi-arid climates, the partial shading from solar panels can reduce evapotranspiration, preserve soil moisture, and even extend growing seasons for shade-tolerant crops. In high-temperature regions, the ambient cooling effects from crops can, in turn, lower PV panel temperatures, potentially improving energy efficiency. In grazing applications, sheep and other livestock help control weeds under panels, providing dual-income opportunities for landowners. A techno-ecological approach encourages designers, farmers, and engineers to co-optimize across multiple dimensions: energy yield, food productivity, water efficiency, biodiversity, and cultural values. Research by Pascaris et al. (2021a) highlights that agrivoltaic projects with strong community involvement and benefit-sharing mechanisms tend to receive greater public support and deliver better outcomes across equity, ecosystem services, and land stewardship. Ultimately, agrivoltaics challenges us to move beyond siloed thinking. By designing systems that embrace complexity and seek synergies rather than trade-offs, we move closer to regenerative models of land use that respect both planetary limits and human livelihoods.

edges of fields, or in the unused corners of center pivot irrigation systems. Because PV panels have become much more affordable and durable over the last decade, and because the sun provides energy free of charge, solar irrigation presents a promising solution to reduce costs and emissions while improving energy access for farmers.

One of the key advantages of solar irrigation is its modular scalability and adaptability. It can be customized for smallholder farmers managing a few acres or for large-scale commercial operations. Systems can be designed as either surface pumps or submersible pumps, depending on the water table depth, and can be paired with drip or sprinkler irrigation to maximize water efficiency. In regions where electricity is expensive or unreliable, solar irrigation can stabilize water access and reduce fuel expenditures—making farming more climate-resilient and economically viable. Additionally, because solar-powered systems require minimal water to cool or operate compared to other energy sources, they align well with water conservation efforts. When combined with water-efficient irrigation practices, such as sensor-triggered irrigation or soil moisture monitoring, solar irrigation can reduce overall water withdrawals even as crop yields improve.

Figure 14.4 Agrivoltaics integrate agriculture and solar production, like this solar array installed over rice plants. Source: Σ64/Wikimedia Commons / CC BY 3.0.

However, solar irrigation is not without challenges. A major concern is the potential for over-extraction of groundwater. Because solar energy is essentially free once installed, and the marginal cost of pumping water approaches zero, farmers may be tempted to run their irrigation systems more often or at higher volumes. In places with limited or declining groundwater supplies, this could accelerate aquifer depletion. Policymakers and planners must consider governance frameworks, metering, or community water-sharing systems to ensure that solar irrigation does not lead to unsustainable water use. Moreover, solar irrigation is often promoted as a development intervention, including by agencies like the United Nations Food and Agriculture Organization (FAO), but the success of these programs depends heavily on technical support, maintenance, and appropriate financing models. Ensuring equitable access, especially for smallholder and women farmers, will be critical if solar irrigation is to be a truly inclusive solution for sustainable agriculture and climate resilience.

Highlights/Takeaways

- Food systems are major drivers of environmental change. The industrialization of food production—through synthetic fertilizers, mechanization, globalized supply chains, and energy-intensive processing—has dramatically increased the sector's carbon and water footprints. Agriculture alone accounts for roughly 25% of global GHG emissions, while also being one of the largest sources of water use and pollution through irrigation, fertilizer runoff, and livestock waste.

- Water and energy are deeply embedded in food production and vice versa. Modern agriculture depends on water for irrigation and energy for fertilizers, pumping, processing, and transportation. At the same time, energy systems require vast amounts of water, particularly for cooling, refining, and fuel extraction—highlighting how resource use in one domain can create vulnerabilities or stress in another.
- Emerging technologies like solar irrigation and agrivoltaics offer promising synergies.

  Innovations that combine renewable energy with agricultural production—such as solar-powered water pumps or co-located solar panels and crop fields—can improve efficiency, lower emissions, and create new forms of land-use optimization. However, these technologies must be carefully managed to avoid unintended consequences, such as over-pumping groundwater or competing land priorities.
- Wastewater and food waste are critical nodes for resource recovery and GHG reduction. Municipal wastewater systems consume large amounts of energy but also present opportunities for energy generation through anaerobic digestion and heat recovery. Similarly, food waste management through composting or biodigesters can reduce methane emissions while recovering nutrients and energy—highlighting the importance of closing loops within the FEWS nexus.

Questions for Review

1. Food and fiber production are major users of energy and generators of GHG emissions. What are some examples of ways that energy and greenhouse gasses can be reduced in these systems?
2. The potential for multiple uses of land or win-win benefits from conservation means that bringing multiple interdisciplinary perspectives together to problem-solve is critical. If these are such good ideas, what in your estimation is the reason there are not more widespread examples of this idea in practice?

Exercise 14.1

Inventory the energy and water inputs for a particular agricultural system. What are the options to replace the fuels with GHG emissions? Is there any potential for on-site energy generation from land or from waste?

Exercise 14.2

Tyner et al. (2010) updated the Searchinger et al. and Fargione et al. studies on land-use change (LUC) from corn ethanol and found LUC from grassland to corn conversion to be closer to 22 $MgCO_2$/ha. (1) How does that change the time to pay back biofuel debt for corn ethanol grown on grasslands? Estimate the volume of ethanol sold in California in 2025 from the resources listed on canvas. (2) How much LUC carbon debt is attributed to ethanol if 50% comes from corn, 25% from sugarcane, and 25% from cellulosic prairie biomass. If California wanted to replace all gasoline + ethanol used in the state with E85, with an ethanol mix of 50% from corn, 25% from sugarcane, and 25% from cellulosic prairie biomass. How much land would this require? Assume the gasoline causes zero LUC. Excluding land use change, what is the carbon intensity of E85 fuel with 50% of the ethanol from corn grown in California (Dry Mill; wet distillers grains with solubles; 80% natural gas [NG]; 20% Biomass) and 50% of the ethanol from Brazilian sugarcane (with average production process, mechanized harvesting, and electricity co-product credit). What percentage reduction in carbon intensity does this have compared to E10, where all the ethanol is processed in a wet mill using coal? Estimate the volume of diesel sold in California in 2013 from the resources listed on canvas. How much LUC carbon debt is attributed to this biofuel if 50% comes from soy, 25% from palm, and 25% from waste vegetable oil (WVO, assume no LUC for this)? If California wanted to replace all diesel with locally produced 50% soy and 50% WVO, how much land would this require? Excluding land use change, what is the carbon intensity of B100 fuel where the biodiesel blend is 25% biodiesel from waste oils requiring cooking, 25% biodiesel from waste oil not requiring cooking, 25% from tallow using the lower range, and 25% is Midwest soybean converted to renewable diesel?

Resources for this Exercise

Fargione 2008. Land Clearing and the Biofuel Carbon Debt. Science. http://www.sciencemag.org/content/319/5867/1235.short (Links to an external site.)

LCFS Lookup table of Carbon intensities of transportation fuels. www.arb.ca.gov/fuels/lcfs/121409lcfs_lutables.pdf

Expanded Lookup table here: www.arb.ca.gov/fuels/lcfs/lu_tables_11282012.pdf

Fuel Energy densities

http://www.ipst.gatech.edu/faculty/ragauskas_art/technical_reviews/Energy%20Basics.pdf

Statistics and Data on Ethanol and E85 as Transportation Fuels:

http://energyalmanac.ca.gov/transportation/ethanol.html

Biodiesel fact sheet:
www.energy.ca.gov/2005publications/CEC-600-2005-028/CEC-600-2005-028-FS.PDF

References

Buck, H.J. (2019). Challenges and opportunities of bioenergy with carbon capture and storage (BECCS) for communities. *Current Sustainable/Renewable Energy Reports* 6: 124–130.

Durga, N., Schmitter, P., Ringler, C. et al. (2024). Barriers to the uptake of solar-powered irrigation by smallholder farmers in sub-Saharan Africa: a review. *Energy Strategy Reviews* 51: 101294.

Fargione, J., Hill, J., Tilman, D. et al. (2008). Land clearing and the biofuel carbon debt. *Science* 319 (5867): 1235–1238.

Grubert, E. (2022). Yellow, red, and brown energy: leveraging water footprinting concepts for decarbonizing energy systems. *Environment, Development and Sustainability* https://doi.org/10.1007/s10668-022-02760-2.

Hernandez, R.R., Armstrong, A., Burney, J. et al. (2019). Techno–ecological synergies of solar energy for global sustainability. *Nature Sustainability* 2 (7): 560–568.

IPCC. (2022). "*Climate Change 2022: Impacts, Adaptation, and Vulnerability. Contribution of Working Group II to the Sixth Assessment Report of the Intergovernmental Panel on Climate Change.*" Cambridge University Press. Cambridge University Press, Cambridge, UK and New York, NY, USA. https://www.ipcc.ch/report/sixth-assessment-report-working-group-ii.

Pascaris, A.S., Schelly, C., Burnham, L., and Pearce, J.M. (2021a). Integrating solar energy with agriculture: industry perspectives on the market, community, and socio-political dimensions of Agrivoltaics. *Energy Research & Social Science* 75: 102023.

Pascaris, A. S, Schelly, C., Rouleau, M., and Pearce, J. M. (2021b). *Do Agrivoltaics Improve Public Support for Solar Photovoltaic Development? Survey Says: Yes!* Michigan Technological University, Houghton, MI (United States).

Pimentel, D. (2019). *Handbook of Energy Utilization in Agriculture*. CRC press.

Pimentel, D. and Pimentel, M.H. (2007). *Food, Energy, and Society*. CRC press.

Searchinger, T., Heimlich, R., Houghton, R.A. et al. (2008). Use of US croplands for biofuels increases greenhouse gases through emissions from land-use change. *Science* 319 (5867): 1238–1240.

Smil, V. (2000). Energy in the twentieth century: resources, conversions, costs, uses, and consequences. *Annual Review of Energy and the Environment* 25 (1): 21–51.

Tyner, W.E., Taheripour, F., Zhuang, Q. et al. (2010). Land use changes and consequent CO_2 emissions due to US corn ethanol production: A comprehensive analysis. *Department of Agricultural Economics, Purdue University* 1–90.

Further Reading

Favi, S.G., Adamou, R., Godjo, T. et al. (2024). Agrivoltaic systems offer symbiotic benefits across the water-energy-food-environment nexus in West Africa: a systematic review. *Energy Research & Social Science* 117: 103737.

Verma, N., Talwar, P., Upadhyay, A. et al. (2024). Food-energy-water nexus in compliance with sustainable development goals for integrating and managing the core environmental verticals for sustainable energy and circular economy. *Science of the Total Environment* 930: 172649.

Chapter 15

Energy Justice

"If we do not change our models and approaches to development, we are doomed to replicate the injustices that we see in our current energy system."

—Shalanda Baker

Decisions regarding energy choices have historically led to significant and disproportionate impacts on communities. The concept of energy justice has emerged as an essential component to integrate into today's energy and climate policies. But often it's still more in name than in practice. So, for many change-makers, the question is how to facilitate energy transitions that materially benefit communities. This means impacting communities in terms of employment and community investment, public health, and education, among other key factors, to make healthy communities the norm. Energy is an essential resource for conducting the activities of everyday life, making technology and resource studies important research areas, as energy transition implies large-scale changes, and this work could and will identify new challenges facing communities and the environment. Implicit in the idea of energy justice is the notion that people should in general both receive the benefits of energy and also not be unequally exposed whether in the home or the workplace. However, there are different interpretations of energy justice, which we will explore in this chapter. The chapter will also highlight some of the key research findings from scholars in this field

Figure 15.1 Energy justice for many activists means moving away from fossil fuels. Source: Stephen Hass / Flickr.

and identify important themes for the future. This topic needs to be approached with humility and care, with attention to listening carefully to the most vulnerable and disaffected. There is nothing intrinsic to the idea of energy justice that can be handed down. The "just" in just transitions should be an open conversation between impacted and benefited communities (Figure 15.1).

Critical Environmental Justice

Environmental injustices are prevalent all around us, often hidden from the everyday activities of most people or do not garner the headlines they should. But we, through our energy and transportation infrastructures, especially cause environmental injustices. Inspired by the work of David Pellow and Andy Szasz, this chapter aims to think about environmental inequality through a critical environmental justice (EJ) lens. In a pivotal paper for my own normative thinking around justice and the environment, Professor Pellow wrote several research papers on "environmental inequality formation" (Pellow 2000). In these papers, the research showed the importance of social movements in the fight for better lives, but also the challenges of EJ problems that do not have the popular news appeal, thus garnering less media attention. Examining injustice involves shedding light on questions of distributional justice (who gets what), recognitional justice (whose lives matter), and procedural justice (whose voice counts) (Jenkins et al. 2016). Critical EJ goes beyond "where we live, work, learn, and play"

Box 15.1 The Environmental Justice Dimensions of the Exide Lead Battery Recycling Plant

The recovery and recycling of materials are key components of a more circular economy. However, materials recovery and recycling facilities can expose communities to heavy metals and other air emissions. Advocates promising a closed-loop system for lead highlight lead recycling as a model for a circular economy. In the industrialized world, nearly 98% lead is recycled, primarily due to the use of lead-acid batteries in automobiles. Auto service stations serve as an ideal node in the reverse logistics that makes recycling possible. However, the case of Exide battery recycling in Southern California raises concerns about the feasibility of circular economies involving toxic heavy metals. This situation is one reason why circular economy principles often focus on the elimination of hazardous materials as important for sustainability and justice. The community continues to seek environmental justice for the cleanup from the company and regulators at the state's Department of Toxic Substances Control.

and adds in the spaces from which the commodities that populate these places originated and where they end up (Box 15.1).

Many EJ studies investigate the geographical distribution of environmental harms and their associated public health consequences (highways, incinerators, and hazardous waste dumps being, among the other things). Early works in the EJ research canon include detailing the community opposition to a PCB landfill in a minority Warren County, North Carolina (Bullard 1994). A major point made by one of the leading EJ scholars, Robert Bullard, is that the costs of environmental harms are disproportionately borne by particular communities, while the benefits are diffused throughout a region or the country. Congressman Walter Fauntroy (D-DC) requested that the U.S. Government Accountability Office investigate waste sites in relation to race and income. The findings revealed that three out of four hazardous waste landfills were located in neighborhoods primarily inhabited by poor and African American residents (Williams 1999). The environmental justice movement has many elements that emerged from the civil rights movement, particularly with relation to organized religion. The 1987 report by the United Church of Christ's Commission for Racial Justice entitled "Toxic Wastes and Race in the U.S." Additionally, EO 12898 aims to address EJ concerns in minority and low-income populations through federal actions.

Pellow's work on recycling in Chicago indicates that most stakeholders viewed the plant as a means to reduce waste and create jobs in an area where such opportunities are scarce. Szasz and Meuser point out the importance of understanding the historical dynamics surrounding the formation of environmental injustices. They also point to the imagery around which environmental problems were identified and point to how some impacts are more visible and other are not (Box 15.2). In the case of Silicon

Box 15.2 Bucket Brigades Bring Citizen Science to Fenceline Communities

Some communities have organized themselves to tackle pollution issues in already overburdened fenceline communities. Around the world, there are examples of communities organizing against big infrastructure projects that pollute their air and water. Many groups fighting against power plants and other polluting industrial facilities have used citizen science to collect data about the levels of pollution in their environment. One example is the Louisiana Bucket Brigade, which organized local residents to collect water samples near some of the most polluting facilities in the United States (Adapted from Ottinger 2010).

Valley, which has become home to the largest collection of Superfund sites in the world and a source of occupational illness for many of its workers, "the term high technology evokes images of clean rooms and workers in bunny suits, utterly different from the smokestacks and stockyards of the older industrial city" (Szasz and Meuser 2000).

What is EJ? This question deserves a careful response, as interpretations may vary, even if subtly. Writ broadly, environmental justice is concerned with unequal distribution of risks, particularly with vulnerable populations.

What then causes environmental injustice from an empirical perspective, as in, how did we get here? What are the mechanisms to fix it?

Two additional aspects of the broader theme of justice worth critically exploring are distributional justice and procedural fairness. Procedural justice is based on the idea that fairness arises from adequate public participation, consultation, inclusion, and consent in the decision-making process. An infamous case of poor public notice regarding the siting of industrial facilities is found in Kettleman City, California, where the largest hazardous waste facility west of the Mississippi River—the Kettleman Hills Hazardous Waste landfill—was proposed in a community that is primarily Latino and immigrant. Public notices required by law were posted in English, a language not spoken by most community members, resulting in minimal public involvement in the project advancement. As a result, the community was not offered an opportunity to weigh in. The case was challenged in the courts, where the ruling favored the community but only mandated that future developments provide notices in languages reflective of the community's composition. Since the facility began operations, numerous controversial research reports have emerged, claiming connections between the toxic waste facility in Kettleman City and a rise in birth defects, making it an important case study to look at the intersection of critical race theory and contested illness (Richter 2018). So, the struggle to find justice for the Kettleman City community continues to this day, as questions remain about the sources of illness and what can be done to protect these communities from industrial pollution. One wonders what the outcomes would have been with greater participation in the siting process.

It is essential to avoid conflicts between the deployment of clean and renewable energy and the well-being of communities and landscapes (Ko et al. 2011). Environmental pressures from renewable energy sources are already emerging from land use changes, extractive industries, waste, and labor forces occupying sites of extraction and manufacturing facilities in new places across the world. For example, there are instances where solar parks in India cut off land access for the landless, suggesting that some voices go unheard in the evolving energy landscape (Yenneti and Day 2015). Similarly, cases from Southern Mexico suggest that a lack of democratic processes could lead to further enclosure of communal lands, loss of sovereignty, and the uneven accrual of privately held benefits (Avila-Calero 2018).

Case studies of large-scale solar energy in the Mayan region of Yucatan, Mexico, have led to procedural injustices where local communities have little input into energy landscape developments and few benefits from them (Avila-Calero 2018; Barragan-Contreras 2022). By failing to prioritize community participation, consent, and direct benefits, common development approaches could reproduce the severe aspects of the decarbonization divide that we are already witnessing globally (Sovacool et al. 2020). This points to a need for more research strategies that incorporate or are led by community voices and perspectives (Box 15.3).

Finally, what do we mean by energy justice? How does it differ from EJ? Energy justice scholars have written extensively on what energy justice means in various

Box 15.3 Participatory Action Research or Community-Based Participatory Research

One key tool for identifying environmental justice issues is participatory action research. This approach sees the researcher as an ally of the community, looking for something of mutual benefit rather than just passively observing. Participatory Action Research teams literally get involved in the subjects they study, to varying degrees. Another variation is Community-Based Participatory Research, which is a systematic mode of inquiry that directly involves the community affected by the issue of study. A community-based research project can help better identify the spatial distribution and cumulative impacts of critical hotspots that cause community environmental health impacts. So, how do we empower these ideas? If we care about environmental injustice, what actions should we take to limit its impact and reach? If we are being true to ideas of justice, how do we ensure that we understand the perspectives we claim to value, as well as what the community truly wants and needs?

contexts, and some provide us with a typology of what they mean or what they do (Jenkins et al. 2016). In the next section, we start to explore the many dimensions of energy justice by highlighting the importance of energy access.

Energy Poverty

Nearly 3 billion people on the planet use less electricity than a typical American refrigerator. Additionally, about 1 billion people do not have access to clean cooking fuels, resulting in high exposures to indoor air pollution. As of 2020, the number of people on the planet without access to reliable electricity could be as high as 3.5 billion (Ayaburi et al. 2020). The numbers are truly astounding. The extent to which people in 2025 lack access to reliable modern energy services is possibly the most stark example of the depths of global inequality. These communities are also more likely to be exposed to natural disasters (Sovacool et al. 2016; Wisner et al. 2004). The United Nations' Sustainable Development Goal Seven aims to correct this by promoting universal access to clean energy by 2030 (Figure 15.2).

Energy access is a severe justice problem in the poorest and most vulnerable parts of the world. Without access to basic energy services, such as electricity and clean cooking fuels, people are exposed to pollution, and it makes it challenging to improve their lives. The dynamics of poverty often exacerbate environmental inequality, as communities seek the best means to make themselves less vulnerable, a topic often explored by anthropologists, sociologists, and human geographers. A just transition means offering opportunities to elevate livelihoods for those always left behind in

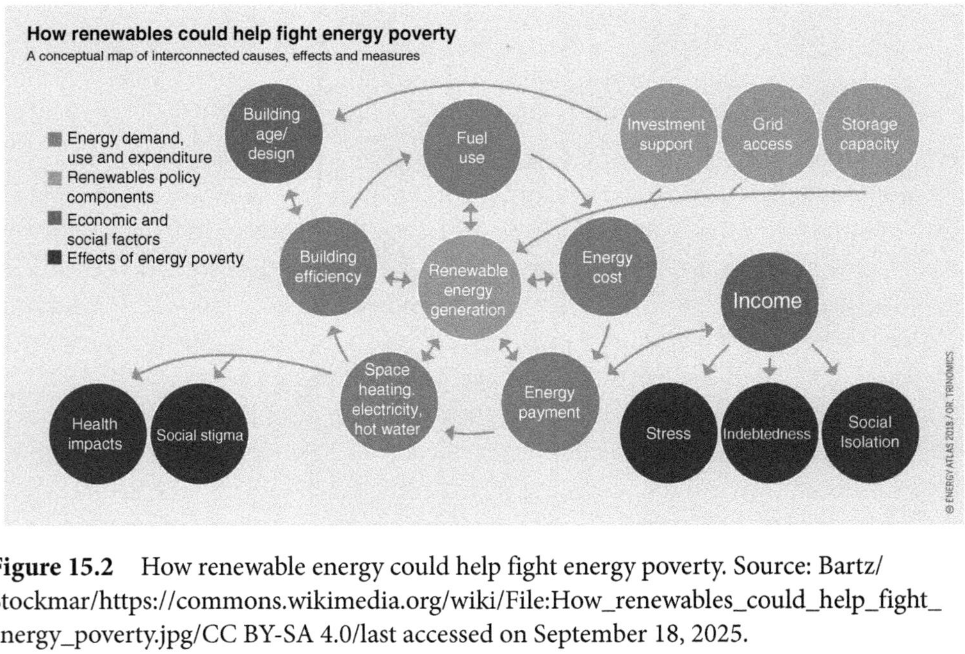

Figure 15.2 How renewable energy could help fight energy poverty. Source: Bartz/Stockmar/https://commons.wikimedia.org/wiki/File:How_renewables_could_help_fight_energy_poverty.jpg/CC BY-SA 4.0/last accessed on September 18, 2025.

poverty by developing the capabilities and social capital of people, households, and communities. The transition to solar power in India has not necessarily led to increased energy access in India, where policies tend to promote renewables for wealthier households at the expense of the poor and vulnerable (Bedi 2022).

Each year, hundreds of people die every year trying to siphon off fuel from broken infrastructures such as failed pipelines or crashed tanker trucks that catch fire and explode. There have been notable accidents in Haiti, Nigeria, and Kenya in recent years. These accidents show the degree to which people will go to access otherwise expensive or difficult-to-get diesel, gasoline, and crude oil, whichever piece of infrastructure comes into contact with people desperate enough to risk their lives trying to obtain it.

How can we work to remove these inequalities of access to reliable energy? What are the best strategies? These are similar questions that have been asked for decades and centuries or more, long before climate change arose as a common concern. Some scholars have suggested fostering better organization through North–South solidarity across the supply chain (Bell 2024). Geographer Siddharth Sareen talks about "solaric solidarities," which refer to transformative ways that solar governance transitions can be grounded not just in technology and economics, but in social justice, equity, and mutual care (Sareen et al. 2023). Unlike top-down electrification models, solaric solidarities foreground *who* participates in, *who* gains from, and *who* bears the risks of solar energy development—particularly in extractive or marginalized regions.

Decolonizing Energy

The global damage caused by the particularly violent and destructive form of colonization that began in the fifteenth century continues to affect us today. This damage is not only biophysical and ecological, but also contributes to the ongoing inequality present in our world. These histories are still with us today materially (in the form of property, where people live, etc.), but also socially and embedded in our knowledge systems. When people discuss the need to decolonize energy, they take on several meanings. In one sense, decolonizing energy means centering the historical legacy of colonialism in our understanding of natural resources and land use conflicts. It also implies benefiting Indigenous peoples or tribes through rural electrification and the general development of energy systems. Alternatively, it could be used as a lens to understand the history of energy development and community impacts on colonized lands, for example, with the development of coal resources in the Navajo Nation (Powell 2018). Myles Lennon, for example, speaks of decolonizing technical expertise to widen the support base for political mobilization and material benefits (Lennon 2017). We also might use the decolonial lens to understand and critique mainstream understandings of related matters of justice (Alexander 2023). These approaches can also inform strategies for addressing past harms (Box 15.4).

Box 15.4 The Case for Climate Reparations as Energy Justice

The concept of reparations—where harmed communities receive material compensation for historical and ongoing injustices—has traditionally been associated with the legacies of slavery, colonization, and systemic racial violence. However, it is now gaining traction in climate and energy justice movements as both a moral imperative and a pragmatic framework for equitable climate action. In the context of global energy transitions, *climate reparations* refer to compensatory investments made by historically high-emitting nations or corporations to support adaptation, mitigation, and resilience-building in communities disproportionately impacted by climate change, particularly in the Global South. The intellectual roots of climate reparations intersect with the principle of *common but differentiated responsibilities and respective capabilities*, a foundational norm in international environmental law enshrined in the 1992 UN Framework Convention on Climate Change (UNFCCC). This principle acknowledges that while all countries share responsibility for addressing climate change, wealthier, industrialized nations bear a greater burden due to their historical emissions and their far greater material and financial capacities. As Olúfẹ́mi Táíwò (2022) argued, reparations should not be seen as mere charity or guilt payments, but rather as structural correction—a redirection of resources and power toward justice-based planetary stewardship.

In energy terms, climate reparations could include debt forgiveness, funding for distributed renewable energy initiatives, capacity-building for community energy projects, technology transfers, and investments in adaptation infrastructure such as flood defenses or drought-resistant agriculture. These actions go beyond generic development aid by explicitly acknowledging responsibilities for harm and centering agency of recipient communities. As Salazar (2023) pointed out, climate reparations are also a strategy for avoiding "climate debt traps," where the poorest countries must borrow money to adapt to a crisis they did not cause—further reinforcing global inequality. Critics argue that reparations are politically infeasible or lack clear mechanisms for implementation. But growing support for frameworks such as the Loss and Damage Fund established under the UNFCCC, along with pressure from frontline movements, Indigenous groups, and youth advocates, suggests that momentum for reparative justice is building. As climate impacts worsen and transitional policies risk reproducing historical injustices, climate reparations present a transformative lens—reimagining the global energy transition not only as a decarbonization effort, but as an opportunity to redress past harm, redistribute power, and regenerate planetary systems in a spirit of solidarity and care (Salazar 2023; Táíwò 2022).

Fenceline Communities Vulnerable to Pollution

Fenceline communities are those that are the most exposed to environmental pollution. As Healy et al. pointed out, there may be embodied environmental injustices in these global commodity chains. Early EJ research explorations of the topic looked at the question of EJ as solely a spatial problem, looking at pollution exposures, plumes, burdens, and other potential risks such as fires and accidents. This led to questions about how these situations arose: Did industrial facilities precede the establishment of residential neighborhoods, or did minorities move into areas already affected by pollution? (Pastor et al. 2001).

The second consideration is justice for the fenceline communities who live, breathe, and depend on water impacted by the fossil fuel economy (Box 15.5). These injustices are widespread and global, spanning the poorer parts of affluent parts of the world—from the Niger Delta to the Port of Richmond, California. These extractive and processing industries are usually located in the least developed and most marginalized communities on Earth (Watts 2001). These dynamics are reproduced in the marginalized parts of wealthier countries as well, suggesting that the distribution of

Box 15.5 The EJ Contradictions of California

Despite its great wealth and public infrastructures, urban areas in California have many communities struggling with poor health outcomes, particularly in environmental justice and fenceline communities located near energy infrastructures (Morello-Frosch et al. 2001; Morello-Frosch et al. 2001). In Northern California, neighborhoods such as East Oakland in Alameda County remain heavily impacted by industrial stationary and mobile sources of air pollution located near homes, schools, recreation centers, and churches. A pointed example of a polluting industry plaguing this community is the American Brass & Iron Foundry (AB&I). Researchers from Communities for a Better Environment, working in a participatory context with the community, found that homes located within the immediate AB&I vicinity suffered from airborne particulate matter (PM 2.5) levels over 50 $\mu g/m^3$, which exceeds California's air quality standards (Communities for a Better Environment 2023). These issues are exacerbated by the lack of greenspace, which is unequally distributed in East Oakland by neighborhood racial characteristics (Kibel 2007). East Oakland's childhood asthma hospitalization rates are 150–200% higher than the county as a whole, and life expectancy in East Oakland for the years 2000–2003 was almost seven years lower than the county average (Alameda Public Health Department 2006). East Oakland is a prime example of cumulative impacts stemming from overexposure to toxic pollution and underexposure to spaces provisioning ecosystem services, resulting in poor community health outcomes. In the Los Angeles region, major polluting facilities are located in neighborhoods that have lower incomes and higher concentrations of racial minorities. The region has some of the highest levels of particulate pollution and smog in the country. In Southeast Los Angeles County, several polluting industries such as chrome plating; hazardous waste treatment, storage, and disposal facilities; and other major toxic emitters combine with a high concentration of mobile sources of pollution to create a toxic hot spot. Community groups have taken the lead in advocating for restoration and justice in these communities, driving efforts to reclaim soil contaminated by the Exide Technologies Battery Recycling Facility in Vernon, California, which was recently closed due to community demand organized by CBE. At the request of California EPA, the California Department of Public Health's Childhood Lead Poisoning Prevention Branch Studies found that children living within 1 mile of the Exide facility had significantly higher blood lead levels compared to children living over a mile away (2016).

environmental harms occurs across a riskscape (Morello-Frosch, (2001); Pastor et al. 2001). Air pollution exposures, groundwater contamination, and living near hazardous waste facilities connected to the energy economy disproportionately expose the poor to public health and occupational risks. Moreover, these fenceline communities may not benefit significantly from these facilities. A just transition means addressing challenges of EJ and ecosystem disservices that affect these communities.

In the context of metals and minerals mining related to the energy transition, it is well-known in the United States that mining often occurs in and around Native American communities. This pattern is also evident in South and Central America, Australia, and China. Treatment of natural resource issues that pertain to Indigenous communities should be familiar with Convention 169 of the International Labor Organization, which guarantees the rights of tribal peoples (International Labor Organization 1989).

Giving ownership of the energy transition process to local communities can have far-reaching consequences for energy justice. We see case examples around the world. In recent years Morocco has seen varying degrees of success regarding large-scale solar energy power plants, and it is hoped that these lessons learned from these projects will inform future initiatives and community benefits offered to local groups.

Labor

The phrase "just transition" implies that technological and social change occur unevenly. While some core ideas go much deeper, the concept of a just transition, often used to describe energy transitions or social responses to climate change, reaches back several decades, starting in some accounts with the identification of the issue initially with labor unions and other worker-led groups (Stevis and Felli 2020). Labor and justice are central themes of recent publications on just transitions. Many narratives are anchored in a moral framework or a transactional one that focuses on the political viability of a change to energy systems without regard to the outfalls. A common just transition refrain has been to plan for a situation to ensure that workers are reemployed or are provided social safety nets from climate policies. Just transition for workers in the coal industry is a common theme, although the contours of that debate are complicated by other socio-ecological dynamics, community history, and personal identities. In some coal-dependent communities, jobs in the solar industry could replace some of the lost income and jobs. One case study of India's most coal-dependent states such as Chhattisgarh (22.2%), Odisha (19.8%), Jharkhand (18.5%), and Madhya Pradesh (16.3%) suggests that these communities will need some kind of retraining or safety net to avoid the worst impacts and to replace the energy jobs lost during the transition away from coal, more than 50% of new capacity additions for solar and wind would need to focus on these four states (Malik and Bertram 2022).

Justice is also invoked for those who will lose employment in the fossil fuel industries, particularly in the western United States, where coal facilities are closing (Haggerty et al. 2018). The reemployment or creation of some new opportunities in transitioning regions are areas of widespread focus because the loss of high-wage jobs is real, and some further suggest it is an ethical imperative to ensure workers who may be employed in an industry driven to change by public policy. The argument presumes that public policy interventions prematurely end an industry, and, therefore, the public is obligated to ensure or help workers to find new employment. For this latter reason, it seems, these just transition approaches favor workers in incumbent industries. Given the initial emphasis on public policy and speculative industrial ventures, workers in, for example, solar industries could also warrant support in jobs that might be lost from new ventures or bankruptcies in new industries. This raises some practical questions about whether it might be simpler to have programs be economy-wide rather than targeted at industries, unless there are specific dimensions that could be drawn upon for fund processes, such as ratepayer funds from utility commissions to support the recovery of workers in the electricity industries.

One study of employment structure and wages in Germany found that jobs in the renewables industries paid a small premium over jobs in non-renewable sectors (Antoni et al. 2015).

The so-called high-road labor policies could encourage the support of good quality, higher-paying jobs rather than those that just rely on a laissez-faire or hands-off approach that assumes these things work themselves out in the market without much difference in cost (Mayfield and Jenkins 2021).

Just Transition

The decarbonization of the global economy is rearranging labor markets and altering geographies and worker flows. The "just transition" concept calls attention to justice issues that arise in the transition from fossil fuels to a low-carbon economy (McCauley and Heffron 2018). Discussions, such as writing, scholarship, and policy, around just transitions encompass justice considerations around several core themes (Newell and Mulvaney 2013). One important theme is justice for the fenceline communities who live, breathe, and depend on water impacted by the fossil fuel economy and the "riskscapes" of toxic landscapes (Morello-Frosch et al. 2001). These injustices are widespread and global, spanning both the poorer areas of affluent parts of the world and the least developed and most marginalized communities on Earth.

The just transition can also be described as a means to gain public acceptance of a low-carbon transition by emphasizing new opportunities in emerging industries. A just transition for workers in the coal industry, for example, is a common theme in policy and practice, where workers are given compensation or retraining to better navigate evolving circumstances. However, simply designing policies to manage just transitions is complicated by questions around who benefits from these programs and who is left out.

The call for a just transition goes back several decades, with the issue initially identified by labor unions and other worker-led groups (Stevis and Felli 2020). In the 1990s, claims that regulations in the United States and Europe offshored petrochemical industries to the developing world led to calls for a just transition for those workers in the form of retraining, compensation, or, more broadly, some plans to ensure that workers are reemployed or provided social safety nets. Today, the concept is widely used in geography, political science, sociology, and by activists and policymakers to describe industries impacted by climate policies. Specific attention is paid to fossil fuel producers and workers, particularly with coal power plant and coal mine closures that are likely to result in significant industrial decomposition and employment loss. This is an area of widespread focus, with some arguing that it is an ethical imperative because of its nexus with public policy. The argument presumes that public policy interventions prematurely end an industry's financial viability and undermine its workforce (and unions in some cases), and therefore the public is obligated to ensure or help workers to continue to find work. For this latter reason, it seems that these approaches favor workers in incumbent industries, particularly the coal industry, because of case studies of targeted policies or plant closures. However, given the initial emphasis on public policy, workers in renewable energy industries could also warrant support in jobs that might be lost from new ventures or bankruptcies. For example, this could include public investments in new photovoltaics manufacturing start-ups that fail. Are they worthy of a similar safety net? It is arguably simpler to have training and reskilling programs be economy-wide rather than targeted at employees from specific industries.

While the United States lacks a nationwide just transition policy, several states have developed policies that align with the ideals of a just transition. On the other hand, many elements of the just transition away from coal in China, where coal accounts for more than 60% of primary energy use, come from the top-down initiatives, as part of China's 5-year planning process. Europe has developed a just transition policy that aims to invest in countries and regions impacted by the transition to low-carbon energy to prepare them for future jobs.

Justice in Just Transitions

Energy justice is gaining prominence and being integrated into formal policies, led by a groundswell of activism that helped elevate the topic. The U.S. Department of Energy now has a dedicated position to understand and develop policy around ensuring it. In 2021, Shalanda Baker was named Secretarial Advisor on Equity and Deputy Director for Energy Justice at the U.S. Department of Energy, where she lead many initiatives aimed at empowering vulnerable communities across the United States.

The varieties of justice invoked in just transition deserve a long conversation in parallel. Justice for whom? Justice for which species or multi-species? How is justice operationalized and put into practice? Philosophers, social scientists, and others from the arts and humanities have speculated on ideas of justice, suggesting how complicated the concept is, even before adding the content of the themes explored in this chapter. The following section focuses less on these questions about the justifications and varieties of justice, and more on the empirical observations of broad EJ themes along solar power commodity chains.

Why should justice be at the center of decarbonization strategies? Some might say getting to decarbonization as quickly as possible would have the greatest effect, so low cost should be at the core. There are moral and ethical reasons that suggest that the answer is yes. Prioritizing EJ can lead to better overall outcomes, even if that is not always guaranteed. Public policies aimed at benefiting the poor do not need to be so instrumental. They are good reasons for doing what is right. An explicit focus on justice can bring attention to lessons to be learned for a better transition to renewable energy, one that aims to improve both people's lives and environmental quality. How then can we also integrate ecosystem services or the cultural resource dimensions of EJ? As we move to a low-carbon energy system that ideally minimizes resource extraction and environmental impact, emphasizing justice as a key principle will not only help us achieve environmental and climate goals, but also social and household-level benefits.

There are several varieties of justice invoked in calls for a just energy transition. Some refer explicitly to fossil fuels. But justice for and according to whom? Philosophers, social scientists, and others from the arts and humanities have long speculated on ideas of justice and its complexity. A rough typology of EJ considerations for energy transitions must focus on themes related to the distribution of environmental harms and benefits, procedural issues related to participation, prior and informed consent with Indigenous groups, the quality of stakeholder involvements, and recognitional issues related to unanswered questions from the past that may be critical to restoring just social relations with fenceline communities and Indigenous groups (Marion Suiseeya et al. 2022). Other considerations of justice refer to raising questions of redistribution and recognition as the connections between impacted communities are framed around social group identities instead of class, raising challenges for social organizing (Fraser 2020).

Energy decision-makers are increasingly finding EJ at the center of consideration of the issues they work on (Sovacool et al. 2016). Whether it's the climate impacts and the distribution of impacts versus historical benefits, or refinery pollution from the new clean energy economy, there will continue to be a core interest in addressing the unequal impacts of the use of and pollution from energy.

Highlights/Takeaways

- Energy justice centers the people most impacted by energy systems.
  Whether through pollution, extraction, or exclusion, many communities—especially low-income, Indigenous, or racialized populations—have historically borne the greatest burdens of energy systems while receiving few of the benefits. Energy justice demands a restructuring of energy governance to prioritize fair access to clean, affordable energy, equitable participation in decision-making, and the redress of historical harms.
- Just transitions must address labor, land, and livelihoods. As societies move away from fossil fuels, ensuring that workers and communities dependent on legacy energy industries are not left behind is critical. A just transition involves not only reemployment and reskilling programs but also community participation, recognition of cultural ties to land, and investment in new, locally beneficial energy infrastructures.
- Procedural and recognitional justice matter as much as outcomes. Environmental injustices are often rooted not just in unequal exposure to harm, but in the failure to engage and recognize marginalized communities in planning and policy. Cases like Kettleman City and solar developments in India and Mexico illustrate how a lack of participation, consent, or recognition can deepen injustice even in seemingly sustainable projects.
- Climate reparations and decolonizing energy offer transformative visions for a more equitable and sustainable relationship between society, energy, and the environment. Beyond mitigation, energy justice movements advocate for reparative policies that account for colonial histories, disproportionate emissions, and current inequalities. Concepts such as *solaric solidarities* and *climate reparations* suggest pathways toward redistributing power, enabling collective self-determination, and centering care and equity in global energy transitions.

Questions for Review

1. Energy justice as a core theme in any energy transition. It emphasizes the importance of ensuring access to modern energy sources, addressing injustices present in the fossil fuel economy, and ensuring justice in the emerging renewable energy technologies at the vanguard of energy transitions.
2. Justice can mean one or more of several dimensions including distributive, procedural, reparative, restorative, and there are even more typologies developed by people thinking about these questions. Make a list of the different types of environmental justice and give one example of each, with a solution for how the injustices might better level the scales.

Exercise 15.1

What kinds of environmental justice issues are most common in your region? Identify a local environmental justice issue and try to identify the main actors such as the companies, organizations, and government agencies and what role they play in addressing the issue.

Exercise 15.2

Think about an issue that might have environmental justice impacts on a community, and think about how you might engage them in a research project. What might you propose to do? How might you help? What can you offer to the impacted community?

Exercise 15.3

Design a cookstove or solar electrification program that addresses energy poverty issues in an underdeveloped context. How would you finance the stoves? How would you verify their use or their benefits? What safeguards for program success would you consider? Spend an hour researching the issues surrounding technology adoption in energy poverty contexts and design a program that utilizes best practices.

References

Alameda County Public Health Department. (2006). Alameda County Health Status Report 2006. Community Assessment, Planning and Evaluation Unit.

Alexander, D. (2023). Decolonizing energy justice: identity politics, attachment to land, and warfare. *Progress in Human Geography* 47 (2): 348–350.

Antoni, M., Janser, M., and Lehmer, F. (2015). The hidden winners of renewable energy promotion: insights into sector-specific wage differentials. *Energy Policy* 86: 595–613.

Avila-Calero, S. (2018). Environmental justice and the expanding geography of wind power conflicts. *Sustainability Science* 13 (3): 599–616.

Ayaburi, J., Bazilian, M., Kincer, J., and Moss, T. (2020). Measuring "Reasonably Reliable" access to electricity services. *The Electricity Journal* 33 (7): 106828.

Barragan-Contreras, S.J. (2022). Procedural injustices in large-scale solar energy: a case study in the Mayan region of Yucatan, Mexico. *Journal of Environmental Policy & Planning* 24 (4): 375–390.

Bedi, H.P. (2022). Solar power for some? Energy transition injustices in Kerala, India. *Environment and Planning E: Nature and Space* 5 (3): 1146–1163.

Bell, K. (2024). International solidarity for a De-colonised just transition: electric vehicles and lithium in Mexico and Europe. *Contemporary Social Science* 19 (1–3): 66–85. https://doi.org/10.1080/21582041.2024.2302615.

Bullard, Robert D. 1994. *Unequal Protection: Environmental Justice and Communities of Color.*

Communities for a Better Environment. (2023). Clean Up at ABI. https://www.cbecal.org/initiative/clean-up-abi/ accessed December 29, 2025).

Fraser, N. (2020). From redistribution to recognition?: dilemmas of justice in a 'post-socialist' age. In: *The New Social Theory Reader* (ed. S. Seidman and J. Alexander). Routledge.

Haggerty, J.H., Kroepsch, A.C., Walsh, K.B. et al. (2018). Geographies of impact and the impacts of geography: unconventional oil and gas in the American West. *The Extractive Industries and Society* 5 (4): 619–633.

International Labor Organization (1989). "Indigenous and tribal peoples convention." https://www.ilo.org/dyn/normlex/en/f?p=NORMLEXPUB:55:0::NO::P55_TYPE,P55_LANG,P55_DOCUMENT,P55_NODE:REV,en,C169,/Document.

Jenkins, K., McCauley, D., Heffron, R. et al. (2016). Energy justice: a conceptual review. *Energy Research & Social Science* 11: 174–182.

Kibel, P.S. (2007). The People Down the Hill: Parks Equity in San Francisco's East Bay. *Golden Gate University Environmental Law Journal* 1 (2): 331–404.

Ko, Y., Schubert, D.K., and Hester, R.T. (2011). A conflict of greens: green development versus habitat preservation–the case of Incheon, South Korea. *Environment* 53 (3): 3–17.

Lennon, M. (2017). Decolonizing energy: black lives matter and Technoscientific expertise amid solar transitions. *Energy Research & Social Science* 30: 18–27.

Malik, A. and Bertram, C. (2022). Solar energy as an early just transition opportunity for coal-bearing states in India. *Environmental Research Letters* 17 (3): 034011.

Mayfield, E. and Jenkins, J. (2021). Influence of high road labor policies and practices on renewable energy costs, decarbonization pathways, and labor outcomes. *Environmental Research Letters* 16 (12): 124012.

McCauley, D. and Heffron, R. (2018). Just transition: Integrating climate, energy and environmental justice. *Energy Policy* 119: 1–7.

Morello-Frosch, R., Pastor, M., and Sadd, J. (2001). Environmental justice and Southern California's 'Riskscape' the distribution of air toxics exposures and health risks among diverse communities. *Urban Affairs Review* 36 (4): 551–578.

Newell, P. and Mulvaney, D. (2013). The Political Economy of the 'Just Transition.'. *The Geographical Journal* 179 (2): 132–140.

Ottinger, G. (2010). Buckets of resistance: standards and the effectiveness of citizen science. *Science, Technology, & Human Values* 35 (2): 244–270.

Pastor, M., Sadd, J., and Hipp, J. (2001). Which came first? Toxic facilities, minority move-in, and environmental justice. *Journal of Urban Affairs* 23 (1): 1–21.

Pellow, D.N. (2000). Environmental inequality formation: toward a theory of environmental injustice. *American Behavioral Scientist* 43 (4): 581–601.

Powell, D.E. (2018). *Landscapes of Power: Politics of Energy in the Navajo Nation*. Duke University Press.

Richter, L. (2018). Constructing insignificance: critical race perspectives on institutional failure in environmental justice communities. *Environmental Sociology* 4 (1): 107–121.

Salazar, A. (2023). "The case for climate reparations in the United States." Roosevelt Institute, April 4. https://rooseveltinstitute.org/publications/the-case-for-climate-reparations-in-the-united-states.

Sareen, S., Sorman, A.H., Stock, R. et al. (2023). Solidaric Solarities: governance principles for transforming solar power relations. *Progress in Environmental Geography* 2 (3): 143–165.

Sovacool, B.K., Heffron, R.J., McCauley, D., and Goldthau, A. (2016). Energy decisions reframed as justice and ethical concerns. *Nature Energy* 1 (5): 1–6.

Sovacool, B.K., Hook, A., Martiskainen, M. et al. (2020). The decarbonisation divide: contextualizing landscapes of low-carbon exploitation and toxicity in Africa. *Global Environmental Change* 60: 102028.

Stevis, D. and Felli, R. (2020). Planetary just transition? How inclusive and how just. *Earth System Governance* 6: 100065.

Suiseeya, M., Kimberly, R., O'connell, M.G. et al. (2022). Waking from paralysis: revitalizing conceptions of climate knowledge and justice for more effective climate action. *The Annals of the American Academy of Political and Social Science* 700 (1): 166–182.

Szasz, A. and Meuser, M. (2000). Unintended, inexorable: the production of environmental inequalities in Santa Clara County, California. *American Behavioral Scientist* 43 (4): 602–632.

Táíwò, Olúfẹ́mi O. (2022). *Reconsidering Reparations*. Oxford University Press.

Williams, R.W. (1999). Environmental injustice in America and its politics of scale. *Political Geography* 18 (1): 49–73.

Wisner, B., Blaikie, P., Cannon, T., and Davis, I. (2004). *At Risk: Natural Hazards, People's Vulnerability and Disasters*. Routledge.

Yenneti, K. and Day, R. (2015). Procedural (in) justice in the implementation of solar energy: The case of Charanaka solar park, Gujarat, India. *Energy Policy* 86: 664–673.

Further Reading

Bell, K. (2021). *Diversity and Inclusion in Environmentalism*. Abingdon: Routledge.

Gilio-Whitaker, D. (2019). *As Long as Grass Grows: The Indigenous Fight for Environmental Justice, from Colonization to Standing Rock*. Boston: Beacon Press.

Lennon, M. (2021). Energy transitions in a time of intersecting precarities: from reductive environmentalism to antiracist praxis. *Energy Research & Social Science* 73: 101930.

Schlosberg, D. (2007). *Defining Environmental Justice: Theories, Movements, and Nature.* OUP Oxford.

Sovacool, B.K., Newell, P., Carley, S., and Fanzo, J. (2022). Equity, technological innovation and sustainable behaviour in a low-carbon future. *Nature Human Behaviour* 6: 326–337.

Glossary

Abiotic factors Nonliving chemical or physical conditions (temperature, wind, solar radiation, moisture) that influence energy systems and ecological processes.

AC (alternating current) Electric current that reverses direction periodically, enabling efficient long-distance power transmission; today used in most grid systems.

Access to energy The ability of individuals or communities to obtain reliable, affordable electricity or clean cooking fuels; lack of access remains a major global inequality shaping health and livelihoods.

Acid mine drainage Polluted water created when sulfide-rich rocks are exposed during mining and generate acidic runoff carrying heavy metals into waterways.

Adaptation Adjustments in systems or practices to reduce harm from climate impacts or energy scarcity (e.g., grid hardening, cooling system upgrades).

Adiabatic compression/expansion A thermodynamic process without heat exchange; explains how heat pumps move heat (compression warms refrigerant; expansion cools it).

Agrivoltaics Dual land use strategy in which solar panels are integrated with crops or grazing, improving land productivity while generating renewable energy.

Air-source heat pump A device that transfers heat between indoor spaces and outdoor air using electricity and refrigerants, enabling efficient heating and cooling.

Albedo The reflectivity of a surface; relevant for solar energy performance and climate feedbacks when land changes alter reflectivity.

Algae biofuels Fuels derived from micro- or macroalgae through biochemical or thermochemical processing; considered an alternative to land-intensive terrestrial biofuel crops.

Aluminum A lightweight, conductive, corrosion-resistant metal produced mainly from bauxite; essential for EVs, solar frames, and transmission hardware; highly energy-intensive to produce but easily recycled.

Anaerobic digestion Microbial breakdown of organic material without oxygen, producing biogas (methane + CO_2) and digestate; widely used in wastewater and agricultural systems.

Anthracite The highest-carbon grade of coal with high energy density and few impurities, formed under extreme heat and pressure.

Aquifer Subsurface water-bearing formation; may be stressed by groundwater withdrawals for lithium brine extraction, oil and gas development, road construction, cooling, or irrigation.

Artisanal and small-scale mining (ASM) Labor-intensive mining conducted with limited mechanization; in cobalt supply chains, ASM supports livelihoods but poses safety and child labor concerns.

Automobility The socio-technical system built around private automobile use, shaping infrastructure, urban planning, energy demand, and social identity.

Avoided cost of electricity A regulatory term referring to the cost a utility avoids by purchasing power from a qualifying facility instead of generating it itself.

Baseload power Power sources capable of providing continuous electricity (e.g., geothermal, hydropower, nuclear) independent of daily or seasonal variation.

Battery chemistries Specific combinations of anode, cathode, and electrolyte materials (e.g., LFP, NMC) that determine performance, safety, cost, and energy density.

Battery energy storage system (BESS) Battery infrastructure that stores electricity for later use, improving grid resilience and enabling higher penetration of variable renewables.

Becquerel (Bq) The SI unit of radioactivity equal to one nuclear disintegration per second.

Biodiesel A liquid biofuel produced from vegetable oils, used cooking oils, or animal fats through transesterification; often blended with petroleum diesel.

Bioenergy Energy from biological material (plants, residues, wastes), including combustion, biogas, and advanced fuels; it can be highly land- and water-intensive.

Biofuel Any fuel derived from biomass, including ethanol, biodiesel, biogas, and renewable diesel.

Biogenic methane Methane produced by biological processes such as anaerobic digestion of wastewater, manure, or food waste.

Biomass Organic material used for energy, including wood, crop residues, algae, and waste streams.

Biomethane Upgraded biogas purified to pipeline-grade methane; often injected into gas grids as a renewable substitute for fossil gas.

Bioplastics Plastics made from biological feedstocks rather than fossil hydrocarbons; seen as alternatives for reducing petrochemical dependence.

Blackouts Complete loss of electricity service in an area due to grid failures, extreme weather, or resource shortages.

Black-boxing A concept from science and technology studies describing how complex socio-technical systems are treated as simple "inputs converted to outputs," obscuring internal dynamics.

Blast furnace High-temperature industrial furnace using coke to reduce iron ore into molten iron for steel production.

Blue water footprint The volume of fresh water withdrawn and consumed in producing a good or service.

Cadmium telluride (CdTe) A thin-film photovoltaic semiconductor using cadmium and tellurium usually paired with cadmium sulfide.

Carbon capture and storage (CCS) Processes capturing CO_2 from industrial sources or power plants and storing it underground, often via pipelines.

Carbon dioxide pipeline A pipeline transporting compressed CO_2 for enhanced oil recovery or sequestration.

Carbon intensity Greenhouse gas emissions per unit of energy or fuel produced; used to compare fossil fuels and biofuels.

Carbon lock-in The persistence of high-carbon systems due to sunk costs, institutional inertia, and infrastructure built for fossil fuels.

Cement A binding material produced by heating limestone, releasing CO_2 through both fuel use and chemical reactions; responsible for ~7–8% of global GHG emissions.

Centrifugal pump A pump commonly used in irrigation or water delivery powered by electricity or diesel; increasingly paired with solar PV for agriculture.

Chain reaction (nuclear) A self-sustaining sequence of nuclear fission events producing heat for electricity generation.

Child labor Hazardous labor by minors, documented in cobalt and other mineral supply chains in the DRC and elsewhere; see ASM entry.

Circular economy A design approach seeking to eliminate waste by reusing, repairing, and recycling materials, contrasting with "take–make–waste" systems.

Clean cookstove Low-emission stoves designed to reduce indoor air pollution relative to biomass open-fire cooking.

Clean firm power Continuous, dispatchable, and low-carbon electricity sources such as nuclear or geothermal.

Climate justice The principle that those least responsible for emissions often face the gravest climate harms, linking climate action to equity and reparations.

Cobalt A metal used in many lithium-ion battery cathodes; over 70% is mined in the DRC, often under conditions involving human rights concerns.

Coal A fossil fuel formed from ancient plant matter over millions of years; historically central to industrialization and still widely used for electricity and steel.

Coal seam A geological layer containing coal deposits; varies in depth, quality, and mining difficulty.

Combustion Burning solid, liquid, or gaseous fuels to generate to release heat and motion, electricity, or other industrial activities.

Community microgrid A localized electricity network capable of operating with or without the main grid, often designed for resilience and community control.

Concentrated solar power (CSP) Solar technology using mirrors or lenses to concentrate sunlight into heat to produce steam for electricity.

Concrete A composite made from cement, sand, gravel, and water; ubiquitous in construction but carbon-intensive through cement production.

Critical energy studies A scholarly approach examining how power, inequality, and socio-ecological relations shape energy systems and transitions.

Critical minerals Metals essential to clean energy technologies, lithium, cobalt, nickel, and rare earths; often sourced from high-impact or geopolitically sensitive regions.

Critical realism A philosophical approach that distinguishes between observable events and the deeper social and material structures that produce them. It argues that concepts such as "energy systems" or "transitions" are abstractions that must be unpacked to reveal underlying causal relations, power dynamics, and socio-ecological processes.

Curie (Ci) A unit of radioactivity equal to 37 billion disintegrations per second.

Dams Large structures built to store water and generate hydropower; they can alter river ecosystems, displace communities, and trap sediments while providing flood control and electricity.

Decarbonization The process of reducing carbon emissions from energy, industry, transportation, and buildings through electrification, renewable energy, efficiency, and carbon management.

Decentralized energy Energy generation located close to where it is used (e.g., rooftop solar, microgrids), reducing transmission needs and enhancing resilience.

Deep sea mining Extraction of minerals from the ocean floor; proposed as a future source of rare metals but raises ecological and geopolitical concerns.

Demand response Programs encouraging consumers to shift or reduce electricity use during peak times to support grid stability.

Desalination Energy-intensive process that removes salt from seawater; raises equity concerns when benefits and harms are unevenly distributed.

Diesel fuel A petroleum distillate used in engines, backup generators, and freight transport; emits particulate matter, NOx, and greenhouse gases.

Direct current (DC) Electric current flowing in one direction; historically used in early grids and now essential in electronics, batteries, EVs, and long-distance HVDC lines.

Direct lithium extraction (DLE) Technologies extracting lithium directly from brines without large evaporation ponds, potentially reducing land and water impacts.

Distribution grid The lower-voltage network delivering electricity from substations to homes, businesses, and local infrastructure.

District heating Community-scale thermal energy systems that distribute hot water or steam from a centralized source.

Dredging Excavation of sediments for port maintenance, often necessary for supertankers and fuel transport but harmful to ecosystems.

Dry cooling power plants Cooling method using air instead of water, reducing water use but lowering thermal efficiency.

Electricity grid A massive interconnected system of generators, transmission lines, substations, and distribution networks supplying power.

Electrification Replacing fossil-fueled systems with electric-compatible technologies (e.g., heat pumps, EVs), enabling cleaner energy as grids decarbonize.

Electric vehicles (EV) Vehicles powered by electricity stored in batteries, offering high efficiency and zero tailpipe emissions.

Electrolysis A process using electricity to split water into hydrogen and oxygen; key for producing green hydrogen.

Embodied energy Total energy used in the extraction, manufacturing, transport, and installation of a material or technology.

Energy density Energy content per unit mass or volume; fossil fuels have high energy density, complicating substitutions in aviation and shipping.

Energy efficiency Using less energy to achieve the same service, often via insulation, efficient appliances, or process improvements.

Energy inequity Unequal distribution of energy burdens and benefits across communities, often along racial, class, or colonial lines.

Energy payback time The time required for an energy system (e.g., PV panel) to generate the amount of energy used to produce it.

Energy poverty Lack of access to reliable, affordable energy services like electricity and clean cooking fuels; affects billions of people globally.

Energy security The reliability, accessibility, and affordability of energy resources; geopolitically shaped by imports, infrastructure, and supply disruptions.

Energy transition Structural shifts in energy systems, such as from coal and oil to renewables, shaped by economics, politics, and technology.

Enhanced oil recovery (EOR) Injecting CO_2 or other fluids into oil reservoirs to increase extraction of remaining oil. Sequesters CO_2 but also results in increased oil production.

Evapotranspiration Combined water loss from soil evaporation and plant transpiration; central concept in green water footprints.

Fenceline community Residents living near industrial facilities (e.g., refineries, power plants) who face disproportionate pollution exposure.

Fertilizer (synthetic) Nitrogen, phosphorus, and potassium compounds produced using fossil fuels (especially ammonia via natural gas), enabling high-yield agriculture but contributing to pollution.

Flaring Controlled burning of excess natural gas at oil and gas facilities, reducing methane release but emitting CO_2 and air pollutants.
Floatovoltaics Solar panels installed on water bodies to reduce land use, lower evaporation, and improve PV efficiency through water cooling.
Food-energy-water (FEW) nexus The interconnected system in which food production, water management, and energy use interact and influence one another.
Forestry residues Wood waste from forest thinning, harvesting, or sawmills used as biomass fuel.
Fossil fuels Energy-dense carbon-based fuels (coal, oil, natural gas) formed over millions of years; major sources of climate and air pollution.
Fuel economy The efficiency with which vehicles convert fuel into motion; regulated through standards such as Corporate Average Fuel Economy (CAFE) standards in the United States.
Fugitive methane Methane leaked unintentionally from extraction, pipelines, or compressors; a potent greenhouse gas with high near-term warming potential.
Geothermal energy Heat from Earth's interior used for electricity or direct heating; considered a firm, low-carbon resource in suitable regions.
Glass Material made from quartz, limestone, and soda ash; energy-intensive to produce and crucial for buildings, vehicles, and solar panels.
Global supply chain Worldwide networks linking extraction, refining, manufacturing, and distribution of energy technologies and materials.
Greenhouse gas (GHG) Gases like CO_2, CH_4, and N_2O that trap heat in the atmosphere and drive climate change.
Green hydrogen Hydrogen produced via electrolysis using renewable electricity, proposed for decarbonizing industry and heavy transport.
Green water footprint Rainwater consumed through crop evapotranspiration in agricultural production.
Grid congestion Transmission bottlenecks that prevent electricity from moving from low-cost generators to high-demand regions.
Grid defection A scenario in which customers leave the utility grid entirely, relying on rooftop solar and batteries.
Grid resilience The ability of the electricity system to withstand and recover from disruptions such as storms, fires, or equipment failures.
Ground-source heat pump A heating and cooling technology that exchanges heat with underground soils or bedrock for high efficiency.
Habitat fragmentation Breaking up contiguous ecosystems through infrastructure like roads or solar farms, harming wildlife movement and biodiversity.
Halogens (plastics manufacturing) Elements such as chlorine and fluorine used in the production of many plastics and polymer additives. Halogens are important in materials such as polyvinyl chloride (PVC), fluoropolymers, flame retardants, and refrigerants, but their use can raise environmental and health concerns because some halogenated compounds are persistent, toxic, or difficult to dispose of safely.

Hard-to-abate sectors Industries like cement, steel, and chemicals that are difficult to decarbonize due to high heat or process emissions.

Hazardous waste Toxic or harmful by-products from mining, manufacturing, or energy production requiring careful management.

Heat island effect Urban warming caused by built surfaces absorbing solar radiation, increasing cooling demand.

Heat pump water heater An efficient device that uses heat pump technology to heat water, reducing household fossil fuel use.

Heating degree days A weather-based metric indicating how much heating is needed to maintain indoor comfort.

High-voltage transmission Long-distance electricity lines operating at high voltages to minimize losses and move bulk power.

Household energy burden The proportion of household income spent on energy bills; disproportionately high for low-income renters.

Industrial heat High-temperature heat (often above 400°C) required in industries such as steel, cement, glass, and chemicals; one of the hardest forms of energy to decarbonize.

Import dependence Reliance on foreign mineral supplies for technologies such as batteries and electronics; the United States is 50–100% import dependent for many critical minerals.

Indoor air pollution Pollution from burning biomass, kerosene, or gas indoors; a major global health burden especially in regions lacking clean cooking fuels.

Induced seismicity Earthquakes triggered by human activities such as geothermal fluid injection, wastewater disposal, or mining.

Industrial ecology A field that studies material and energy flows in industry to design systems that mimic ecological cycles and reduce waste.

Industrial symbiosis When one industry's waste becomes another's input, reducing resource extraction and emissions (e.g., fly ash used in cement and waste heat used in community heating districts).

Inequality (environmental) Unequal exposure to pollution, unequal access to clean energy, and unequal participation in decision-making across communities.

Internal combustion engine (ICE) An engine that burns gasoline or diesel, converting only ~20–30% of energy into motion while producing major air pollution.

International supply chains The globally distributed processes of mining, refining, manufacturing, transport, and disposal associated with energy systems.

Irrigation energy Electricity or diesel used to pump and move water for agriculture; a major energy demand in water-stressed regions.

Jet fuel A highly energy-dense hydrocarbon fuel essential for aviation; difficult to replace due to weight and volume constraints.

Just transition A framework ensuring workers and communities are supported during decarbonization through fairness in employment, participation, and benefits.

Kilowatt vs. kilowatt-hour A kilowatt is a unit of power (rate of energy use), while a kilowatt-hour is a unit of energy consumed over time.

Kinetic energy (hydropower or wind power) The energy of moving fluid used to turn turbines and generate electricity.

Knowledge coproduction Collaborative creation of knowledge between researchers and communities, essential for participatory energy justice approaches.

Land dispossession Loss of access to land caused by energy projects such as solar parks, dams, or mining, disproportionately affecting marginalized communities.

Land use change Conversion of landscapes for energy, agriculture, or industry; a major source of emissions and biodiversity loss.

Land use conflict Tension over how land is allocated among energy production, farming, conservation, and cultural uses.

Landfills Sites where organic waste is disposed of and decomposes, releasing methane; increasingly tapped for biogas or biomethane production.

Leaded gasoline Gasoline containing tetraethyl lead, used historically to prevent engine knock and later phased out due to health hazards.

Levelized cost of energy (LCOE) The lifetime cost per unit of energy from a technology; used to compare different energy sources.

Life cycle assessment (LCA) An evaluation of environmental impacts from extraction through disposal of a product or system.

Lignin/cellulose Organic polymers in plants; precursors to coal under geological pressure and key to many biofuel pathways.

Lignite A low-grade, high-moisture coal with lower energy density, often burned near where it is mined. Sometimes called brown coal.

Liquefied natural gas (LNG) Natural gas cooled to −162°C to reduce volume for shipping; requires energy-intensive liquefaction and regasification.

Lithium A lightweight metal essential for most lithium-ion batteries; extracted from hard rock or brines with high water use.

Lithium-ion battery Rechargeable battery system using lithium compounds in the cathode; critical for EVs and storage.

Lithium triangle Region spanning Argentina, Bolivia, and Chile containing vast lithium-rich brine deposits. Some suggest eschewing this name because of connotations invoked.

Macroelectrification Large-scale electrification of national energy systems to support renewable integration and end-use electrification.

Methane A potent greenhouse gas with 80 times the warming effect of CO_2 over 20 years; shorter over longer time horizons because of oxidation. It's emitted from leaks, agriculture, and landfills.

Methanol A liquid fuel produced from biomass, gasification, or industrial waste gases; used as a fuel additive or chemical feedstock. It can be made from renewable or fossil feedstocks.

Microgrid A localized electricity system capable of islanding from the main grid, improving resilience and local control.

Mineral intensity The amount of mineral input required for an energy technology.

Mining waste, tailings, or overburden Material left after extraction of valuable minerals, often far exceeding the volume of useful ore.

Mobility justice The recognition that transportation systems produce unequal mobility opportunities, affecting access to jobs, safety, and emissions burdens.

Modularity Designing products so parts can be easily replaced, reused, repaired, or recycled, reducing waste.

Molten salt storage A thermal storage medium used in concentrated solar power systems to retain heat for later electricity generation.

Monoculture bioenergy Large-scale cultivation of a single crop (e.g., sugarcane or palm oil) for fuel, often linked to biodiversity loss and land conflicts.

Mooring (offshore energy) Anchoring systems for offshore wind turbines or floating solar installations.

MTBE Methyl tert-butyl ether, a gasoline oxygenate later phased out due to groundwater contamination.

Municipal wastewater treatment Energy-intensive facilities that treat sewage; some can recover energy via biogas and heat.

Naphtha A liquid hydrocarbon coproduct of fossil and renewable diesel production that can be blended into transportation fuels.

Nationally Determined Contributions (NDCs) Country-level climate commitments under the Paris Agreement outlining planned emissions reductions and adaptation actions.

Natural gas A fossil fuel primarily composed of methane used for heating, electricity, and industrial processes; associated with leakage, groundwater contamination risks, and global methane emissions.

Natural gas liquids (NGLs) Hydrocarbons like propane, butane, and ethane produced with natural gas, widely used in petrochemicals.

Net metering A policy allowing solar customers to receive credit for excess power sent to the grid.

Net zero A state where greenhouse gas emissions are balanced by removals through sinks or carbon capture; central to climate policy.

Nickel A metal used in many high-energy-density lithium-ion batteries; typically mined with low-grade ores.

Nitrogen cycle The biogeochemical cycling of nitrogen through soil, plants, and the atmosphere; heavily altered by fertilizer use and biofuel production.

Nitrogen dioxide (NO_2) A combustion pollutant commonly produced indoors by gas stoves and outdoors by vehicles and power plants.

Nitrous oxide (N_2O) A potent greenhouse gas released from fertilizer use and some industrial processes.

Nuclear fission The splitting of uranium atoms to release heat for electricity; the basis of most nuclear reactors.

Nuclear fuel cycle The sequence from uranium mining and milling to enrichment, reactor use, and waste management.

Nuclear waste Radioactive materials produced during nuclear generation; includes spent fuel requiring secure long-term storage.

Nuclear weapons complex Facilities historically built for plutonium and uranium production, often leaving long-term contamination burdens.

Ocean energy Renewable energy sources derived from waves, tides, and ocean currents.

Off-grid systems Independent energy systems not connected to centralized grids, commonly used in rural electrification.

Offshore drilling Extraction of oil and gas from beneath the seafloor using rigs and platforms; carries risks of spills and blowouts.

Offshore wind Wind turbines located in marine environments taking advantage of higher and steadier wind speeds.

Oil refinery Industrial complex converting crude oil into gasoline, diesel, jet fuel, and petrochemicals; major source of air pollution for nearby communities.

Oil spill Accidental release of petroleum into the environment, often causing long-term ecological and economic damage.

Oil supertanker Enormous ships capable of transporting millions of barrels of crude oil across oceans; critical nodes in global oil infrastructure.

Overburden Soil and rock removed to access ore deposits; often stored as waste piles that can cause flooding or contamination. See mining wastes.

Over-the-fence rule A regulation (e.g., in California) preventing private parties from sharing electricity across property lines without being regulated as utilities.

Ozone (ground level) A harmful pollutant formed from reactions between NOx and VOCs; contributes to smog and respiratory illness.

Passive cooling Building design strategies such as shading, ventilation, and thermal mass that reduce the need for mechanical cooling.

Peak oil The theorized point at which global oil production would reach a maximum before declining; largely overtaken by technological and economic factors.

Peaker plant Power plants that operate during periods of high electricity demand, often using fossil fuels with high emissions.

Perovskite solar cell An emerging photovoltaic technology with high efficiency potential; may be stacked atop silicon cells.

Petrochemicals Chemicals derived from oil and natural gas used to create plastics, fertilizers, solvents, and synthetic materials.

Petroleum A liquid fossil fuel extracted from underground reservoirs and refined into transportation fuels, lubricants, and chemical products.

Photovoltaics (PV) Technology that converts sunlight into electricity using semiconductor materials.

Pipeline Infrastructure transporting oil, gas, CO_2, hydrogen, or other fluids; raises safety, environmental, and justice concerns.

Plastics Petrochemical-derived materials used extensively in consumer goods, industry, and clean energy devices (e.g., batteries).

Pleistocene energy regime Humanity's early reliance on biomass, muscle power, and fire before fossil fuels.

Pollution haven effect The movement of polluting industries to regions with weaker environmental laws.

Power density Energy output per unit area; fossil fuels and nuclear have higher power densities than wind or solar.

Power plant Facility that converts primary energy into electricity using turbines, reactors, generators, or photovoltaic arrays.

Power purchase agreement (PPA) Contract through which energy buyers purchase electricity from renewable projects at preset prices.

Radiation (ionizing) High-energy particles or waves emitted during radioactive decay; includes alpha, beta, and gamma radiation with varying health risks.

Radionuclide An atom with an unstable nucleus that undergoes radioactive decay.

Rare earth elements (REEs) A group of 17 metals critical for magnets, wind turbines, and EV motors; dominated by Chinese processing.

Recycled water Treated wastewater reused for irrigation, industrial processes, or potable applications; typically less energy-intensive than desalination.

Redlining The discriminatory practice that segregated communities and shaped unequal access to transportation, housing, and urban infrastructure.

Refinery flaring Combustion of excess hydrocarbon gases at refineries to prevent explosions, releasing pollutants that affect nearby communities.

Renewable diesel A hydrotreated fuel chemically similar to petroleum diesel but produced from waste oils, fats, or vegetable oils.

Renewable energy Energy from naturally replenished sources such as solar, wind, geothermal, and hydropower.

Renewable portfolio standard (RPS) Policy requiring a portion of electricity generation to come from renewable sources.

Reservoir (hydropower) Stored water used to generate electricity through controlled releases across turbines.

Resource curse The economic and political instability associated with heavy dependence on extractive energy resources.

Reverse logistics (recycling) Collection and recovery of materials (e.g., lead-acid batteries, PV modules) for reuse or recycling.

Rooftop solar Decentralized PV systems installed on buildings, reducing land-use impacts and often improving grid resilience. See net metering.

Rural electrification Extending modern energy access to rural communities, historically central to development strategies worldwide.

Sand Made from quartz, it is a major input for concrete, glass, and PV manufacturing; high global demand contributes to resource pressures and mining impacts.

Sea-level rise Rising ocean levels driven by climate change; threatens coastal energy infrastructure and settlements.

Seismic hazard Earthquake risks exacerbated by mining, dam reservoirs, geothermal operations, and wastewater injection.

Silicon tetrachloride A toxic by-product of polysilicon refining that must be recycled or safely handled in PV manufacturing.

Slag A metallurgical by-product used in construction materials and clinker substitution, reducing cement-sector emissions.

Slash (forest bioenergy) Residual woody debris from logging or thinning used as biomass fuel.

Smart grid Digitally enhanced grid using sensors, automation, and distributed resources to improve reliability and flexibility.

Smelter Facility where metal ores are heated and chemically reduced to produce refined metals; often pollution-intensive.

Solar farm Utility-scale photovoltaic or CSP installation generating bulk solar electricity, often with land-use and ecological implications.

Solar irrigation Using PV-powered pumps to deliver water for agriculture, reducing diesel use and emissions.

Solar module The assembled PV unit containing cells, glass, aluminum, polymers, and wiring to convert sunlight into electricity.

Solar-quality quartz High-purity quartz required for silicon and PV glass production.

Solid-state battery Next-generation battery design replacing liquid electrolytes with solid materials for safety and performance gains.

Specialty chemicals High-value petrochemicals used in coatings, solvents, electronics, and PV manufacturing; energy- and emissions-intensive to produce.

Spent nuclear fuel Used nuclear fuel rods that remain highly radioactive and require secure long-term storage.

Stakeholder engagement Inclusion of community members, workers, and other affected parties in energy planning and decisions, central to justice.

Steam turbine Device converting high-pressure steam into mechanical rotation for electricity generation in nuclear, geothermal, and fossil plants.

Stranded assets Energy infrastructure or resources that lose economic value due to policy, market shifts, or climate constraints.

Sugarcane ethanol A major biofuel produced predominantly in Brazil; lower processing energy than corn but linked to labor and land-use concerns.

Supply chain The sequence of extraction, processing, manufacturing, transport, use, and disposal that defines products of technologies and their impacts.

Sustainable aviation fuel (SAF) Lower-carbon jet fuel alternatives derived from waste oils, biomass, or synthetic pathways; limited by feedstock constraints.

Tailings Waste material left after ore processing, often stored in large impoundments that pose contamination and safety risks. See mining waste.

Thermal pollution Heating of water bodies from power plant cooling discharges, harming aquatic ecosystems.

Thermal power plant Facility producing electricity by combusting fuels or using other sources of heat to create steam; includes coal, natural gas, biomass, and nuclear plants.

Thin-film PV Solar technology using very thin semiconductor layers (CdTe, CIGS, and perovskites) that require fewer materials but may involve toxic metals. See perovskite solar cell.

Tidal power Electricity generated from predictable ocean tidal flows driven by gravitational forces.

Transmission lines High-voltage infrastructure transporting electricity over long distances.

Transpiration Water vapor release by plants; part of the green water footprint and agricultural water demand.

Trichlorosilane A reactive gas used in polysilicon refining for PV manufacturing.

Truck freight emissions Pollution from diesel-powered freight movement, concentrated near logistics hubs and impacting fenceline communities.

Turbine Rotating machinery that converts kinetic energy of steam, wind, or water into mechanical power for electricity.

Underground storage Subsurface facilities storing natural gas for seasonal demand management; vulnerable to leakage.

Uranium A radioactive element mined for nuclear fuel, with major reserves in Canada, Australia, Namibia, and the United States.

Uranium enrichment The process of increasing the proportion of fissile U-235 in uranium to make reactor fuel.

Urban heat Elevated temperatures in cities due to built environments absorbing solar energy.

Urban mining Recovery of valuable materials (e.g., metals, electronics) from buildings or waste streams instead of natural deposits.

Vapor compression cycle The refrigeration principle underlying heat pumps and air conditioners.

Vehicle-to-grid (V2G) Technology enabling EVs to supply electricity back to the grid during peak demand.

Vertical integration When a single firm owns generation, transmission, and distribution; historically common in electricity systems.

Virtual power plant (VPP) Aggregated distributed resources (solar, batteries, smart loads) coordinated to provide grid services like a conventional power plant.

Volatile organic compounds (VOCs) Air pollutants emitted from fuels, solvents, and industrial processes contributing to ozone formation.

Voltaic pile The first practical battery, invented by Alessandro Volta, enabling early electrochemical power experiments.

Wastewater treatment energy The substantial electricity required to pump, aerate, and treat sewage; facilities are exploring energy recovery to reduce costs.

Wave energy Electricity generated from the kinetic motion of ocean waves.

Weatherization Home upgrades (insulation, air sealing) that reduce heating and cooling energy demand, often underfunded in low-income communities.

Well-to-wheels emissions Total greenhouse gas emissions across the fuel life cycle from extraction to final use in vehicles. See LCA.

Wildfire risk Increasing danger to transmission lines and power plants from climate-driven heat and drought conditions.

Wind farm A cluster of wind turbines generating utility-scale renewable electricity.

X-ray (radiation) Penetrating electromagnetic radiation produced during radioactive decay or in medical/industrial equipment; requires shielding.

Zero-emission vehicle (ZEV) Vehicle producing no tailpipe emissions, typically EVs or fuel-cell vehicles.

Zero-net energy building A building that generates as much energy annually (e.g., via rooftop PV) as it consumes through high efficiency and electrification.

Zoning (land/energy) Land-use regulations affecting where energy projects such as solar farms and transmission lines can be sited.

Index